THE ROLE OF CLIMATE CHANGE IN GLOBAL ECONOMIC GOVERNANCE

The Role of Climate Change in Global Economic Governance

BRADLY J. CONDON AND TAPEN SINHA

OXFORD
UNIVERSITY PRESS

OXFORD
UNIVERSITY PRESS

Great Clarendon Street, Oxford, OX2 6DP,
United Kingdom

Oxford University Press is a department of the University of Oxford. It furthers the University's objective of excellence in research, scholarship, and education by publishing worldwide. Oxford is a registered trade mark of Oxford University Press in the UK and in certain other countries

First Edition published in 2013

Impression: 1

Published in the United States of America by Oxford University Press
198 Madison Avenue, New York, NY 10016, United States of America

British Library Cataloguing in Publication Data
Data available

Library of Congress Control Number: 2013940568

ISBN 978–0–19–965455–0

Printed in Great Britain by
CPI Group (UK) Ltd, Croydon, CR0 4YY

We dedicate this book to the future generations
who will bear the brunt of the effects of climate change.

Foreword

Climate change caused by greenhouse gas emissions presents an enormously difficult problem to humanity. Unlike the first truly global problem of chlorofluorocarbon refrigerant emissions, which was resolved at little economic cost, reduction of greenhouse emissions will not be cheap. Greenhouse gas reduction is a classic "tragedy of the commons" problem with two nasty adjuncts. From the point of view of the individual or nation, my activities producing emissions profit me, but these emissions will *in the future* hurt everyone *in an uneven way*. The future aspect raises the question, "What does the present owe the future?" The uneven aspect raises the selfish issue of "My country can mitigate and adapt to the consequences. Why should I care about what happens elsewhere?" In this context, can humanity find a way to cooperate to reduce greenhouse gas emissions?

In order to attack this international economic problem, it is necessary to know the existing international rules and institutions and to explore how they can be used to address the problem or altered if they are obstructing its solution. The relation between the present international landscape and addressing climate change is the subject of this book.

Condon and Sinha have produced the definitive treatise on all legal and economic issues that might be encountered in the context of climate change and international relations. A significant part of the book explores the relationship between local, national, and international efforts to reduce global carbon dioxide emissions and the existing international system governing trade. This book carefully explores the international legal issues that could be encountered when individual countries make efforts to reduce their own carbon emissions. Any country, state, locality, or individual proposing an emissions reduction scheme would be well advised to read it.

The scope of this book goes well beyond emissions control and international economic regulation. Chapter 6 is a lucid exposition of the application of economic theory to modeling the effects of climate change in the international context. For several different reasons, the effect of climate change will affect different countries in different ways. Thus all of the Maldives and major portions of Bangladesh are very likely to go underwater. In countries like India, that have a large fraction of the population dependent on subsistence farming, a small change in climate can have a devastating effect. In countries that have many people who are very poor, a small negative economic shift can create great suffering. Condon and Sinha combine all such considerations to create a susceptibility index for countries, which can be applied to prioritize international aid for mitigation efforts.

This book (Chapter 7) describes how the finance industry can combine with government to make cap-and-trade markets giving a historical account of how the US SO2 market and the European CO2 market have functioned. Any mandatory emissions reduction scheme, of necessity, affects different industries and regions disproportionately. In the SO2 market case, cap-and-trade proved highly successful in reducing emissions, but after more than a decade, the Environmental Protection Agency introduced a change that caused the cost of permits to escalate with the result that several states successfully sued. The result of that lawsuit was that the price of permits went to near zero effectively ending the program, but the beneficial changes it produced persist.

Emissions reductions and mitigation will be expensive. The poor in less developed countries will be some of the most affected adversely by climate change; the rich countries will be called upon to help them adapt. Chapter 8 describes the existing international mechanisms for distributing such funds, but at present the available funding is inadequate. Condon and Sinha propose that international efforts can be financed in great measure by global elimination of fossil fuel subsidies. If developed countries eliminated these subsidies, nearly USD 58 billion would be freed yearly to develop lower emission energy sources while simultaneously significantly reducing emissions through higher fossil fuel prices. Many less developed countries also subsidize fossil fuel consumption, and most of these subsidies do not benefit their poor; the money thus freed will help but will not be enough.

There are a number of people zealously devoted to combating the emissions causing climate change. Before they can be effective, they must understand the existing institutions of global economic governance in order to use them or change them. Their self-education can be greatly helped by study of *The Role of Climate Change in Global Economic Governance*.

Robert F. Curl
Pitzer-Schlumberger Professor of Natural Sciences Emeritus
University Professor Emeritus, Rice University
Nobel Prize in Chemistry in 1996

Preface

On January 11, 2013, the US Global Change Research Program released the National Climate Assessment Development Advisory Committee Draft Climate Assessment Report. The Committee coordinates federal research on changes in the global environment and their implications for society. Thirteen US government departments and agencies participate in the Research Program. The Executive Office of the President of the United States oversees this program. The Draft Report concludes that the evidence for a changing climate has strengthened considerably since its last Climate Assessment Report of 2009. Most significantly, the Draft Report concludes that observed climatic changes are having wide-ranging impacts in every region of the United States and most sectors of its economy. Climate change is no longer a future threat. It is happening now. This Draft Report changes the tone of the climate change debate significantly. In an appendix to Chapter 1, we reproduce the Draft Report's Introduction, which is entitled "Letter to the American People." The Joint US–China Statement on Climate Change is also significant, and notes "the overwhelming scientific consensus about *anthropogenic* climate change and its worsening impacts."[1]

Climate change represents an unprecedented global challenge. Because climate change raises scientific, economic, financial, social, political, and legal issues, it requires interdisciplinary research. This book analyzes the challenges that climate change poses for global economic governance, integrating economic, financial, and legal perspectives. Our research reveals that several policies are misguided and some are just plain backwards.

The division of countries into developed and developing is too simplistic to address the complex issues that arise from climate change. If one is to categorize countries according to their capacity for mitigation and adaptation, or according to their vulnerability to climate change, then one must use a more sophisticated categorization. We propose one.

The role of intellectual property rights with respect to technology dissemination should be analyzed according to different types of technologies. They may create obstacles to technology dissemination where plant varieties are concerned, which has implications for the capacity of subsistence farmers to adapt to a more variable climate. However, they are less likely to create obstacles to the diffusion of clean energy technologies. Nevertheless, many developing countries insist that intellectual property rights are an obstacle to the adoption of clean energy technologies, while other developing countries agree to stricter intellectual property rights for new plant varieties. These policy stances are backwards.

[1] Joint US–China Statement on Climate Change, Media Note, Office of the Spokesperson, Washington, D.C., April 13, 2013 <http://www.state.gov/r/pa/prs/ps/2013/04/207465.htm> (accessed April 15, 2013).

Developing countries insist that developed countries pay for climate finance. However, subsidies for fossil fuels in developing countries are four times the amount of climate financing that they seek for mitigation and adaptation actions. In WTO subsidies law, the legality of clean energy subsidies is uncertain. At the same time, WTO subsidies law permits Members to apply countervailing duties on imports of clean energy technologies, which raises their cost. Lowering the cost of fossil fuels, while increasing the cost of clean energy technologies, are also backward policies.

These policy incongruencies need to be addressed if multilateral climate change negotiations are to advance towards effective solutions. We hope that this book will contribute to that end.

PowerPoint slides to accompany this book are available at <http://cdei.itam.mx/medios_digitales/educacion.php#materiales>.

Bradly J. Condon
Tapen Sinha

Mexico City
April 15, 2013

Acknowledgements

We are very grateful to Nobel Laureate Dr. Robert F. Curl for his encouragement of our research and, in particular, for his kind and generous foreword for this book. There is no one who is better qualified than Bob to write this Foreword and to evaluate the value of our contribution to climate change research.

The authors gratefully acknowledge the help, support, and encouragement of the following people: Luis Yahir Acosta Pérez, M. Fernanda Alcalá Durand, Sankarshan Basu, Alyssa Benedict, Rebecca Benedict, Perla Buenrostro, Françoise Carner, Jorge Cerdio, Ashok Chaudhuri, Cathy Condon, Kip Condon, Fernando de Mateo, ISICal73 Group, Jean Paul Dutoit, Arturo Fernández, Alejandro Hernández, Diego Hernández, Georgina Jarquín, Rajeeva Karandikar, Stephen Kay, Dorotea López, César Martinelli, Felipe Muñoz, Héctor Murguía Holguín, Gabriela Moya, Mark Nordman, María Pereyra, Beatriz Rumbos, Dipendra Sinha, Sheena Sinha, Todd Wetmore, Wendy Williamson, and "El Grupete": Carlos Bernal, Gabriela Rodríguez, and Alejandra Sierra.

We gratefully acknowledge the support of the WTO Chairs Program and the many individuals at the WTO who have contributed to the success of Mexico's WTO Chair and the research for this book: Hakim Ben Hammouda, Pablo Bentes, Jorge Castro, Fatima Chaudhri, Bridget Chilala, Mireille Cossy, Fernanda Garza, Alejandro Jara, Pascal Lamy, Patrick Low, Hamid Mamdouh, Jose-Antonio Monteiro, María Pereyra, Ricardo Ramírez, Verónica Scerbo, Karsten Steinfatt, Gerardo Thielen, Héctor Torres, Raymundo Valdes, and Wendy Williamson. We also thank Vlasta Macku and Diana Tussie for their support of Mexico's WTO Chair research program.

Chapter 3 is based on a paper prepared for the International Conference on the Future of the WTO, Asian Center for WTO & International Health Law and Policy, College of Law, National Taiwan University, 24 May 2009. We thank the conference participants for their helpful comments on an earlier draft, particularly those of Victoria Donaldson, Lothar Ehring, Chang-fa Lo, Mitsuo Matsushita, Bryan Mercurio, David Morgan, and Marie-Isabelle Pellan. This paper was later published as Bradly J. Condon, "Climate Change and Unresolved Issues in WTO Law" (2009) 12 Journal of International Economic Law 895. I thank the anonymous reviewers of that article for their insightful comments. Chapter 3 has been extensively modified in light of more recent WTO jurisprudence.

Chapter 5 is reprinted from Journal of World Trade 47:4 (2013) 867–892, with permission of Kluwer Law International. We gratefully acknowledge the permission of the Journal of World Trade to use this work in this book. We also thank the participants in the conference on International Economic Law and the Environment for their helpful comments on an earlier version of this research, particularly Luis Yahir Acosta Pérez, José Manuel Álvarez, Luis Ángel Madrid, and Ramón

Eduardo Guacaneme. Parts of Chapters 6 and 7 were presented at a conference at the Indian Institute of Management, Bangalore. The hospitality of the Chennai Mathematical Institute is gratefully acknowledged.

We thank the Instituto Tecnológico Autónomo de México and the Asociación Mexicana de Cultura AC for their generous support of our research. We are grateful to our publisher, Oxford University Press, for seeing the value in this topic. We deeply appreciate the diligent work of the editors at OUP, in particular Merel Alstein, Anthony Hinton, Ela Kotkowska, Emma Hawes, John Louth, and Vengatakrishnan Subramaniam. We gratefully acknowledge the helpful comments of the three anonymous peer reviewers on earlier drafts of this book.

Contents

List of Figures

List of Tables

List of Abbreviations

2005 Guidelines	Guidelines for Research and Development Expenditures (Canada)
ARP	Acid Rain Program (US)
AUD	Australian Dollar
BCSC	British Columbia Supreme Court
BITs	Bilateral Investment Treaties
BRICS	Brazil, Russia, India, China, and South Africa
CAD	Canadian Dollar
CAIR	Clean Air Interstate Rule (US)
CARB	California Air Resources Board
CBD	Convention on Biological Diversity
CCS	Carbon-dioxide capture and storage
CCX	Chicago Carbon Exchange
CDM	Clean development mechanism
CDR	Common but differentiated responsibilities
CDVI	Climate disaster vulnerability index
CERs	Certified Emission Reductions
CITES	Convention on International Trade in Endangered Species of Wild Fauna and Flora
CODEX STAN	1 1985 General Standard for the Labeling of Prepackaged Foods
COP	Conference of Parties (UNFCCC)
CO2	Carbon dioxide
CO2e	Carbon dioxide equivalent
CH4	Tetrahydridocarbon (Methane)
CSAPR	Cross-State Air Pollution Rule (US)
CSI	Climate Sensitivity Index
DICE model	Dynamic Integrated Climate Economy model
E&T	Education and Training
EC	European Communities
ECJ	European Court of Justice
EEC	European Economic Community
EIA	Environmental impact assessment
EPA	Environmental Protection Agency (US)
EU	European Union
EUA	European Union Allowance
EU ETS	European Union Emissions Trading System
FAO	Food and Agriculture Organization
FIT	Feed-in tariff
FTA	Free Trade Agreement
GATS	General Agreement on Trade in Services
GATT	General Agreement on Tariffs and Trade
GDP	Gross Domestic Product
GHG	Greenhouse gas
GM	Genetically modified

GMO	Genetically modified organism
HS	Harmonized System of Tariff Classification
ICJ	International Court of Justice
ICSID	International Centre for Settlement of Investment Disputes
IIA	International Investment Agreements
ILC	International Law Commission
IMF	International Monetary Fund
IPCC	Intergovernmental Panel on Climate Change
IPR	Intellectual property right
JI	Joint Implementation
MEA	Multilateral environmental agreement
MFN	Most-favored nation
MIGA	Multilateral Investment Guarantee Agency
MTBE	Methyl tertiary-butyl ether
NAAEC	North American Agreement on Environmental Cooperation
NAFTA	North American Free Trade Agreement
NBP	US NOx Budget Trading Program
NOAA	US National Oceanic and Atmospheric Administration
N2O	Nitrous oxide
OECD	Organisation for Economic Co-operation and Development
PCB	Polychlorinated biphenyl
ppm	parts per million
PPM	Processing and production methods
PPP	Polluter-pays principle
R&D	Research and development
REDD	Reduced Emissions from Deforestation and Degradation
RGGI	Regional Greenhouse Gas Initiative (US)
RICE model	Regional Integrated model of Climate and the Economy model
SCM Agreement	Agreement on Subsidies and Countervailing Measures
SO2	Sulfur dioxide
SPS Agreement	Agreement on Sanitary and Phytosanitary Measures
TBT Agreement	Agreement on Technical Barriers to Trade
TEEB Report	The Economics of Ecosystems and Biodiversity Report
TRIMS	Agreement on Trade-Related Investment Measures
TRIPS Agreement	Agreement on Trade-Related Aspects of Intellectual Property Rights
UK	United Kingdom
UN	United Nations
UNDP	United Nations Development Program
UNFCCC	United Nations Framework Convention on Climate Change
UNTS	United Nations Treaty Series
UPOV	International Convention for the Protection of New Varieties of Plants
US	United States
USD	United States Dollar
WTO	World Trade Organization

Table of Cases

WTO APPELLATE BODY REPORTS

WTO PANEL REPORTS

WTO CASES FOR WHICH NO REPORT YET ISSUED

GATT PANEL REPORTS

INTERNATIONAL CENTER FOR THE SETTLEMENT OF INVESTMENT DISPUTES CASES

ICSID CASES IN PROGRESS

NAFTA CHAPTER TWENTY CASES

INTERNATIONAL COURT OF JUSTICE CASES

OTHER INTERNATIONAL CASES

NATIONAL COURT CASES

Table of Legislation and International Agreements

WORLD TRADE ORGANIZATION

OTHER TREATIES AND INTERNATIONAL DOCUMENTS

1

The Climate Change Policy Puzzle

1.1 Introduction

Climate change is a complex problem that requires solutions on many fronts. Policy responses to climate change can be divided into two broad categories—mitigation and adaptation. Mitigation seeks to reduce greenhouse gas (GHG) emissions in order to limit the degree of climate change that will occur in the future. Adaptation seeks to respond to the effects of climate change that have occurred or that will occur as a result of the GHG emissions that have entered the atmosphere. This book will analyze these issues against the backdrop of two central themes: technology diffusion (rather than transfer, which implies donations or transfer of property rights) and unilateral responses to multilateral negotiation failure.

Policy responses to climate change can be divided along jurisdictional lines—subnational, national, and international; and unilateral, bilateral, regional, and multilateral. Examples of subnational measures include changes to zoning by-laws to account for expanded flood zones or changing the choice of trees planted on city streets to account for higher temperatures and changing rainfall, both of which are adaptation measures. Examples of national measures include higher energy efficiency requirements for housing construction, eliminating subsidies on fossil fuel consumption and taxing fossil fuel consumption, all of which are mitigation measures. Measures taken by the European Union are examples of regional measures. The United Nations Framework Convention on Climate Change (UNFCCC) and the Kyoto Protocol are examples of international measures.

To these "jurisdictional" categories we also could add individual actions, such as hanging the laundry to dry in the sun instead of using a dryer (mitigation), taking public transportation or walking instead of driving a car (mitigation), and planting drought-tolerant plants in the garden to reduce watering requirements (adaptation). Governments can create incentives for such individual actions, by increasing the cost of electricity, gasoline, and water; implementing watering restrictions for gardens; improving public transportation; making cities more walkable; and implementing public education campaigns. Thus, government policy has a role to play even in the realm of the individual choices that we make.

Unilateral measures refer to measures taken without reaching international agreement on the appropriate policy response. They can be taken, and have implications, at the subnational, national, or international level. If Chicago changes its street trees from maples to magnolias, its actions do not require international agreement and do not risk violating the international obligations of the United States. Any legal implications are likely to be strictly local. If New York changes the building code and land-use zoning in some of the areas flooded by Hurricane Sandy, there could be international legal implications for the US federal government if these changes are made in a way that violates the rights of foreign investors under international investment agreements (for example, if the rezoning is tantamount to expropriation without adequate compensation). When the Province of Ontario introduces renewable energy subsidies, they may be inconsistent with Canada's World Trade Organization (WTO) obligations. When the European Union applies its Aviation Directive to airline emissions that occur outside its territory, it creates an incentive to improve progress in multilateral negotiations on the regulation of airline emissions, but also increases its litigation risk. Thus, unilateral measures may be taken to address local or global concerns and may be used to create incentives for multilateral action. They may be consistent with international obligations or not, depending on the circumstances of each case. Regardless of whether climate change measures are unilateral, regional, or multilateral, they may be implemented or applied in a manner that is consistent with international economic law, or not.

In a multipolar world, it is difficult to reach multilateral agreements with respect to climate change, intellectual property rights, and technology transfer; protections for international investors; or international trade in goods and services. In the case of the WTO, deadlock at the multilateral level has led to unilateral, bilateral, and regional policy responses. The same may happen with respect to climate change negotiations, unless some unanticipated catalyst will permit a breakthrough in multilateral negotiations. However, catalysts that trigger new regulation may produce unilateral responses, rather than multilateral responses.

Weather events, such as Hurricane Sandy, are potential catalysts. The issue is not whether climate change caused Hurricane Sandy, but whether it will serve as a catalyst for unilateral or multilateral action on climate change. It is not uncommon for a trigger event to serve as a catalyst for regulation. When the event gets the attention of voters it also gets the attention of politicians.

The 2013 Draft Report may serve as a catalyst in the United States. It represents a significant change in tone with respect to the discussion of climate change:

> Climate change is already affecting the American people. Certain types of weather events have become more frequent and/or intense, including heat waves, heavy downpours, and, in some regions, floods and droughts. Sea level is rising, oceans are becoming more acidic, and glaciers and arctic sea ice are melting. These changes are part of the pattern of global climate change, which is primarily driven by human activity.[1]

Indeed, the Joint US–China Statement on Climate Change, which was issued three months later, indicates that recent weather events may also lead to increased bilateral cooperation between the world's two largest GHG emitters:

[1] National Climate Assessment Development Advisory Committee, *Draft Climate Assessment Report* (January 11, 2013) 3 <http://ncadac.globalchange.gov/> (accessed January 12, 2013).

The two countries took special note of the overwhelming scientific consensus about anthropogenic climate change and its worsening impacts, including the sharp rise in global average temperatures over the past century, the alarming acidification of our oceans, the rapid loss of Arctic sea ice, and the striking incidence of extreme weather events occurring all over the world.[2]

Like weather events, unilateral measures can serve as catalysts for multilateral action on climate change, by prompting the affected economic actors to pressure their governments to seek a solution, through litigation or negotiation. For example, when litigation by private actors against the EU Aviation Directive in the European Court of Justice (ECJ) failed, the EU measure had the effect of prompting other countries to make a greater effort to negotiate a multilateral agreement on airline emissions.

In addition to unilateral and multilateral actions, policy responses may be taken at the bilateral or regional level. However, these categories are not mutually exclusive. The EU Aviation Directive is a unilateral measure with respect to non-EU members, but it was taken at the regional level of the European Union. An example of a bilateral/regional measure is the agreement between the EU and Australia to fully link their emissions trading systems with the mutual recognition of carbon units between their two cap and trade systems.[3] Bilateral or regional actions need not be formalized in an international agreement. For example, Canada is likely to follow the lead of the United States on climate change. Bilateral and regional initiatives, like unilateral initiatives, may serve as a stepping stone to multilateral initiatives.

Regardless of the shape that climate change law and policy negotiations take—unilateral, bilateral, regional, or multilateral responses—they will need either to comply with or to modify existing international economic realities, with respect to international economic law, global models of economic governance, and economic and financial realities.

Rather than seek to predict how climate negotiations might succeed, this book analyzes the manner in which these other factors constrain possible policy outcomes. Existing international economic law places limitations on the right of national and sub-national governments to regulate to address climate change. Given the current difficulty in reaching multilateral agreements, we assume for the most part that countries will have to develop climate change policy and law within the constraints of the existing legal, economic, and financial framework. We will only suggest efforts to negotiate changes to that framework where we identify particularly egregious lacunae in the existing framework.

[2] Joint US–China Statement on Climate Change, Media Note, Office of the Spokesperson, Washington, D.C., April 13, 2013 <http://www.state.gov/r/pa/prs/ps/2013/04/207465.htm> (accessed April 15, 2013).

[3] Australia and European Commission agree on pathway towards fully linking Emissions Trading systems, EU Doc. IP/12/916, 28/08/2012 <http://europa.eu/rapid/press-release_IP-12-916_en.htm> (accessed October 3, 2012).

This chapter begins with a brief overview of the science of climate change, and its implications for the environment, human health, and the global economy, to demonstrate the severity and urgency of this environmental problem. We then examine how the evolving political and economic context has paralyzed multilateral negotiations, both with respect to climate change mitigation and adaptation and with respect to multilateral trade negotiations at the WTO. We argue that the shifting fortunes of developed and emerging economies have altered the dynamics of global governance, and will continue to do so for the foreseeable future. As a result, we are unlikely to break the multilateral logjam through negotiation alone. Rather, unilateral action by countries and regions will be necessary to create incentives to address climate change.

This chapter also examines regulatory capture. Regulatory capture at the national level has translated into regulatory capture of negotiating positions, contributing to multilateral negotiation paralysis. Regulatory capture also risks distorting unilateral measures in unacceptable ways. Thus, while we advocate unilateral means to achieve multilateral ends, unilateral measures should be taken within a range of acceptable actions, both in terms of their consistency with international law and their economic viability.

Finally, this chapter introduces the themes that we will examine in subsequent chapters.

1.2 Scientific Foundations and Potential Effects

Climate change is referred to as changes in the statistical distribution of weather patterns in the long run (over decades to millions of years). The most comprehensive study is "Climate Change 2001: The Scientific Basis" by the Intergovernmental Panel on Climate Change (IPCC). There are additional studies that examine in detail what is happening in specific developed countries. Two such comprehensive studies are the National Oceanic and Atmospheric Administration (NOAA) Report and the 2013 *Draft Climate Assessment Report.*[4]

There has been a gradual change of the surface temperature over the past century. This change has been well documented. The mean temperature changed globally (Northern Hemisphere) in January and July between 1880 and 2011 (see Figures 1.1a and 1.1b). The scale is Celsius multiplied by 100. Thus, both in the coldest month and the hottest month of the year, there have been steady rises in temperature of about 1°C over the course of the past century. This change of just one degree may not sound like much, but there is clear scientific evidence that an additional change of temperature of 0.7°C will have dire consequences (see Figure 1.2). In the United States, average temperature has increased by about

[4] Thomas R. Karl, Jerry M. Melillo, and Thomas C. Peterson (eds.), *Global Climate Change Impacts in the United States* (Cambridge University Press, Cambridge 2009); *Draft Climate Assessment Report.*

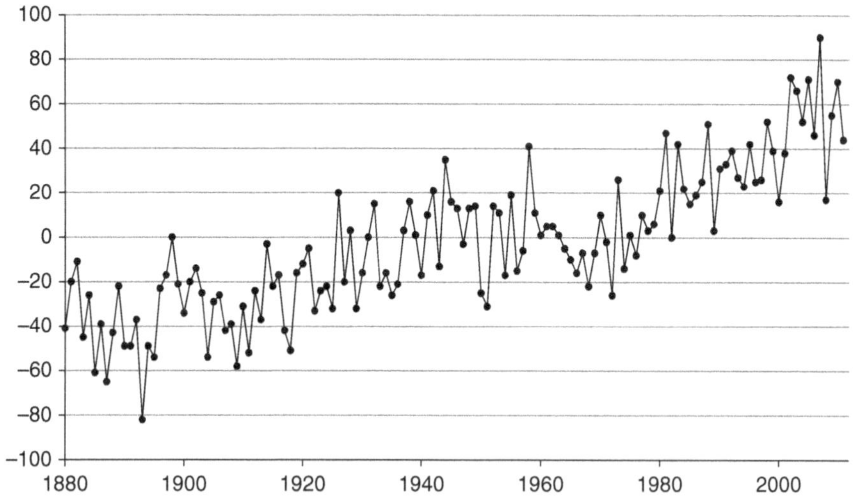

Figure 1.1a Temperature anomaly 1880–2011 (January)

Note: The scale on the vertical axis is measured in 1/100 of degrees Celsius.

Source: <http://data.giss.nasa.gov/gistemp/tabledata/GLB.Ts+dSST.txt>

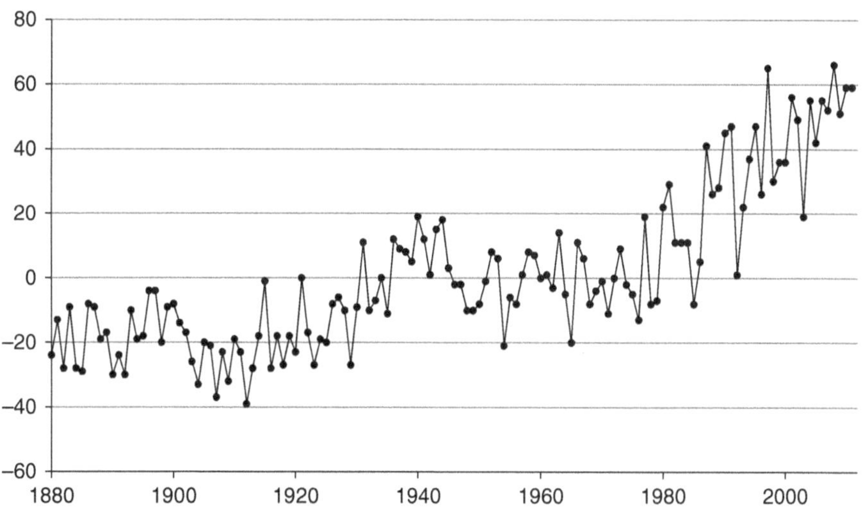

Figure 1.1b Temperature anomaly 1880–2011 (July)

Note: The scale on the vertical axis is measured in 1/100 of degrees Celsius.

Source: <http://data.giss.nasa.gov/gistemp/tabledata/GLB.Ts+dSST.txt>

1.5°F since record keeping began in 1895; more than 80 percent of this increase has occurred since 1980.[5]

There are clear links between human activities and the rise in temperature, principally through the emission of CO2 and other greenhouse gases (such as N2O, SO2, and CH4). The 2007 Fourth Assessment Report compiled by the IPCC (AR4) stated that "changes in atmospheric concentrations of greenhouse gases and aerosols, land cover and solar radiation alter the energy balance of the climate system." It concluded that "increases in anthropogenic greenhouse gas concentrations are *very likely* to have caused most of the increases in global average temperatures since the mid-20th century."[6] The 2013 *Draft Climate Assessment Report* states: "Much of the climate change of the past 50 years is primarily due to human activities."[7]

What impact will it have on hunger, water shortage, flooding, and diseases associated with rising temperature? The answer depends on the rise in temperature. Even if actions are taken now to reduce CO2 emissions, the emissions will at least double by the end of 2100. This will produce a rise of mean temperature of 3°C (with a 95 percent band of between 2 to 4.5°C).[8] In most cases, the number of people affected in each dimension mentioned above will depend on the temperature in a *nonlinear* way. All of that is demonstrated in Figure 1.2 below. We summarize here what Figure 1.2 exactly says. On the vertical left axis, we plot the impact on the number of people affected in terms of (1) hunger, (2) malaria, and (3) flooding as a result of the temperature rise. (1) Rising temperature will cause more frequent droughts mostly in Africa and some other pockets such as Latin America and Asia. This impact will be felt starting at a temperature rise above 1.7°C. It will negatively affect agriculture in Australia. However, it will not increase hunger there, since Australia is already developed. In Africa, especially in Sub-Saharan Africa, the population increase is already outstripping the increase in food production. It will get much worse in the current century.[9] (2) Malaria will reach new places. This process has already begun.[10] (3) Climate change will result in rising sea levels. Investigations have predicted a rise in sea level in 2100 of between 0.5 meters and 1.4 meters.[11] It will also lead to an increase in flooding of river basins.[12] In addition, it will lead to more frequent flooding in other parts of the

[5] *Draft Climate Assessment Report* 25.

[6] IPCC Report (2007) 5 <http://www.ipcc.ch/pdf/assessment-report/ar4/syr/ar4_syr_spm.pdf> (accessed October 23, 2012).

[7] *Draft Climate Assessment Report* 25.

[8] Reto Knutti and Gabriele C. Hegerl, "The Equilibrium Sensitivity of the Earth's Temperature to Radiation Changes" (2008) 1 Nature Geoscience 735.

[9] Michael Herrmann, *Food Security and Agricultural Development* (UNCTAD Report No. 196, UNCTAD, Geneva 2009) 14, Fig. 5.

[10] Simon I. Hay et al., "Climate Change and the Resurgence of Malaria in the East African Highlands" (2002) 415 Nature 903.

[11] Stefan Rahmstorf, "A Semi-Empirical Approach to Projecting Future Sea-Level Rise" (2007) 315 Science 368.

[12] M.J. Booij, "Impact of Climate Change on River Flooding Assessed with Different Spatial Model Resolutions" (2005) 303 Journal of Hydrology 176.

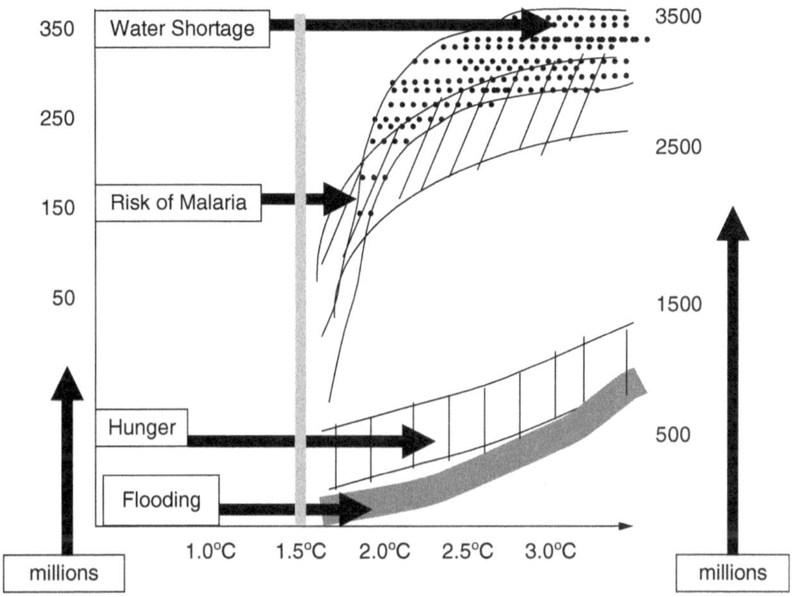

Figure 1.2 Hunger, malaria, flooding, and water shortage in 2080

Note: Flooding, Hunger, and Malaria risks are on the left scale and Water Shortage risk on the right scale.

Source: Adapted from N.W. Arnell et al, "The consequences of CO2 stabilisation for the impacts of climate change" (2002) 53 Climatic Change 413.

world.[13] On the right hand side vertical axis, we plot the impact on water shortage. While hunger, malaria, and flooding will affect tens of millions, the water shortage problem will affect hundreds of millions by 2080.

There will be more droughts in more places than before, more storms, more hurricanes, as well as more inundations. While these events will have an environmental impact, they also will have economic effects with implications for the financial industry. There will be irreversible losses on coral seas, and coastal land will be lost forever. It will lead to a change in the comparative attractiveness of cities where most people live. In the developing countries, cities with large, informally living populations will suffer. On the other hand, in developed countries, some cities that today are less desirable could become more desirable in the future.[14] This will have an economic impact: first, there will be a redistribution of wealth within countries. We know from the past that such redistributions can lead to regional tensions and civil wars. Thus, the manner in which the environmental impact of climate change is distributed among and within countries will also influence the

[13] Jens H. Christensen and Ole B. Christensen, "Severe Summertime Flooding in Europe" (2002) 421 Nature 805.

[14] Matthew Kahn, *Climatopolis: How Our Cities will Thrive in Our Hotter Future* (Basic Books, New York 2010).

political environment. That in turn will affect the laws and policies that take shape in international economic law.

1.3 Political Economy of Climate Change

Four big questions asked about climate change are: (1) how much global warming and climate change will occur; (2) how bad will it get; (3) when will all this occur; and (4) what should be done about it. Nicholas Stern provided a detailed analysis of the economic impact by taking into account future generations in a well-defined economic model.[15] His model posits all the future generations with very *low* discounting (the utilitarian approach). The model has been criticized by a number of economists on various grounds. Weitzman starts with a different approach. He argues that the standard cost–benefit analysis done in economics breaks down while analyzing climate change because there is too much uncertainty about too many parameters of any climate change model (deep structural uncertainty).[16] We will analyze these models in Chapter 6, along with that of Nordhaus, from a political economy perspective—the feasibility of implementing these changes given the real world constraints.[17]

UNFCCC Article 15(3) requires the parties to make every effort to reach agreement on any proposed amendment to the Convention by consensus. If all efforts at consensus have been exhausted, and no agreement reached, the amendment can be adopted by a three-fourths majority vote of the parties present and voting at the meeting, as a last resort. Similarly, WTO decisions are taken by consensus. While the WTO Agreement provides for majority voting, the practice to date has been to take all decisions by consensus. In both cases, a subset of parties can enter into a plurilateral agreement (in the case of the WTO) and a protocol (in the case of the UNFCCC). Multilateral negotiations have been stalled in recent years with respect to the UNFCCC and the WTO Doha Round, in part because of consensus-based decision-making.

As WTO Deputy Director Alejandro Jara has observed, this negotiation paralysis must be considered in the global economic context, which has been characterized by financial crises, economic deceleration and uncertainty. It also reflects a new global reality, in which the rise of major emerging economies has led the traditional economic powers to demand greater concessions and commitments from the emerging economies. These changes in the economic balance are likely to continue, making consensus increasingly difficult to achieve. In turn, the lack of multilateral agreement generates further uncertainty, with respect to both the UNFCCC and the WTO negotiations. In addition, the failure to advance in the UNFCCC

[15] Nicholas Stern, *The Economics of Climate Change: The Stern Review* (Cambridge University Press, Cambridge 2007).

[16] Martin L. Weitzman, "Insurance for a Warming Planet" (2010) 467 Nature 784.

[17] William D. Nordhaus, *A Question of Balance: Weighing the Options on Global Warming Policies* (Yale University Press, New Haven 2008).

negotiations prevents progress at the WTO with respect to the compatibility of multilateral and unilateral policy responses to climate change. Moreover, the climate of uncertainty leads to risk aversion, making it less likely that financing of green technologies will come from the private sector. However, public financing of green technologies will be uneven across countries and could lead to serious market distortions that will create further political and economic instability. As a result, it is important to identify policy issues and options and ways to overcome negotiation obstacles.[18]

One proposal, with respect to WTO negotiations, is to make negotiations less ambitious, by abandoning the rule that "Nothing is agreed until everything is agreed," and to abandon decision-making by consensus.[19] There are precedents for this approach at the WTO, with negotiations to liberalize specific sectors, such as information technology products[20] and basic telecommunication services.[21] With this approach, once enough Members are on board to cover 90 percent of trade in the sector, the agreement is finalized. The most-favored-nation rule extends concessions to all WTO Members and the resulting agreement is left open for other Members to join. A similar approach could be taken with UNFCCC negotiations, by seeking agreement among the countries that account for the overwhelming majority of emissions, and by leaving it open for other countries to join. Less ambition and more progress may work better than the more idealistic consensus-based approach and the negotiation paralysis at the WTO and UNFCCC.

International politics plays a big role in responses to climate change, just as it does with respect to international economic governance. In both cases, the divergence is not simply between the perspectives of developed and the developing countries. Rather, it depends on how the interests of countries are aligned, which vary with the specific issue. Thus, there may be not just a North–South divide, but North–North and South–South divides as well. International organizations can play a key role in bringing all the parties to the table. The Economics of Ecosystems and Biodiversity (TEEB) Report, released in October 2010, points to some common ground emerging between the developed and the developing countries.[22] However, policy responses that work in developed countries will not necessarily work in developing countries, for example when the effectiveness of those policy responses depends on the maturity of financial markets or the robustness of national regulatory institutions. Thus, differences in levels of economic development influence the design of global solutions both at the negotiation stage and at the implementation stage.

[18] Alejandro Jara, "El Medio Ambiente y la OMC" (2012) 2 Revista de Derecho Económico Internacional 5.

[19] "Goodbye Doha, hello Bali," The Economist (September 6, 2012) <http://www.economist.com/node/21562196> (accessed March 15, 2013).

[20] Information Technology Agreement <http://www.wto.org/english/tratop_e/inftec_e/inftec_e.htm> (accessed October 23, 2012).

[21] Results of the basic telecommunication negotiations <http://www.wto.org/english/tratop_e/serv_e/telecom_e/telecom_results_e.htm> (accessed October 23, 2012).

[22] TEEB, *Mainstreaming the Economics of Nature: A Synthesis of the Approach* (2010) <http://www.teebweb.org/publications/teeb-study-reports/synthesis/> (accessed March 15, 2013).

Another consideration is the manner in which international economic and environmental policies are formulated at the national level. For example, the processes and actors differ in China, the European Union, and the United States. In the United States, the private sector has a very significant role in the formulation and implementation of policies that affect their interests. In the European Union, while the private sector has an important role, civil society also has influence, which alters the policy outcomes with respect to the intersection of trade and environment. In China, the politburo determines policy. Thus, the interests of the actors and the processes have to be taken into account in determining the parameters of possible policies, with respect to both international economic relations and international environmental policy.[23] As Bruce Yandle described it, the process of regulating environmental protection can involve Baptist-and-bootlegger coalitions, in which different interest groups favor the same regulatory outcome for different reasons. Just as Baptists and bootleggers were both in favor of alcohol prohibition, domestic industry and environmentalists can both be in favor of environmentally motivated trade restrictions.[24]

Climate change regulation—whether based on regional and multilateral agreements or taken unilaterally—can be designed or implemented in a manner that benefits domestic industry and discriminates against imports or foreign investors. In subsequent chapters, we will examine such measures and the implications of their inconsistency with international trade agreements and international investment agreements. In addition to potential inconsistency with international economic law, regulatory capture of climate change regulation implies a preference for less efficient producers of adaptation and mitigation technology, which raises the cost of addressing climate change. In addition to trade and investment barriers, it is common to structure foreign aid in a manner that favors domestic industry, regardless of whether it is the best or most cost-effective technology.

National regulatory capture can lead to *de facto* international regulatory capture when the national government in question has disproportionate influence in international organizations, such as the World Bank, or disproportionate bargaining power in international negotiations, such as trade negotiations at the WTO or in regional trade agreements. The source of the influence or bargaining power may include the percentage of world trade of a country, the size of its market, its spending power with respect to foreign aid, the size and percentage of its monetary contributions to international organizations, and its ability to facilitate membership in a desirable economic group (such as the European Union or the WTO). The degree of bargaining power or influence for the same national government can vary with the context. For example, the bargaining power of the United States is greater in free trade negotiations with small developing countries than it is in the WTO context, where the bargaining power of the United States is offset by the bargaining

[23] Aluisio de Lima-Campos, "Políticas de Comercio y Medio Ambiente: En Busca de un Alineamiento" (2012) 2 Revista de Derecho Económico Internacional 35.

[24] Bruce Yandle, *The Political Limits of Environmental Regulation: Tracking the Unicorn* (Praeger, New York 1989).

power of the European Union or other coalitions of influential countries. Where an industry succeeds in capturing the regulatory process of one or more national governments, that regulatory capture can lead to the indirect capture of international regulatory processes, leading to international rules that favor the industry at the expense of global welfare.

Climate change regulation provides evidence of regulatory capture. For example, in the United States, during 2001–2008, the Senators and Congressmen who were elected from the states with big support from the energy lobby were powerful members of the committees that oversaw environmental regulation. This has led to the weakening of many laws (such as the laws governing Superfund sites) and to the failure to ratify international agreements (such as the Kyoto Protocol). These activities have not gone unchallenged. The State attorneys-general challenged in court the Bush administration's proposal to relax environmental standards for new plants or upgrades of industrial facilities. State attorneys-general have moved to the forefront of social regulatory activism, ahead of state legislatures, piggybacking on the popular product liability movement and their financial success in the tobacco cases.[25]

There is also an equity issue involved when we examine developed versus developing countries. Specifically, there is a clear difference between the developed and the developing countries for the following reason. The countries that were already developed by the end of the twentieth century had undertaken development with scant regard to the environment in the nineteenth and the twentieth centuries. Thus, there is a perception among (now) developing countries that these countries could achieve development relatively "cheaply."[26] It is perceived that the (now) developed countries are trying to impose restrictions on the development paths of the developing countries by asking them to industrialize with cleaner technologies that are more costly. The developing countries are not willing to bear that cost. Until recently, the developing countries have not mounted any unified effort to deal with these issues. But now there are signs that they are.[27]

1.4 Mitigation, Adaptation, and Global Governance

It is useful to think about climate change and global governance in terms of adaptation and mitigation. Climate change mitigation seeks to reduce the rate and magnitude of climate change by reducing greenhouse gas emissions or

[25] Paul Teske, "The New Role and Politics of State Regulation" (2004) 27 Regulation <http://www.cato.org/pubs/regulation/regv27n3/v27n3-1.pdf> (accessed March 15, 2013).

[26] Ross Garnaut et al., "Emissions in the Platinum Age: the implications of rapid development for climate-change mitigation" (2008) 24 Oxford Review of Economic Policy 377.

[27] Mike Goldblatt and Julie Middleton, "Climate Policy Coherence and Institutional Coordination: Clarifying Institutional Responsibilities in South Africa," Basic-Project Paper 13 (September 2007) <http://www.basic-project.net/data/final/Paper13South%20Africa%20Climate%20Policy%20Coherence%20and%20Instituti%85.pdf> (accessed March 15, 2013).

enhancing the absorption of carbon or carbon dioxide from the atmosphere by "sinks," such as oceans or forests. Adaptation seeks to diminish the negative impacts of climate change by increasing the ability of humans or ecosystems to cope with the changes. Mitigation and adaptation deal with different aspects of the risks imposed by climate change, but they are interrelated. Since mitigation would prevent the worst-case scenario from occurring, it increases the chances that the remaining climate risks can be successfully managed through adaptation. The benefits of mitigation are global, but the cost of reductions is primarily local. In contrast, adaptation costs and benefits are both primarily local. Mitigation benefits are long-term, whereas adaptation benefits are short- and medium-term. Because of these differences, national and international climate change response efforts have followed separate mitigation and adaptation paths, with a major focus on mitigation. However, some climate change impacts are unavoidable now, so adaptation efforts have begun. Nevertheless, the technological and financial capacity of countries to adapt to climate change differs significantly, as does their vulnerability to the effects of climate change.[28] These differences among countries are part of the reason for incorporating the principle of common but differentiated responsibilities into the UNFCCC.[29]

Incentives to take mitigation and adaptation measures are asymmetrical, since the costs and benefits of these measures are not the same for all countries.[30] Developing countries may insist that developed countries take responsibility for the cost of mitigation and adaptation. Regardless, at some point, some developing countries will have to focus on adaptation at the national level to address climate-related catastrophes, with or without the help of the developed countries. Some developed countries might find adaptation less costly than mitigation and just leave the most vulnerable to fend for themselves, absent significant spillovers, such as climate refugees. However, discussing the costs and benefits of mitigation and adaptation in terms of developed and developing countries ignores the global nature of the problem, the potential seriousness of the consequences and the great variation among countries in their vulnerability to, and their capacity to adapt to, the effects of climate change.

In the context of climate change, global governance means risk governance and international processes, institutions, and legal regimes.[31] Risk management requires identifying and assessing risks, then finding ways to mitigate and adapt. Scientists identify and assess risks. Mitigation and adaptation have financial and economic implications. Global governance also requires the coordination of legal regulation in many areas, which involves both international environmental law and

[28] WTO and UNEP, *Trade and Climate Change* (WTO Secretariat, Geneva 2009) 24–6.

[29] Christopher D. Stone, "Common but Differentiated Responsibilities in International Law" (2004) 98 AJIL 276.

[30] Hanh H. Danga, Axel Michaelowa, and Dao D. Tuan, "Synergy of Adaptation and Mitigation Strategies in the Context of Sustainable Development: The Case of Vietnam" (2003) 3 Climate Policy (Supplement 1) S81.

[31] Carolyn Deere-Birkbeck, "Global Governance in the Context of Climate Change: The Challenges of Increasingly Complex Risk Parameters" (2009) 85 International Affairs 1173.

international economic law. In turn, legal regulation implies international negotiation, avoiding regulatory capture and coordination between global institutions.

There is a synergy between mitigation and adaptation, since the former facilitates the latter. For example, reinforcing dykes in anticipation of rising sea levels and storm surges is a precautionary measure and an adaptation measure.

1.5 Technology Transfer for Mitigation and Adaptation

Debates between developing and developed countries over access to technology to mitigate or adapt to climate change tend to overlook the importance of plant varieties. Climate change will increase the importance of the development of new plant varieties that can adapt to changing climactic conditions. Chapter 5 compares intellectual property rights (IPRs) for plant varieties in the WTO Agreement on Trade-Related Aspects of Intellectual Property Rights (TRIPS), the International Convention for the Protection of New Varieties of Plants (UPOV Convention), and the Convention on Biological Diversity. It concludes that TRIPS Article 27.3(b) provides an appropriate degree of flexibility regarding the policy options available to facilitate adaptation to climate change.

Clean energy technologies are often cited as an example of the kind of technology that needs to be developed and transferred internationally in order to combat climate change. In international debates regarding the effect of IPRs on the transfer of environmental technologies, developing countries often draw upon the experience regarding pharmaceutical patents. However, IPRs play a different role in the renewable energy industries than they do in the pharmaceutical sector, and seem less likely to create barriers to technology access.[32] New plant varieties represent another important technology that developing countries, in particular, will need in order to adapt to the effects of climate change.[33] The applicable intellectual property laws and the technology transfer issues are different for biotechnologies such as plant varieties, where IPRs may create barriers to access that are similar to the pharmaceutical sector. Thus, it is not possible to analyze the subject of intellectual property rights and international technology transfer in a generalized manner. The analysis must be done according to specific categories of technology. In the case of clean energy technologies, the availability of competing technologies will diminish the impact of IPRs on their cost. In the case of new plant varieties, where a technology has no or few substitutes, or IPRs are concentrated in the hands of relatively few firms, IPRs will increase costs due to monopoly pricing power.

[32] John H. Barton, "Intellectual Property and Access to Clean Energy Technologies in Developing Countries: An Analysis of Solar Photovoltaic, Biofuel and Wind Technologies" (2007) ICTSD Programme on Trade and Environment, Issue Paper No. 2 <http://ictsd.org/i/publications/3354/> (accessed March 15, 2013).

[33] The transfer of agricultural technology for adaptation to climate change is discussed in Intergovernmental Panel on Climate Change, *Methodological and Technological Issues in Technology Transfer*, ch. 11 <http://www.grida.no/publications/other/ipcc_sr/?src=/climate/ipcc/tectran/index.htm> (accessed April 1, 2012).

In general, developing countries are more vulnerable to the impacts of climate change on agriculture and subsistence farmers. The impact of climate change on agriculture will affect developing countries disproportionately, for several reasons. Climate change will have a greater impact on the viability of traditional plant varieties in developing countries in the tropics than in developed countries in temperate zones. This means that the need for genetically modified organism (GMO) seeds will be greater in developing countries. However, a larger percentage of the population depends on agriculture in developing countries (50 percent in India, for example) and the poorest in developing countries depend on subsistence agriculture. Their poverty means that they rely on collecting seeds from traditional plant varieties to sow future crops. However, climate change will make these varieties increasingly untenable. As production in these varieties decline, developing countries will require greater access to GMO varieties that can raise yields and adapt to climate change. However, the vast majority of the rights to these GMO varieties belong to corporations in the United States, Europe, and Japan. The percentage of GMO crops owned by Monsanto alone is: 91 percent of soy, 97 percent of maize, 63 percent of cotton, and 59 percent of canola. Thus, the food security of developing countries will depend on access to technology from developed countries, in particular private companies from developed countries. This will give a company like Monsanto tremendous bargaining power over developing country governments.[34] The six large multinational companies (DuPont, BASF, Monsanto, Syngenta, Bayer and Dow) not only own the GMO crops, but they have patented 77 percent of all "climate ready crop genes" during June 2008 and June 2010.[35]

Innovations in plant breeding play an important role in a number of public objectives, such as food security, environment, sustainability, and transitions in the rural economy. Farmers and growers have an interest in competition in the seed market. However, stronger intellectual property rights have combined with technological developments to produce increasing consolidation among breeding companies. Strengthening intellectual property rights may conflict with development objectives. Strengthening intellectual property rights, by contributing to a decreasing diversity in breeding companies, also threatens innovation in plant breeding.[36] These negative impacts of intellectual property rights will become more apparent as climate change begins to have a greater impact on crop yields. The combination of

[34] Peter Drahos and Michael Blakeney (eds.), *IP in Biodiversity and Agriculture: Regulating the Biosphere* (Sweet & Maxwell, London 2001); Carliene Brenner, *Integrating Biotechnology in Agriculture: Incentives, Constraints, and Country Experiences* (OECD, Paris 1996); OECD-FAO, *Agricultural Outlook 2008–2017* (OECD-FAO, Paris 2008); Michael Blakeney, *Intellectual Property Rights and Food Security* (CABI, Wallingford, MA 2009); OECD, *The Bioeconomy to 2030: Designing a Policy Agenda* (OECD, Paris 2009); OECD, *Biotechnologies in Agricultures and Related Natural Resources to 2015* (OECD, Paris 2009).

[35] ETC Group, "Gene Giants Stockpile Patents on 'Climate-ready' Crops in Bid to become 'Biomassters'" (2010) <http://www.etcgroup.org/sites/www.etcgroup.org/files/publication/pdf_file/Genegiants2011_0.pdf> (accessed March 15, 2013).

[36] Niels Louwaars et al., "Breeding Business: The Future of Plant Breeding in the Light of Developments in Patent Rights and Plant Breeder's Rights" (2009) Netherlands Ministry of Agriculture, Nature and Food Quality (LNV), Wageningen, Centre for Genetic Resources (CGN) <http://ssrn.com/abstract=1720088> (accessed March 2, 2012).

climate change, increasing intellectual property rights for new plant varieties and growing demand for food supplies raises serious concerns regarding affordable access to new biotechnologies and food staples in developing countries. Technological developments may also require a reassessment of policies regarding intellectual property rights for new plant varieties. Genetic use restriction technology or terminator genes render the harvested seed sterile. This technology prevents farmers from replanting saved seed and thereby consolidates the seed companies' monopoly. When applied to seeds that are protected by intellectual property rights, this technology not only prevents infringement of intellectual property rights and farmers' reuse of seeds, but also ensures the continuation of the monopoly beyond the life of any patent or breeder's rights.[37] The advent of terminator genes raises the issue of whether, rather than strengthening intellectual property rights, governments should focus on strengthening competition laws to enhance access to new plant varieties that can adapt to climate change. Indeed, as Debra M. Strauss observes, the application of IPRs to biotechnology may undermine the public interest in the security of a global food supply.[38]

Some have proposed the negotiation of an agreement on intellectual property rights on technologies necessary for mitigation efforts in developing countries, based on the WTO decision on compulsory licensing of pharmaceuticals.[39] The benefit of this approach is that the negotiation could take place separately from the negotiation of the Doha Round. Others have argued that pharmaceuticals and environmental technologies are too different to adopt the same approach in the two areas, but have focused mainly on clean energy technologies.[40] It is true that, for environmental technologies such as those that address ozone depletion, the nature of international business strategies differs significantly from those in the pharmaceutical sector.[41] However, the extent to which solutions in one field of technology can be applied to another must take into account the specific legal and economic issues that arise in each area of technology.

In spite of the different implications IPRs may have in different technologies, the debate over IPRs and technology transfer tends to divide along North–South lines,

[37] Enrico Bonadio, "Crop Breeding and Intellectual Property in the Global Village" (2007) 29(5) European Intellectual Property Review 167.

[38] Debra M. Strauss, "The Application of TRIPS to GMOs: International Intellectual Property Rights and Biotechnology" (2009) 45 Stanford Journal of International Law 287, 291.

[39] H.A.C. Prasad and J.S. Kochher, "Climate Change and India—Some Major Issues and Policy Implications" (2009) Working Paper No. 2/2009-DEA, Department of Economic Affairs, Ministry of Finance, Government of India, March 2009 <http://finmin.nic.in/WorkingPaper/Working%20paper%20Climate%20Change.pdf> (accessed March 12, 2012).

[40] Jérôme de Meeûs, "Patent Rules under Scrutiny in the United Nations Negotiations on Climate Change and Technology Transfer: Analysis and Observations," Paper presented at the 6th Annual Conference of the European Policy for Intellectual Property (EPIP) Association Fine-Tuning IPR Debates, Brussels, September 8, 2011.

[41] Jayashree Watal, "Case Study 3 India: The Issue of Technology Transfer in the Context of the Montreal Protocol" in Veena Jha and Ulrich Hoffman (eds), *Achieving Objectives of Multilateral Environmental Agreements: A Package of Trade Measures and Positive Measures* (United Nations Conference on Trade and Development UNCTAD/ITCD/TED/6) 45 <http://www.unctad.org/en/docs/itcdted6_en.pdf> (accessed March 15, 2013).

whether with respect to clean energy technologies, new plant varieties, or other environmental technologies. In climate change negotiations, developing countries push for financing to acquire technology or relaxation of IPRs to lower the cost of acquiring technology, whereas developed countries tend to defend IPRs. This reflects the concentration of technologies and IPRs in developed countries and the prediction that there will be disproportionate impacts of climate change on developing countries. However, the role that IPRs play in technology transfer in the more scientifically advanced developing nations will be significantly different in poorer, less scientifically advanced nations.[42]

As technologies mature and IPRs expire, their cost will go down. However, newer technologies may be more effective and, hence, more desirable. Many of the relevant technologies are owned by private interests in developed countries, not governments, and may be acquired by private interests in developing countries. Private firms are not likely to transfer technology in a way that would hamper their competitiveness and would be concerned about creating competitors among the recipients of the technologies.

Once we recognize that the environment at any level is a public good, we understand that the same problems of pricing public goods apply with full force. The environmental externality is negative (that is, my action causes harm to others with whom I have no direct market transactions). Therefore, solutions through private markets will not yield socially optimal results. There have been attempts to produce solutions along the line argued by Ronald Coase. More recent literature (propelled by David Levine and Michele Boldrin) has argued that the only solution to the problem of intellectual monopoly is to get rid of it entirely. Thus, patent rights on innovative technological solutions to the problem of climate change can be mitigated through *not* granting patents (intellectual monopoly) at every turn. Levine and Boldrin have shown that the more global the market is the less there is a case for an intellectual monopoly.[43]

Striking the right balance between incentives to invent new technologies and affordable access to those technologies is a key issue in addressing climate change. The theoretical foundation for IPRs in general, and patents in particular, lies in the economic argument that these monopoly rights are the *sine qua non* of innovation. In essence, this argument states that, without patents, inventions would cease, making the issue of affordable access to technology a moot point. The same theoretical argument underpins the intellectual property protection provided by breeders' rights or patents for new plant varieties. We have demonstrated elsewhere that this argument weakens considerably when it comes to incentives to invent treatments for poor country diseases.[44] Biotechnology with respect to plant var-

[42] Barton, "Intellectual Property and Access to Clean Energy Technologies in Developing Countries."

[43] Michele Boldrin and David K. Levine, "IP and Market Size" <http://www.micheleboldrin.com/Papers/scalenew.pdf> (accessed March 21, 2012).

[44] Bradly J. Condon and Tapen Sinha, "Global Diseases, Global Patents and Differential Treatment in WTO Law: Criteria for Suspending Patent Obligations in Developing Countries" (2005) 26 Northwestern Journal of International Law and Business 1.

ieties is akin to pharmaceuticals for HIV/AIDS, since demand for the technology to grow weather- and plague-resistant plant varieties is global, and there is sufficient purchasing power of the farmers in *developed* countries to create adequate incentives to invent the technologies. Similar issues arise regarding the effect of monopoly rights on affordable access to the technology.

With respect to clean energy technologies, while IPRs may have a negative impact on innovation, competition, and affordable access, we argue that the focus of the debate should not be on IPRs for clean energy technologies. First, achieving reforms to the international intellectual property regime is likely to prove difficult. Second, IPRs do not represent the main obstacle to innovation, competition and affordable access for clean energy technologies. For example, the fact that the United States has applied countervailing duties on imports of solar panels from China indicates that IPRs are not a sufficient barrier to competition in this sector. Otherwise, countervailing duties would not be necessary to protect the US solar panel industry from Chinese competition.

With respect to clean energy technologies, the debate on IPRs is misplaced and distracts from the real issues: creating incentives for and removing obstacles to clean energy development and dissemination. In particular, as we argue in Chapter 8, fossil fuel subsidies need to be eliminated and that source of financing reallocated to clean energy technologies. This may require reforms to WTO subsidies law. Countries also need to remove barriers to trade in clean energy technologies, rather than erect such barriers. This, too, may require reforms to WTO subsidies law, although reforms to national countervailing duty law could be undertaken without reforming WTO law. WTO law does not require countries to apply countervailing duties; it merely permits this practice as long as it is done in accordance with the requirements of the SCM Agreement. Dissemination of clean energy technologies can also be facilitated by removing barriers to foreign investment and international trade in services. For these reasons, our analysis of clean energy technology transfer focuses on these issues, rather than on IPRs.

1.6 Themes in Subsequent Chapters

In Chapter 2, we examine a series of international environmental principles that enjoy broad acceptance, as evidenced by their inclusion in the1992 Rio Declaration on Environment and Development. We analyze how these principles find expression in international climate change law and WTO law, in order to determine the legal parameters that unilateral measures must respect to be consistent with both international legal regimes. This analysis shows that there is no inherent inconsistency between the climate change regime and WTO law. This discussion will inform more detailed analyses in subsequent chapters of international economic law (international trade law, international investment law, and international intellectual property law).

Governments have the right to regulate matters that fall within their jurisdiction. They have agreed to restrict that right in international treaties on trade, investment, and intellectual property rights. Thus, the right to regulate is constrained to the extent that the conduct in question is specifically covered by international treaty obligations[45] or other sources of international legal obligations. Therefore, to determine the scope of the right to regulate requires the interpretation of those international treaty obligations to determine their scope and meaning. This task has become increasingly complex with the proliferation of international treaties, many of which have not been judicially interpreted. In addition, these international treaties must be interpreted in light of each other's content, in order to avoid conflicting interpretations. The complexity of this legal analysis represents an additional obstacle to the effective regulation of climate change, particularly since the relevant treaties that regulate international economic law were not drafted with climate change in mind.

The legal constraints require climate change policies to be implemented within the limits imposed by international obligations regarding trade and foreign investment. Compliance with the WTO Agreement on Technical Barriers to Trade and the General Agreement on Tariffs and Trade (GATT) will require careful design of measures such as standards, carbon labels, carbon taxes, and border tax adjustments. Similarly, governments will need to respect the provisions of international investment agreements to avoid the cost of compensating foreign investors. We address these issues in Chapters 3 and 4.

Technological innovation will have to play a central role in addressing climate change mitigation and adaption. We do not focus on the policies needed to spur such technological innovation nor the scientific challenges. Rather, we focus on what is needed to overcome obstacles to technological diffusion, which are legal, economic, financial, and political impediments or constraints. Impediments refer to obstacles that may foreclose certain policy options. Constraints refer to parameters within which policies must be designed and implemented.

The legal impediments affect policies regarding intellectual property and subsidies. With respect to intellectual property, there is no one-size-fits-all solution; the solution depends on the nature of the technology in question. This means that flexibility is important regarding the policy options that countries have at their disposal. However, a growing number of free trade agreements have introduced "TRIPS plus" provisions that reduce flexibility in this regard, which may affect the affordability of some climate change technologies. We analyze the potential effects of the TRIPS Agreement and the UPOV Convention on access to climate resistant plant varieties in Chapter 5.

Chapter 6 applies the concepts of environmental economics to the problem of climate change. Greenhouse gas emissions create a negative economic externality, by imposing the costs of climate change on society. Climate change regulation seeks to internalize this cost by adjusting the market price for carbon emissions. Unilateral measures represent only a partial solution because they will not achieve the desired level of emissions reductions. However, they can still be used to

[45] Robert Jennings and Arthur Watts (eds.), *Oppenheim's International Law* (9th edn., Longman, London 1996) vol. 1, 931–2.

create incentives for multilateral action, which is what is needed to effectively address the risks associated with climate change. Chapter 6 examines the macroeconomic models used to calculate those risks.

International treaty obligations are not the only constraints on climate change regulation. Governments also face financial constraints, which will also influence the design of climate change regulation. Chapter 7 analyzes the role of financial markets in addressing climate change, examining the lessons learned from the SO2 market and the EU emissions trading system to consider how CO2 mitigation will work in the future. Chapter 7 also considers the role of insurance in climate change adaptation. The catastrophic risk markets in the reinsurance industry give us a market-based way to address the adaptation problem.

Existing WTO law also restricts the use of subsidies and permits the application of countervailing measures to subsidized imports, both of which have implications for investments in technology development and dissemination. We examine climate finance in Chapter 8 and consider sources of financing and constraints on financing mechanisms in WTO subsidies law.

Appendix to Chapter 1
Climate Change and the American People

Climate change, once considered an issue for a distant future, has moved firmly into the present. This report of the National Climate Assessment and Development Advisory Committee concludes that the evidence for a changing climate has strengthened considerably since the last National Climate Assessment report, written in 2009. Many more impacts of human-caused climate change have now been observed. Corn producers in Iowa, oyster growers in Washington State, and maple syrup producers in Vermont have observed changes in their local climate that are outside of their experience. So, too, have coastal planners from Florida to Maine, water managers in the arid Southwest and parts of the Southeast, and Native Americans on tribal lands across the nation.

Americans are noticing changes all around them. Summers are longer and hotter, and periods of extreme heat last longer than any living American has ever experienced. Winters are generally shorter and warmer. Rain comes in heavier downpours, though in many regions there are longer dry spells in between.

Other changes are even more dramatic. Residents of some coastal cities see their streets flood more regularly during storms and high tides. Inland cities near large rivers also experience more flooding, especially in the Midwest and Northeast. Hotter and drier weather and earlier snow melt mean that wildfires in the West start earlier in the year, last later into the fall, threaten more homes, cause more evacuations, and burn more acreage. In Alaska, the summer sea ice that once protected the coasts has receded, and fall storms now cause more erosion and damage that is severe enough that some communities are already facing relocation.

Scientists studying climate change confirm that these observations are consistent with Earth's climatic trends. Long-term, independent records from weather stations, satellites, ocean buoys, tide gauges, and many other data sources all confirm the fact that our nation, like the rest of the world, is warming, precipitation patterns are changing, sea level is rising, and some types of extreme weather events are increasing. These and other observed climatic changes are having wide-ranging impacts in every region of our country and most sectors of our economy. Some of these changes can be beneficial, such as longer growing seasons in many regions and a longer shipping season on the Great Lakes. But many more have already proven to be detrimental, largely because society and its infrastructure were designed for the climate of the past, not for the rapidly changing climate of the present or the future.

This National Climate Assessment collects, integrates, and assesses observations and research from around the country, helping to show what is actually happening and what it means for people's lives, livelihoods, and future. This report includes analyses of impacts on seven selected sectors: human health, water, energy, transportation, agriculture, forests, and ecosystems and biodiversity. This report additionally focuses on the interactions among several sectors at the national level. It also assesses key impacts on the regions of the US: Northeast, Southeast and Caribbean, Midwest, Great Plains, Southwest, Northwest, Alaska and the Arctic, Hawaii and the Pacific Islands; as well as coastal areas, oceans, and marine resources. Finally, this report is the first to explicitly assess the current state of adaptation, mitigation, and decision support activities.

Climate change presents a major challenge for society. This report and the sustained assessment process that is being developed represent steps forward in advancing our understanding of that challenge and its far-reaching implications for our nation and the world.[46]

[46] "Letter to the American People," Draft Climate Assessment Report of the National Climate Assessment Development Advisory Committee (January, 11 2013) <http://ncadac.globalchange.gov/download/NCAJan11-2013-publicreviewdraft-letter.pdf> (accessed January 12, 2013).

2

International Legal Framework

2.1 Introduction

This chapter will analyze the regulation of climate change within the existing international legal framework. There are three main sources of international climate change law: treaties, customary international law, and general principles of law.[1] In addition, international jurisprudence has become an increasingly important source of international law and plays an important role in the interpretation of treaties, the recognition of customary international law, and the clarification of general principles of law. International climate change regulation involves many different fields of international law, including international environmental law and international economic law; it also involves rules of customary international law, but it is not always clear which rules have attained the status of custom. International climate change regulation further involves general principles of law that may be vague or that may appear to be irreconcilable. Thus, while many bemoan the lack of progress towards international agreements (treaties) on how to address climate change, the reality is that there are multiple sources of international climate change regulation and this makes the process of interpreting and applying international climate change law particularly complex.

Our approach in this chapter is to analyze the existing international agreements on climate change, principally the UNFCCC, according to the customary rules of interpretation of treaties set out in the Vienna Convention on the Law of Treaties.[2] However, this does not mean that treaty law is the only source of law being analyzed. The UNFCCC contains a lengthy preamble, sets out an objective, incorporates a series of principles, and establishes obligations. The preamble, objective, and principles do not establish binding obligations, but they are relevant to the interpretation of the obligations set out in the treaty. In addition to the text of the treaty, the interpretative context must be considered, including agreements and instruments relating to the treaty. In addition, the treaty interpreter must take into account any subsequent agreement or subsequent practice regarding the interpretation of the treaty or the application of its provisions and the relevant

[1] Statute of the International Court of Justice, Art. 38 <http://www.icj-cij.org/documents/?p1=4&p2=2&p3=0> (accessed April 19, 2013).

[2] *Vienna Convention on the Law of Treaties*, opened for signature May 23, 1969, 1155 UNTS 331 (entered into force January 27, 1980) Arts. 31–3.

rules of international law. Finally, we can use supplementary means of interpretation, including the preparatory work of the treaty and the circumstances of its conclusion, and consider how the treaty terms are expressed in different languages. In other words, we must consider what the UNFCCC says, what it does not say, and what it does or does not say in Arabic, Chinese, English, French, Russian, and Spanish.

Regarding what the UNFCCC does not say, we will consider "agreements" and "instruments" relating to the treaty and "relevant rules of international law." Any agreement that we consider as relating to the treaty must have been made between all the parties in connection with the conclusion of the treaty. Any instrument must have been made by one or more parties in connection with the conclusion of the treaty and accepted by the other parties as an instrument related to the treaty. Any relevant rules of international law must be applicable in the relations between the UNFCCC parties, but not necessarily all of them. The UNFCCC was concluded at the UN Conference on Environment and Development in Rio de Janeiro in 1992, as was the Rio Declaration and Agenda 21.[3] We will consider provisions in the latter documents in our interpretation of the UNFCCC, as related instruments, sources of relevant rules or supplementary means of interpretation, depending on the provision. In addition, the Uruguay Round of Multilateral Trade Negotiations was well under way in 1992, the results of which now apply to 159 States (as of March 2013). Therefore, we will consider WTO law to be a source of relevant rules of international law applicable in the relations between the UNFCCC parties or, at the very least, part of the circumstances of the conclusion of the UNFCCC.

General international law can sometimes be used to fill in the gaps in treaties. However, general principles cannot displace clear treaty language. Moreover, while some principles have been recognized as forming part of customary international law, others have not. A treaty interpreter must take care not to exaggerate the importance of non-binding principles, which can guide actions without obligating governments, and to carefully consider the role that they might play in the interpretative process. At the same time, a treaty interpreter must avoid conflicting interpretations of obligations in different fields of international law, and general principles may prove useful in this regard. We will structure our analysis of the UNFCCC in this chapter according to a series of general principles, since this will facilitate an analysis of the UNFCCC that takes into consideration its relation to other instruments and agreements and notes both what it does and does not say.

Several principles of international environmental law are relevant to climate change. The 1992 Rio Declaration sets out 27 principles regarding sustainable development and related international environmental principles and norms, including the precautionary principle, common but differentiated responsibility (CDR), polluter pays, environmental assessment, responsibility for transboundary environmental damage, and the duty to cooperate to address transboundary and global environmental problems. All are relevant to the international law of climate

[3] *Agenda 21*, in Report of the United Nations Conference on Environment and Development, Rio de Janeiro, UN Doc. A/CONF.151/26/Rev.1 (Vol. 1), Annex II (1992).

change and have a role to play in defining the relationship between international climate change law and various aspects of international economic law that we address in subsequent chapters.

2.2 UN Framework Convention on Climate Change

In 1990, the Ministerial Statement of the Second World Climate Conference recognized the concept of climate change as a common concern of humankind, the principle of equity and the common but differentiated responsibility of countries at different levels of development, the concept of sustainable development, and the precautionary principle. The Ministerial Statement also called for the elaboration of a framework treaty on climate change and the necessary protocols. The Intergovernmental Panel on Climate Change (IPCC) had completed its First Assessment Report for the Conference, which later provided input for the first session of the International Negotiating Committee for a Framework Convention on Climate Change in 1991.[4] The UNFCCC opened for signature at the 1992 UN Conference on Environment and Development in Rio de Janeiro and entered into force in 1994. The 1992 Conference also produced the Rio Declaration, a set of 27 international environmental protection principles.

UNFCCC Article 2 sets out the "ultimate objective" of the Convention, and any related legal instruments that the Conference of the parties may adopt, as follows:

> to achieve, in accordance with the relevant provisions of the Convention, stabilization of greenhouse gas concentrations in the atmosphere at a level that would prevent dangerous anthropogenic interference with the climate system. Such a level should be achieved within a time frame sufficient to allow ecosystems to adapt naturally to climate change, to ensure that food production is not threatened and to enable economic development to proceed in a sustainable manner.

At the UNFCCC Conference of Parties (COP) 18, while the parties reiterated their determination to achieve the ultimate objective, they also reaffirmed that "adaptation must be addressed with the same priority as mitigation" and that "enhanced action and international cooperation on adaptation is urgently required."[5] It now appears increasingly unlikely that the ultimate objective of mitigation will be

[4] UNEP, The Second World Climate Conference <http://unfccc.int/resource/ccsites/senegal/fact/fs221.htm> (accessed August 20, 2012).

[5] The Cancun Agreements: Outcome of the work of the Ad Hoc Working Group on Long-term Cooperative Action under the Convention, 1/CP.16, Report of the Conference of the Parties on its sixteenth session, held in Cancun from 29 November to 10 December 2010, FCCC/CP/2010/7/Add.1, March 15, 2011 <http://unfccc.int/resource/docs/2010/cop16/eng/07a01.pdf#page=2> (accessed December 29, 2012); Decision 1/CP.18, Agreed outcome pursuant to the Bali Action Plan, Report of the Conference of the Parties on its eighteenth session, held in Doha from 26 November to 8 December 2012, Addendum, Part Two: Action taken by the Conference of the Parties at its eighteenth session (Advance version) <http://unfccc.int/resource/docs/2012/cop18/eng/08a01.pdf> (accessed April 4, 2013).

achieved.[6] Nevertheless, the ultimate objective still informs the interpretation of the UNFCCC and any related legal instruments that the Conference of the Parties may adopt. It should also inform the interpretation of international economic law, in particular provisions that set out exceptions for measures that are necessary to protect human, animal and plant life or health and to conserve exhaustible natural resources (including the climate itself).

UNFCCC Article 3 provides that, in their actions to achieve the objective of the Convention and to implement its provisions, the parties shall be guided, *inter alia*, by a series of principles. Article 4 is the source of the UNFCCC's principal obligations. Article 4 sets out commitments, including GHG reductions, financing, and technology transfer. We analyze these principles and commitments in greater detail below, in the broader context of international environmental law and international economic law. Here, we simply provide an overview of the UNFCCC and related instruments.

The UNFCCC divides the parties into three categories, each of which is placed in a different annex to the Convention. Annex I parties include the industrialized countries and countries with economies in transition. Annex II lists the OECD members of Annex I and excludes the countries with economies in transition. Developing countries (non-Annex I) are the third category, and the least developed receive special consideration. Mitigation is the main focus of the UNFCCC and the Kyoto Protocol to the UNFCCC, but UNFCCC Articles 4.1, 4.3, and 4.4 address adaptation, and Article 12 of the Kyoto Protocol sets up an Adaptation Fund generated from a levy on clean development mechanism (CDM) projects to help the most vulnerable countries to meet the costs of adaptation. In addition, United Nations Development Program (UNDP) has developed an Adaptation Policy Framework.[7] Moreover, the parties have increased adaptation activities, with the Cancun Adaptation Framework, an Adaptation Committee, a work program on loss and damage, and a national adaptation plan process for developing countries.[8]

The 1997 Kyoto Protocol set binding targets for 37 industrialized countries and the European Community to reduce GHG emissions. The initial commitments expired in 2012. Kyoto Protocol Article 3(1) expresses the central obligation in these terms:

> The Parties included in Annex I shall, individually or jointly, ensure that their aggregate anthropogenic carbon dioxide equivalent emissions of the greenhouse gases listed in Annex A do not exceed their assigned amounts, calculated pursuant to their quantified emission limitation and reduction commitments inscribed in Annex B and in accordance with the provisions of this Article, with a view to reducing their overall emissions of such gases by at least 5 per cent below 1990 levels in the commitment period 2008 to 2012.

[6] Potsdam Institute for Climate Impact Research and Climate Analytics, *Turn Down the Heat: Why a 4° Warmer World Must be Avoided* (World Bank, Washington 2012).

[7] Fabio Morosini, "Trade and Climate Change: Unveiling the Principle of Common but Differentiated Responsibilities from the WTO Agreements" (2010) 42 George Washington International Law Review 713.

[8] Decision 1/CP.18, Agreed outcome pursuant to the Bali Action Plan.

Kyoto Protocol Article 3(9) provides that commitments for subsequent periods for Annex I parties be considered at the COP by 2006. At the 2012 COP, Australia and European countries extended their Kyoto commitments. The second commitment period runs from 1 January 2013 to 31 December 2020.[9]

Kyoto Protocol Article 3(14) requires each Annex I party to "strive to implement" its commitments in such a way as to minimize adverse social, environmental and economic impacts on developing country parties, and to consider issues such as the establishment of funding, insurance and transfer of technology.

The Kyoto parties can meet their targets through three market-based mechanisms: the carbon emissions trading market, joint implementation, and the clean development mechanism. The CDM allows industrialized countries to invest in projects that reduce emissions in developing countries as an alternative to domestic reductions. While 192 parties ratified the Kyoto Protocol, the United States did not, due to the absence of binding targets for developing countries and concerns regarding its economic impact.[10] Canada withdrew from the Kyoto Protocol, effective 15 December 2012. Japan and the Russian Federation formally notified the secretariat that they do not intend to make commitments after 2012.[11]

The UNFCCC parties hold annual Convention of the Parties, at which progress has been painfully slow. In 2007, in Bali, the Kyoto parties launched the Adaptation Fund, and made progress on technology transfer and reducing emissions from deforestation. In 2009, the Copenhagen Accord made progress on emissions cuts, financing, and verification, but not on technology transfer. It also adopted a new deforestation program. However, it failed to create an international treaty or any binding emissions caps.

In 2010, the Cancun Agreements set emissions mitigation targets for 2020 for both developed and developing countries. The participating countries agreed to keep temperature increases below a global average of 2°C, having noted this goal in Copenhagen. They also established a "Green Climate Fund" to finance mitigation and adaptation to climate change, and new initiatives to protect tropical forests and to transfer clean energy technology to poorer countries. They also adopted the Cancun Adaptation Framework.[12] The most recent COPs took place in Durban

[9] Amendment to the Kyoto Protocol pursuant to its Article 3(9), Report of the Ad Hoc Working Group on Further Commitments for Annex I Parties under the Kyoto Protocol, Draft decision proposed by the President, Conference of the Parties serving as the meeting of the Parties to the Kyoto Protocol, Eighth session, Doha, November 26–December 7, 2012, FCCC/KP/CMP/2012/L.9, December 8, 2012 <http://unfccc.int/resource/docs/2012/cmp8/eng/l09.pdf> (accessed December 28, 2012); Roger Harrabin, "UN climate talks extend Kyoto Protocol, promise compensation," BBC News, December 8, 2012 <http://www.bbc.co.uk/news/science-environment-20653018> (accessed December 28, 2012).

[10] Morosini, "Trade and Climate Change."

[11] Amendment to the Kyoto Protocol pursuant to its Article 3(9).

[12] The Cancun Agreements: Outcome of the work of the Ad Hoc Working Group on Long-term Cooperative Action under the Convention, 1/CP.16, Report of the Conference of the Parties on its sixteenth session, held in Cancun from November 29 to December 10, 2010, FCCC/CP/2010/7/Add.1, March 15, 2011 <http://unfccc.int/resource/docs/2010/cop16/eng/07a01.pdf#page=2> (accessed December 29, 2012); Robert N. Stavins, "Why Cancun Trumped Copenhagen: Warmer Relations on Rising Temperatures," The Christian Science Monitor, 20 December 2010

(2011) and Doha (2012). The Green Climate Fund is to balance the allocation of the resources between adaptation and mitigation activities. As of December 2012 the Green Climate Fund was not yet operational, but had received cumulative contributions of USD 4.298 million and cumulative pledges of USD 4.554 million for the administrative budget of the secretariat in South Korea.[13]

The most notable signs that the UNFCCC parties are not advancing quickly enough to address climate change are the failure to stabilize GHG emissions, the shrinking commitments of industrialized parties under Kyoto, the lack of commitments from major developing countries, and the unilateral measures that some parties have begun to take in response to the failures of the UNFCCC and Kyoto processes.

2.3 Sustainable Development Principle

The 1992 Rio Declaration refers repeatedly to sustainable development, confirming that it is anthropocentric (Principle 1) and requires development "to equitably meet developmental and environmental needs of present and future generations" (Principle 3), integrating environmental protection into the development process (Principle 4), poverty eradication and decreasing economic disparities (Principle 5), changes to production and consumption and appropriate demographic policies (Principle 8), technology development and transfer (Principle 9), an open international economic system (Principle 12), the full participation of women (Principle 20), the creativity, ideals, and courage of the youth of the world (Principle 21), the effective participation of indigenous people and other local communities (Principle 22), environmental protection during wars (Principle 24), peace, development, and environmental protection (Principle 25), and cooperation in the further development of international law (Principle 27).

The UNFCCC does not contain a binding "sustainable development" obligation. Moreover, its take on sustainable development places greater emphasis on economic growth than the Rio Declaration does. The preamble recognizes that all countries, especially developing countries, need access to resources required to achieve "sustainable social and economic development." Article 2 states the treaty's objective as stabilization of greenhouse gas concentrations in the atmosphere at a level and within a time frame sufficient to enable economic development to proceed in a sustainable manner. Article 3(4) sets out the principle that the parties have a right to, and should, promote sustainable development, but also notes that

<http://theenergycollective.com/robertstavins/49196/why-cancun-trumped-copenhagen> (accessed December 29, 2012).

[13] Decision –1/CP.18, Agreed outcome pursuant to the Bali Action Plan para. 72; Decision –6/CP.18, Report of the Green Climate Fund to the Conference of the Parties and guidance to the Green Climate Fund, Report of the Conference of the Parties on its eighteenth session, held in Doha from 26 November to 8 December 2012, Addendum, Part Two: Action taken by the Conference of the Parties at its eighteenth session (Advance version) <http://unfccc.int/resource/docs/2012/cop18/eng/08a01.pdf> (accessed April 4, 2013).

economic development is essential for adopting measures to address climate change. Article 3(5) adds that the parties should cooperate to promote a supportive and open international economic system that would lead to sustainable economic growth and development. Article 4(1)(d) obliges all parties to "promote" sustainable management, and promote and cooperate in the conservation and enhancement, as appropriate, of sinks and reservoirs of all greenhouse gases not controlled by the Montreal Protocol, including biomass, forests, and oceans. This is not a particularly meaningful obligation, since it is too vague to be enforceable. Article 4 (2)(a) allows developed countries and other Annex I countries to take into account the need to maintain strong and sustainable economic growth when taking measures to mitigate climate change. Thus, while the UNFCCC might be considered a means to achieve sustainable development, this concept plays a different role than it does in the Rio Declaration.

Some authors have argued that sustainable development has become accepted as a rule of customary international law that would oblige nations to exploit their resources in a manner that is sustainable.[14] This is a difficult argument to accept given the vagueness and utopian nature of the concept.[15] Moreover, State sovereignty gives countries the right to exploit their own resources as they see fit, subject to their corresponding responsibility to not cause environmental damage beyond their borders, as confirmed by Stockholm Declaration, Principle 21, and Rio Declaration, Principle 2. However, the concept of sustainable development can be taken into account in treaty interpretation, depending on the specific treaty language. Measures taken to address climate change are likely to be litigated at the WTO or the International Court of Justice (ICJ), where sustainable development has arisen in past cases.

In *US—Shrimp*, the WTO Appellate Body referred to sustainable development to support its interpretation of the general exception for measures relating to the conservation of exhaustible natural resources under GATT Article XX(g). The references to environmental protection and sustainable development in the preamble of the WTO Agreement meant that the generic term "natural resources" in Article XX(g) is not "static" in its content but is "evolutionary" and should take into consideration the international community's efforts to protect living natural resources. As a result, the Appellate Body interpreted Article XX(g) as referring to the conservation of exhaustible living as well as non-living natural resources.[16]

In 2001, the WTO Doha Ministerial Declaration "strongly" reaffirmed the commitment to the objective of sustainable development, as stated in the Preamble to the Marrakesh Agreement. The Declaration also emphasizes "that the aims of upholding and safeguarding an open and nondiscriminatory multilateral trading

[14] Virginia Dailey, "Sustainable Development: Reevaluating the Trade vs. Turtles Conflict at the WTO" (2000) 9 Transnational Law and Policy 331.

[15] Philippe Sands, "International Law in the Field of Sustainable Development: Emerging Legal Principles" in Winfried Lang (ed.), *Sustainable Development and International Law* (Springer, Berlin 1995) 58.

[16] Appellate Body Report, *United States—Import Prohibition of Certain Shrimp and Shrimp Products* (*US—Shrimp*), WT/DS58/AB/R, adopted November 6, 1998, paras. 129–31.

system, and acting for the protection of the environment and the promotion of sustainable development can and must be mutually supportive." The Declaration further recognizes that

> under WTO rules no country should be prevented from taking measures for the protection of human, animal or plant life or health, or of the environment at the levels it considers appropriate, subject to the requirement that they are not applied in a manner which would constitute a means of arbitrary or unjustifiable discrimination between countries where the same conditions prevail, or a disguised restriction on international trade, and are otherwise in accordance with the provisions of the WTO Agreements.[17]

The latter statement incorporates language from GATT Article XX(b) and the introductory language of GATT Article XX, as well as other WTO Agreements, all of which we analyze in depth in a later chapter. This Declaration is important for two reasons. First, it is relevant to the interpretation of WTO law. Second, it represents a renewed political commitment to sustainable development, in accordance with WTO rules, whose timing suggests acceptance of the preexisting WTO jurisprudence on this point.

Measures to address climate change are likely to qualify as relating to the conservation of an exhaustible natural resource, taking into account the linkage between climate change and sustainable development and WTO jurisprudence finding that clean air is an exhaustible natural resource.[18] Alternatively, they could be found to constitute measures necessary to protect human, animal, or plant life or health, in a manner similar to the analysis of the Appellate Body in *Brazil—Retreaded Tyres*.[19] Either way, the consistency of such measures with WTO law would still be subject to the requirement that they are not applied in a manner which would constitute a means of arbitrary or unjustifiable discrimination between countries where the same conditions prevail, or a disguised restriction on international trade, and are otherwise in accordance with the provisions of the WTO Agreements.

Disputes regarding the interpretation and application of the UNFCCC are to be submitted to the International Court of Justice or arbitration, in accordance with UNFCCC Article 14(2). In *Pulp Mills on the River Uruguay (Argentina v. Uruguay)*, the ICJ considered the concept of "sustainable development" in the context of a transboundary environmental dispute.[20] In that case, Argentina alleged that Uruguay had breached its obligations under the Statute of the River Uruguay,[21]

[17] Ministerial Declaration, Adopted on November 14, 2001, WT/MIN(01)/DEC/1, para. 6.

[18] Appellate Body Report, *United States—Standards for Reformulated and Conventional Gasoline (US—Gasoline)*, WT/DS2/AB/R, adopted May 20, 1996.

[19] Appellate Body Report, *Brazil—Measures Affecting Imports of Retreaded Tyres* (*Brazil—Retreaded Tyres*), WT/DS332/AB/R, adopted December 17, 2007.

[20] *Pulp Mills on the River Uruguay (Argentina v. Uruguay)*, (2010) (Judgment) <http://www.icj-cij.org/docket/files/135/15877.pdf> (accessed August 20, 2012); *Pulp Mills on the River Uruguay (Argentina v. Uruguay)* (2006) (Provisional Measures) ICJ Reports 2006 113 <http://www.icj-cij.org/docket/files/135/11235.pdf> (accessed August 23, 2012).

[21] United Nations, Treaty Series (UNTS), Vol. 1295, No. I-21425, 340, signed by Argentina and Uruguay at Salto (Uruguay) February 26, 1975, entered into force September 18, 1976.

due to the effects of two pulp mills on the River Uruguay on the quality of the waters of the river and on the areas affected by the river. The ICJ initially characterized the case as one that "highlights the importance of the need to ensure environmental protection of shared natural resources while allowing for sustainable economic development" and stated that "account must be taken of the need to safeguard the continued conservation of the river environment and the rights of economic development of the riparian States."[22] In its judgment, the Court noted that the object and purpose set forth in the Statute is for the parties to achieve "the optimum and rational utilization of the River Uruguay" by means of the "joint machinery" for cooperation. The Court also referred to its judgment in the *Gabčíkovo-Nagymaros* case, in which it noted that the concept of sustainable development refers to the need to reconcile economic development with protection of the environment, but added that "[i]t is for the Parties themselves to find an agreed solution that takes account of the objectives of the Treaty."[23] The Court also referred to "the need to strike a balance between the use of the waters and the protection of the river consistent with the objective of sustainable development" and added that, when any party wished to exercise its right to use the waters of the river, such utilization could not be considered to be equitable and reasonable if the interests of the other riparian State in the shared resource and the environmental protection of the latter were not taken into account. The Court considered that Article 27 of the Statute "embodied this interconnectedness between equitable and reasonable utilization of a shared resource and the balance between economic development and environmental protection that is the essence of sustainable development."[24]

However, in *Pulp Mills on the River Uruguay*, the Court concluded that there was no clear evidence that substances with harmful effects had been introduced into the aquatic environment of the river through the emissions of the mill into the air and that there was no conclusive evidence that Uruguay had not acted with the requisite degree of due diligence or that the discharges of effluent from the mill had had deleterious effects or caused harm to living resources or to the quality of the water or the ecological balance of the river. The Court concluded that Uruguay had breached only its procedural obligations under the Statute, but not any substantive obligations. The Court considered that its finding of wrongful conduct by Uruguay in respect of its procedural obligations per se constituted a measure of satisfaction for Argentina, and ordered no further remedies.[25] This case demonstrates the

22 *Pulp Mills on the River Uruguay (Argentina v. Uruguay)* (2006) para. 80.

23 *Pulp Mills on the River Uruguay (Argentina v. Uruguay)* (2010) paras. 75–6; *Gabčíkovo-Nagymaros Project (Hungary v. Slovakia)* (1997) (Judgment) ICJ Reports 1997 paras. 140–1 <http://www.icj-cij.org/docket/files/92/7375.pdf> (accessed August 23, 2012).

24 *Pulp Mills on the River Uruguay (Argentina v. Uruguay)* (2010) para. 177; Art. 27 stipulates that

> [t]he right of each party to use the waters of the river, within its jurisdiction, for domestic, sanitary, industrial and agricultural purposes shall be exercised without prejudice to the application of the procedure laid down in Articles 7 to 12 when the use is liable to affect the régime of the river or the quality of its waters.

25 *Pulp Mills on the River Uruguay (Argentina v. Uruguay)* (2010) paras. 264–5, 275–6.

importance of conclusive evidence of environmental harm to secure an order for restitution or to pay compensation, which may be difficult to achieve given the current uncertainty regarding the causes and effects of climate change.

The foregoing analysis of the concept of sustainable development suggests that it does not create specific obligations for States with respect to climate change. Moreover, in treaties in which sustainable development forms part of the preamble or is otherwise incorporated into the treaty text, it can influence the interpretation of that treaty. However, this concept cannot be used to alter clearly stated obligations in treaties, whether treaties specifically on climate change or other treaties, such as those in the field of international economic law. Within these limits, the concept of sustainable development is relevant to interpret WTO law and international investment law in cases involving national and international environmental regulations related to climate change, as well as the UNFCCC and its related instruments.

2.4 Precautionary Principle

The essence of the precautionary principle is that States should act to prevent environmental damage without waiting for definitive scientific proof regarding causation. Principle 15 of the Rio Declaration states this principle as follows:

> In order to protect the environment, the precautionary approach shall be widely applied by States according to their capabilities. Where there are threats of serious or irreversible damage, lack of full scientific certainty shall not be used as a reason for postponing cost-effective measures to prevent environmental degradation.

The UNFCCC adopts a similar phrasing of the principle in Article 3(3), albeit in greater detail:

> The Parties should take precautionary measures to anticipate, prevent or minimize the causes of climate change and mitigate its adverse effects. Where there are threats of serious or irreversible damage, lack of full scientific certainty should not be used as a reason for postponing such measures, taking into account that policies and measures to deal with climate change should be cost-effective so as to ensure global benefits at the lowest possible cost. To achieve this, such policies and measures should take into account different socio-economic contexts, be comprehensive, cover all relevant sources, sinks and reservoirs of greenhouse gases and adaptation, and comprise all economic sectors. Efforts to address climate change may be carried out cooperatively by interested Parties.

At COP 18, in the decision regarding loss and damage that climate change causes to developing countries, the parties underlined that "the lack of full scientific certainty should not be used as reason for postponing action." They also decided to establish, at COP 19, institutional arrangements to address loss and damage from climate change in developing countries that are particularly vulnerable to its adverse effects.[26] While this does not create an obligation to compensate vulnerable developing countries for the damage caused by climate change, it is an interesting development nevertheless, since it raises the sensitive issue of liability for damage caused by climate change. However, it would be difficult to allocate responsibility for damage caused by climate change, due to the complexity of proving causation.[27]

In the *Gabčíkovo-Nagymaros* case, Hungary argued that the previously existing obligation not to cause substantive damage to the territory of another State had evolved into a general obligation of prevention of damage pursuant to the "precautionary principle," and that this precluded performance of its treaty with Slovakia. Slovakia argued that no intervening developments in environmental law gave rise to norms that would override the treaty.[28] The ICJ rejected Hungary's argument. The Court found that the Treaty could be terminated only on the limited grounds enumerated in the Vienna Convention, none of which existed in this case.[29] It pointed out that the parties could agree to incorporate newly developed norms of environmental law under the treaty provisions that require them to ensure that the quality of water in the Danube is not impaired and that nature is protected, and to take new environmental norms into consideration during implementation. These evolving provisions recognized the potential necessity to adapt the project and left the treaty open to adapt to emerging norms of international law, through a process of consultation and negotiation. The Court noted the importance of engaging in this process in good faith, in order to address potential environmental risks on a continuous basis. The Court also noted that both parties agreed on the need to take environmental concerns seriously and to take the required precautionary measures. However, they could not agree on the consequences this had for their joint project. The Court suggested third-party involvement to help them to reach an agreement.[30] Thus, in this case, the ICJ did not determine whether the precautionary principle had evolved into a rule of customary international law. Instead, it indicated that the treaty permitted the incorporation of new environmental norms. Thus, this case merely confirms that parties to a treaty are free to incorporate new (environmental) provisions by mutual agreement. The precautionary principle cannot be invoked to excuse performance of a treaty or to terminate a treaty, unless of course the treaty says so.

In *EC—Hormones*, the WTO Appellate Body also rejected the precautionary principle as a defense, this time under the WTO Agreement on the Application of

[26] Draft decision 3/CP.18, Approaches to address loss and damage associated with climate change impacts in developing countries that are particularly vulnerable to the adverse effects of climate change to enhance adaptive capacity, December 8, 2012, Preamble, para. 9, Report of the Conference of the Parties on its eighteenth session, held in Doha from 26 November to 8 December 2012, Addendum, Part Two: Action taken by the Conference of the Parties at its eighteenth session (Advance version) <http://unfccc.int/resource/docs/2012/cop18/eng/08a01.pdf> (accessed April 4, 2013).

[27] See e.g. *Trail Smelter Case (United States v. Canada)* 3 RIAA 1905 (1941); Michael Bowman and Alan Boyle (eds.), *Environmental Damage in International and Comparative Law: Problems of Definition and Valuation* (Oxford University Press, Oxford 2002).

[28] *Gabčíkovo-Nagymaros Project (Hungary v. Slovakia)* (1997) para. 97.

[29] *Gabčíkovo-Nagymaros Project (Hungary v. Slovakia)* (1997) para. 100.

[30] *Gabčíkovo-Nagymaros Project (Hungary v. Slovakia)* (1997) paras. 111–15.

Sanitary and Phytosanitary Measures (the SPS Agreement). It observed that the status of the precautionary principle in international law is the subject of debate among academics, law practitioners, regulators, and judges. It noted that the precautionary principle is regarded by some as having crystallized into a general principle of customary international environmental law, but whether it has been widely accepted as a principle of general or customary international law appears less than clear. It stated that the status of the precautionary principle in international law, at least outside the field of international environmental law, still awaits authoritative formulation.[31] However, it did not explain how a norm could qualify as customary international environmental law and yet not form part of customary international law. Article 38 of the Statute of the International Court of Justice refers to "international custom, as evidence of a general practice accepted as law." Accepting that a norm could crystallize as international custom in just one area of public international law, without doing so in general, would promote the fragmentation of international law and increase the difficulty of ensuring coherence between different areas of international law. With respect to climate change, it is especially important to maintain coherence between international economic law and international environmental law. Indeed, the general acceptance of the concept of sustainable development in both fields is consistent with an intention to avoid divergence in the development of norms.

The *EC—Hormones* case, like the *Gabčíkovo-Nagymaros* case, confirms that the effect of the precautionary principle on a treaty depends on the specific treaty language. The Appellate Body found that the more concrete formulation of aspects of the precautionary principle found in the SPS Agreement provided a more solid basis for the Appellate Body to make its ruling than the defense that the EC put forward based on the precautionary principle itself. In this regard, the Appellate Body noted that the principle has not been written into the SPS Agreement as a ground for justifying SPS measures that are otherwise inconsistent with the obligations of Members set out in specific provisions of that agreement. While it is reflected also in the sixth paragraph of the preamble, the precautionary principle does not, by itself, prevent applying the normal principles of treaty interpretation to the provisions of the SPS Agreement.[32] However, in determining whether a measure was maintained without sufficient scientific evidence, the Appellate Body said that a panel should bear in mind "that responsible, representative governments commonly act from perspectives of prudence and precaution where risks of irreversible, e.g. life-terminating, damage to human health are concerned."[33]

This approach of the Appellate Body is consistent with the view that the leeway that a treaty interpreter has depends on the degree of ambiguity in the provision in

[31] Appellate Body Report, *European Communities—Measures Concerning Meat and Meat Products (Hormones)* (*EC—Hormones*), WT/DS26/AB/R, WT/DS48/AB/R, adopted February 13, 1998, para. 123.

[32] Appellate Body Report, *EC—Hormones* para. 124.

[33] Appellate Body Report, *EC—Hormones* para. 124.

question.[34] However, this view is not limited to international environmental law or international health law. Nor should one read too much into the Appellate Body's observation regarding how responsible governments behave. In the case of climate change, due to the length of time that GHGs remain in the atmosphere, there is a risk of irreversible damage to the environment (at least in the short to medium term), with serious increases in the risks to human health from diseases, malnutrition, and severe weather events. However, this is not sufficient to permit the precautionary principle to influence the interpretation and application of a treaty in the absence of a specific provision that provides greater guidance. As with treaty interpretation in general, the result also depends on the degree of ambiguity in the provision itself. Even if the precautionary principle achieves the status of customary international law, that would only be sufficient for it to influence treaty interpretation as a relevant rule of international law applicable in the relations between the parties, not to override clear treaty provisions. Thus, while the precautionary principle might be used to persuade countries that they should reach international agreements to address effectively the risks from climate change, it is far less important than the actual terms of those agreements.

Nevertheless, the precautionary principle may serve as an additional argument to justify unilateral and multilateral measures taken to mitigate or adapt to climate change measures based on limited or insufficient scientific proof, both in WTO law and in international investment law. In both areas of international economic law, it would be relevant to arguments regarding the nature of the measure. In the case of international investment law, it could support an argument that the measure relates to environmental protection, not investment, and is not subject to the provisions of the international investment agreement. In the case of WTO law, it could be used to support the same argument, that the measure is an environmental measure, and that any discrimination is neither arbitrary nor unjustifiable. These issues will be analyzed in more detail in Chapters 3 and 5.

2.5 Common but Differentiated Responsibilities

The principle of common but differentiated responsibilities is included explicitly in the UNFCCC and is implicit in other multilateral environmental agreements, including the Convention on Biological Diversity and the Montreal Protocol on Substances that Deplete the Ozone Layer. While this principle is not exclusive to international environmental law, it has become the most fertile ground for creating non-uniform international treaty obligations.[35] In international environmental law, the principle of common but differentiated responsibilities means that not every country that faces the common problem shares equally in the common

[34] Bradly J. Condon, *Environmental Sovereignty and the WTO: Trade Sanctions and International Law* (Transnational Publishers, Ardsley 2006) 276–82.

[35] Christopher D. Stone, "Common but Differentiated Responsibilities in International Law" (2004) 98 AJIL 276, 279.

responsibility to fix it.[36] Principle 12 of the Stockholm Declaration calls for international technical and financial assistance to help developing countries with the cost of environmental protection.[37] Rio Declaration Principle 6 provides that "[t]he special situation and needs of developing countries, particularly the least developed and those most environmentally vulnerable, shall be given special priority." Principle 7 of the Rio Declaration expresses this principle in the following terms:

> States shall cooperate in a spirit of global partnership to conserve, protect and restore the health and integrity of the Earth's ecosystem. In view of the different contributions to global environmental degradation, States have common but differentiated responsibilities. The developed countries acknowledge the responsibility that they bear in the international pursuit to sustainable development in view of the pressures their societies place on the global environment and of the technologies and financial resources they command.

UNFCCC Article 3(1) expresses this principle as follows:

> The Parties should protect the climate system for the benefit of present and future generations of humankind, on the basis of equity and in accordance with their common but differentiated responsibilities and respective capabilities. Accordingly, the developed country Parties should take the lead in combating climate change and the adverse effects thereof.

UNFCCC Article 4(1) requires all parties to take into account "their common but differentiated responsibilities and their specific national and regional development priorities, objectives and circumstances" in implementing the reporting and cooperation commitments set out in that provision. The commitments in Article 4(2), to limit GHG emissions and to protect and enhance sinks and reservoirs, only apply to the developed country parties and other parties included in Annex I. The commitments regarding financing and technology transfer in Article 4, paragraphs (3), (4), and (5) only apply to the developed country parties and other developed parties included in Annex II. Article 4(7) conditions the implementation of developing country parties' commitments under the Convention on the effective implementation by developed country parties of their commitments related to financial resources and transfer of technology. Article 4(9) requires the parties to take full account of the specific needs and special situations of the least developed countries in their actions with regard to funding and transfer of technology. Article 4(10) requires the parties to take into consideration in the implementation of the commitments of the Convention the situation of parties, particularly developing country parties, with economies that are vulnerable to the adverse effects of the implementation of measures to respond to climate change.

In public international law, there is no agreement on what CDR means or when it applies, which makes it difficult to consider as a rule of customary international law that applies to all nations regardless of whether they have agreed to a specific

[36] Stone, "Common but Differentiated Responsibilities" 276.

[37] Declaration of the United Nations Conference on the Human Environment (1972) <http://www.unep.org/Documents.Multilingual/Default.asp?documentid=97&articleid=1503> (accessed August 28, 2012).

application in a treaty.[38] Indeed, the responsibility of States to ensure that activities within their jurisdiction do not cause damage to the environment in other States or in areas beyond the limits of their jurisdiction does not permit a lack of mitigation resources to be used as a defense.[39] Moreover, in public international law, differential obligations are the exception.[40]

Are there normative arguments that support differential obligations with respect to climate change? Stone argues that Rio Declaration Principles 6 and 7 advance three arguments in favor of differentiation: (1) needs/vulnerability; (2) contribution to the problem; and (3) financial and technological capacity to resolve the problem. However, he notes that the most needy nations are not necessarily less responsible for environmental damage, nor are those most responsible necessarily those with the greatest financial and technological capacity. He argues that, in the case of climate change, need must be understood as need that is due to a combination of vulnerability and lack of resources to mitigate risks. Assigning responsibility according to the contribution to the environmental damage is justifiable on the same basis as the polluter-pays principle (PPP), but is more problematic when the payment is for past harms, not caused by today's citizens, such as, for example, an immigrant to a developing country who played no part in the developing country's past pollution. Assigning responsibility based on wealth and technology can be viewed as a form of distributive justice, but the pursuit of wealth redistribution should not make poverty a defense against antipollution laws.[41]

The UNFCCC provisions on common but differentiated responsibility seem rather outdated now, only twenty years after the parties agreed to them. So do the Kyoto Protocol commitments. The ensuing growth of the technological capacities, economies, and emissions of countries like China, the reduction in the debts of major developing countries like Mexico, and the financial crises in Japan, the United States, and the European Union have combined to dramatically alter the financial and technological landscape. Robert Stavins points out that the 1997 Kyoto Protocol assigned emission targets only to 40 industrialized countries, leaving more than 140 countries without any such commitments. Little more than a decade later, 50 of those developing countries had higher per capita income than the poorest of the countries with emission-reduction responsibilities under Kyoto.[42] Nevertheless, India and China recently alleged a violation of the CDR principle of the UNFCCC in ordering their airlines not to comply with the EU Aviation Directive.[43] Some consider that disputes over the scope of CDR are a primary cause of the failure to advance sufficiently in UNFCCC negotiations.[44]

[38] Stone, "Common but Differentiated Responsibilities" 281.

[39] Stone, "Common but Differentiated Responsibilities" 282.

[40] Stone, "Common but Differentiated Responsibilities" 282.

[41] Stone, "Common but Differentiated Responsibilities" 290–2; Rio Declaration on Environment and Development (1992) <http://www.unep.org/Documents.Multilingual/Default.asp?documentid=78&articleid=1163> (accessed August 28, 2012).

[42] Stavins, "Why Cancun Trumped Copenhagen."

[43] Joshua Meltzer, "Regulating CO2 Emissions from Aviation in the EU" (2012) 16(27) ASIL Insights <http://www.asil.org/pdfs/insights/insight120831.pdf> (accessed March 15, 2013).

[44] Stone, "Common but Differentiated Responsibilities" 280.

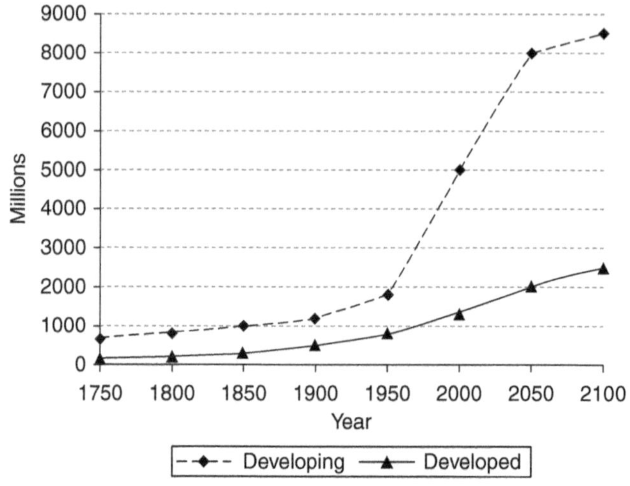

Figure 2.1 World population 1750–2100 AD

As Figure 2.1 shows, the overwhelming majority of the world's population will still be in developing countries by the end of this century, when serious impacts from climate change are expected to be felt. What is notable about Figure 2.1 is that the countries which were in the developed country league in 1800, were the same in 2000 (with one clear exception: Japan). However, this will change substantially in the twenty-first century. Among the countries with more than 100 million population, Brazil, China, India, Indonesia, Mexico, and Russia will join the league of developed countries (along with a group of smaller ones). Some major developing countries have begun to produce both clean energy technologies (China for example) and new plant varieties (Brazil for example). This trend is likely to continue as other countries join the foray.

Using CDR to divide the world into developed and developing countries, and using that categorization as a basis for addressing climate change, whether through adaptation or mitigation, is too simplistic and is not working. Indeed, it has become an obstacle to reaching international agreements to address climate change. Moreover, the financial and technological endowments of countries are not frozen in time. It makes more sense to assign responsibility on a scale, based on objective criteria that are in accordance with the governing principles of international environmental law and determine responsibility for the cost of mitigation and adaptation, respectively. These criteria could be used to create an index of countries that evolves as conditions change, what we shall call the "Climate Sensitivity Index." We develop this index in Chapter 6.

A comparable principle exists in WTO law (special and differential treatment for developing countries). The principle of common but differentiated responsibilities is unlikely to influence the interpretation of WTO law, since it would be overridden by the express inclusion of special and differential treatment provisions. Some argue that special and differential treatment of developing countries is inappropriate in the trade context because it may hamper the economic development that could flow from adopting stricter trade disciplines.[45] However, this argument is not persuasive in the context of other aspects of WTO law, such as intellectual property.[46]

Aside from specific provisions in WTO agreements that provide special and differential treatment, Morosini argues that CDR is implicit in other provisions of the WTO agreements. In *US—Shrimp*, the Appellate Body found that GATT Article XX requires flexible application of measures that takes into account the conditions prevailing in different countries. In *Brazil—Retreaded Tyres*, the Appellate Body took Brazil's level of technological and economic development into consideration in rejecting an alternative approach to environmental protection that the European Union proposed in place of Brazil's trade restriction. Morosini concludes that "the judicial bodies of the WTO will resort to CDR to resolve conflicts between trade, environment, and development."[47] However, this is unlikely to occur, for two reasons. First, CDR cannot override the text of the WTO agreements any more than special and differential treatment or the precautionary principle can. Second, most of the special and differential treatment provisions in the WTO Agreements are not binding in a meaningful way.[48] Thus, both the text and the context of the WTO Agreements indicate that CDR would have a limited role, if any, in the interpretation and application of their provisions.

Nevertheless, Morosini's analysis is useful, because it shows that WTO law probably is not inconsistent with the principle of CDR. Morosini argues that the principle of CDR is part of, or inherent to, the concept of sustainable development, which is recognized in the preamble of the WTO Agreement. Alternatively, if one disagrees that the principle of CDR is inherent in the concept of sustainable development, the principle of CDR should inform the interpretation of the WTO covered agreements based either on a different passage of the preamble to the WTO Agreement ("seeking both to protect and preserve the environment and to enhance the means for doing so in a manner consistent with their respective needs and concerns at different levels of economic development") or on the GATT Article XX chapeau ("between countries where the same conditions prevail"), as interpreted in *US—Shrimp*.[49] It is an interesting argument and could provide support for developing CDR/special and differential treatment criteria along the lines that we

[45] Michael Hart and Bill Dymond, "Special and Differential Treatment and the Doha Development Round" (2003) 37 Journal of World Trade 395.

[46] Bradly J. Condon and Tapen Sinha, "Global Diseases, Global Patents and Differential Treatment in WTO Law: Criteria for Suspending Patent Obligations in Developing Countries" (2005) 26 Northwestern Journal of International Law and Business 1.

[47] Morosini, "Trade and Climate Change" 718.

[48] See e.g. Antidumping Agreement, Art. 15; Panel Report, *United States—Anti-Dumping and Countervailing Measures on Steel Plate from India (US—Steel Plate)*, WT/DS206/R, adopted July 29, 2002, para. 7.110.

[49] Morosini, "Trade and Climate Change" 721–5.

propose above. Nevertheless, in WTO disputes, what will be relevant are the specific circumstances of each case and each country and the specific terms of the relevant WTO provisions, particularly when considered in light of *EC—Tariff Preferences*, in which the Appellate Body found that countries in similar financial and economic circumstances should be treated in a similar fashion under the Enabling Clause.[50]

2.6 Polluter-Pays Principle

The polluter-pays principle is a principle in international environmental law where the polluting parties are made liable to pay for the environmental damages they cause. The objective of this principle is to shift the responsibility from governments to the polluting entities. The PPP is normally implemented through two different policy approaches: (1) the command-and-control approach and (2) the market-based approach. The command-and-control approach seeks to prevent environmental problems by regulating how a company should manage a pollution-generating process. The market-based approach charges fees or taxes to polluters.

Principle 16 of the Rio Declaration expresses this principle as follows:

> National authorities should endeavour to promote the internalization of environmental costs and the use of economic instruments, taking into account the approach that the polluter should, in principle, bear the cost of pollution, with due regard to the public interest and without distorting international trade and investment.

The UNFCCC does not incorporate this principle. Instead, it leaves parties to determine the means by which to implement their obligations. Moreover, the provisions regarding financing for and technology transfer to developing countries, combined with the absence of emissions reduction commitments on the part of developing countries, indicate that developed country governments, rather than private sector polluters, are to pay for mitigation and adaptation, at least in developing countries. Indeed, part of the mandate of the Green Climate Fund is "to provide for effective direct and indirect public and private-sector financing *by* the Green Climate Fund."[51] Nevertheless, COP18 "[u]rges all developed country Parties to scale up climate finance from a wide variety of sources, public and private, bilateral and multilateral, including alternative sources, to the joint goal of mobilizing USD 100 billion per year by 2020."[52] However, seeking financing from private sources is not the same as obliging the private sector to internalize the cost of pollution.

[50] Appellate Body Report, *European Communities—Conditions for the Granting of Tariff Preferences to Developing Countries (EC—Tariff Preferences)*, WT/DS246/AB/R, adopted April 20, 2004; Condon and Sinha, "Global Diseases, Global Patents and Differential Treatment in WTO Law."

[51] Report of the Green Climate Fund to the Conference of the Parties and guidance to the Green Climate Fund para. 7(a). Emphasis added.

[52] Draft decision –/CP.18, Agreed outcome pursuant to the Bali Action Plan para. 66.

One could argue that, by having developed countries pay for their past and current emissions, PPP is implicit. However, this would imply accepting a somewhat distorted version of PPP, because it is governments that pay, not the individuals and companies that actually did the polluting. Moreover, having today's citizens in a developed country pay for pollution that had occurred before their ancestors even immigrated is a further distortion of PPP.[53]

Kyoto Protocol Article 2(1)(a)(v) requires that Annex I parties, in achieving their emissions commitments, take measures "such as" "[p]rogressive reduction or phasing out of market imperfections, fiscal incentives, tax and duty exemptions and subsidies in all greenhouse gas emitting sectors that run counter to the objective of the Convention and application of market instruments." Cap-and-trade programs and carbon taxes are two types of PPP-based measures that have been implemented.

Kyoto Protocol Article 2(1)(a)(v) discourages what might be termed a "pay-the-polluter" approach in developed countries. However, many developed and developing countries still spend significant amounts of money on subsidies for fossil fuels. These subsidies do not appear to be in conflict with the WTO Agreement on Subsidies and Countervailing Measures (SCM Agreement), and there is no legal obstacle to eliminating them. However, the SCM Agreement creates disincentives to subsidize clean energy technology development and permits WTO Members to apply countervailing duties that raise the cost of subsidized technologies when they are exported. The SCM Agreement thus creates obstacles to converting fossil fuel subsidies to clean energy subsidies. In this regard, it runs counter to the objective of the UNFCCC. We analyze the SCM Agreement in Chapter 3.

Clean energy technologies are often cited as an example of the kind of technology that needs to be developed and transferred internationally in order to combat climate change. However, the current multilateral legal framework favors "pay-the-polluter" over "polluter-pays." In Chapter 8, we analyze the reallocation of fossil fuel subsidies to finance climate change adaptation and mitigation.

2.7 Environmental Impact Assessment

Principle 17 of the Rio Declaration addresses environmental impact assessment (EIA) in a manner that suggests it is mandatory:

> Environmental impact assessment, as a national instrument, shall be undertaken for proposed activities that are likely to have a significant adverse impact on the environment and are subject to a decision of a competent national authority.

UNFCCC Article 4(1)(f) obliges *all* parties, taking into account their common but differentiated responsibilities and their specific national and regional development priorities, objectives and circumstances, to:

[53] Stone, "Common but Differentiated Responsibilities" 292.

> Take climate change considerations into account, to the extent feasible, in their relevant social, economic and environmental policies and actions, and employ appropriate methods, for example impact assessments, formulated and determined nationally, with a view to minimizing adverse effects on the economy, on public health and on the quality of the environment, of projects or measures undertaken by them to mitigate or adapt to climate change. . . .

The UNFCCC, unlike the Rio Declaration, does not say *environmental* impact assessments, but addresses instead *adverse effects on* economy, on public health, and on the quality of the environment of the climate change measures themselves. Similarly, Kyoto Article 2(3) is concerned primarily with the adverse effects of climate change measures, especially on developing country parties. Thus, the UNFCCC *lacks* the EIA obligation of the Rio Declaration. However, all of the UNFCCC parties must now be considered bound to conduct EIAs, in the light of recent developments in customary international law.

In *Pulp Mills on the River Uruguay (Argentina v. Uruguay)* the International Court of Justice observed that the practice of EIA "has gained so much acceptance among States that it may now be considered a requirement under general international law to undertake an environmental impact assessment where there is a risk that the proposed industrial activity may have a significant adverse impact in a transboundary context, in particular, on a shared resource."[54] GHG emissions have a significant, adverse, and transboundary impact on the climate, and the climate can be considered a shared resource. However, the Court observed that general international law does not specify the scope and content of an EIA. This qualification makes the obligation less useful.

Nevertheless, general international law, as reflected in draft Article 9(3) of the 2001 draft Articles of the International Law Commission Prevention of Transboundary Harm from Hazardous Activities,[55] requires the State of origin to "take into account the interests of the State likely to be affected in case it decides to authorize the activity to be pursued . . . " when a State requests consultations on how to prevent or minimize the risk of transboundary harm.[56] While this norm is procedural, and does not impose substantive obligations, when a State conducts an EIA with respect to activities that risk causing transboundary environmental harm, it is reasonable to conclude that the scope and content of the EIA would have to take into account the risk posed to other States. This does not entitle the affected State to prevent the activity, but it could affect the liability of the State that proceeds with the activity in question.

With respect to unilateral measures to address climate change, the foregoing norms lead us to conclude that a State has an obligation to take into account the impacts of a unilateral measure on other States in the process of designing the

[54] *Pulp Mills on the River Uruguay (Argentina v. Uruguay)* (2010) para. 204.

[55] International Law Commission, "Prevention of Transboundary Damage from Hazardous Activities" (2001) Yearbook of the International Law Commission Vol. II, Pt. 2 <http://untreaty.un.org/ilc/texts/9_7.htm> (accessed July 28, 2012).

[56] *Pulp Mills on the River Uruguay (Argentina v. Uruguay)* (2010) paras. 152–4.

measure. Depending on the nature of the measure, there also may be an obligation to consult other States at the design stage.[57] In the case of trade measures, the absence or presence of such an assessment may affect the outcome of the analysis in GATT Article XX and in other WTO provisions that incorporate language from that Article.[58]

2.8 Unilateral Measures and State Jurisdiction

There are outstanding issues in international law regarding the extent to which States have jurisdiction to regulate transboundary and global environmental issues unilaterally.

Principle 12 of the Rio Declaration addresses this issue as follows:

> States should cooperate to promote a supportive and open international economic system that would lead to economic growth and sustainable development in all countries, to better address the problems of environmental degradation. Trade policy measures for environmental purposes should not constitute a means of arbitrary or unjustifiable discrimination or a disguised restriction on international trade. *Unilateral actions* to deal with environmental challenges outside the jurisdiction of the importing country should be avoided. Environmental measures addressing transboundary or global environmental problems should, *as far as possible*, be based on an international consensus. (Emphasis added.)

The Rio Declaration thus expresses a preference for international consensus over unilateral actions, but without prohibiting unilateral actions.

UNFCCC Article 3(5) does not rule out the use of unilateral measures, as long as they comply with some of the language of GATT Article XX: "Measures taken to combat climate change, *including unilateral ones*, should not constitute a means of arbitrary or unjustifiable discrimination or a disguised restriction on international trade" (emphasis added). Paragraph (6) of the Doha Ministerial Declaration also implicitly recognizes a right to take unilateral measures, in accordance with GATT Article XX:

> We recognize that under WTO rules no country should be prevented from taking measures for the protection of human, animal or plant life or health, or of the environment *at the levels it considers appropriate*, subject to the requirement that they are not applied in a manner which would constitute a means of arbitrary or unjustifiable discrimination between countries where the same conditions prevail, or a disguised restriction on international trade, and are otherwise in accordance with the provisions of the WTO Agreements. (Emphasis added.)

There are two cases that support the use of unilateral measures to address climate change. In one, the European Court of Justice considered the right of the EU to use

[57] See e.g. WTO Agreement on Technical Barriers to Trade Art. 2.

[58] Ministerial Declaration, Fourth Ministerial Conference, Doha, Qatar, adopted November 14, 2001, WT/MIN(01)/DEC/1, November 20, 2001, para. 6, <http://www.wto.org/english/thewto_e/minist_e/min01_e/mindecl_e.htm> (accessed April 5, 2013).

a unilateral climate change measure in light of the Kyoto Protocol and customary international law. In the other, the WTO Appellate Body considered the right of the United States to use a unilateral environmental protection measure in light of GATT Article XX.

The ECJ case of EU Directive 2008/101 (Aviation Directive) considers the validity of a unilateral climate change measure. Kyoto Article 2(2) requires Annex I parties to "pursue limitation or reduction of emissions of greenhouse gases... from aviation... bunker fuels, working through the International Civil Aviation Organization...." After insufficient progress on limiting emissions from airlines under Kyoto Protocol Article 2(2), the EU acted unilaterally and applied the directive to both EU and non-EU airlines.

China and India prohibited their airlines from complying with the EU Aviation scheme, arguing that it violates the UNFCCC principle of common but differentiated responsibility. Airlines challenged the law before the UK High Court, which referred several legal issues to the ECJ. The ECJ ruled that the Kyoto Protocol does not provide a legal basis for challenging EU action and it does not breach customary international law principles of State sovereignty because it applies only to aircraft that choose to operate in EU airspace.[59]

Treaties concluded by the EU prevail over EU acts, but the provisions of the treaty must appear, as regards their content, to be unconditional and sufficiently precise. The Kyoto Protocol allows certain parties undergoing the process of transition to a market economy a degree of flexibility in the implementation of their commitments. The protocol also allows certain parties to meet their reduction commitments collectively. Finally, the Conference of the Parties is responsible for approving procedures and mechanisms to determine and to address cases of non-compliance with the provisions of the protocol. Thus, in the opinion of the ECJ, even though the Kyoto Protocol imposes quantified greenhouse gas reduction commitments for 2008 to 2012, the parties to the Protocol may comply with their obligations in the manner and at the speed upon which they agree. Article 2(2) of the Kyoto Protocol provides that the parties are to pursue limitation or reduction of emissions of certain greenhouse gases from aviation bunker fuels, working through the International Civil Aviation Organization. The ECJ found that this provision, as regards its content, cannot be considered to be unconditional and sufficiently precise so as to confer on individuals the right to rely on it in legal proceedings in order to contest the validity of the Aviation Directive.[60] A similar conclusion could follow from an analysis of the relationship between the provisions of the Kyoto Protocol and WTO law; the conditionality and imprecision of the

[59] Meltzer, "Regulating CO2 Emissions from Aviation in the EU"; Case C-366/10 *Air Transportation Association of America v. Secretary of State for Energy and Climate Change of the United Kingdom of Great Britain and Northern Ireland*, 21 December 2011 <http://curia.europa.eu/juris/liste.jsf?language=en&num=c-366/10> (accessed March 15, 2013); Council Directive 2008/101/EC, 2009 OJ (L 8/3) (EC) (amending Council Directive 2003/87/EC so as to include aviation activities in the scheme for greenhouse gas emission allowance trading within the community) (Aviation Directive).

[60] Case C-366/10 *Air Transportation* paras. 50–4, 75–8.

former means that it would be unlikely to prevail over more concrete and precise WTO obligations.[61]

When the European Union adopts an act, it is bound to observe international law, including customary international law. The ECJ concluded that three of the principles to which the referring court makes reference are recognized as forming part of customary international law: (1) that each State has complete and exclusive sovereignty over its airspace; (2) that no State may validly purport to subject any part of the high seas to its sovereignty; and (3) freedom to fly over the high seas. However, the ECJ determined that a fourth principle, that aircraft flying over the high seas are subject to the exclusive jurisdiction of the State in which they are registered, did not constitute customary international law and therefore could not affect the validity of the Aviation Directive. With respect to the first three principles, the ECJ had to determine whether the EU had competence to adopt Directive 2008/101, since it applies to aircraft operators of third States whose flights arriving in and departing from the EU are carried out in part over the high seas and over the third States' territory. However, since a principle of customary international law does not have the same degree of precision as a treaty provision, judicial review must be limited to the question whether, in adopting the act in question, the EU institutions made manifest errors of assessment concerning the conditions for applying those principles. Moreover, the Aviation Directive had to be interpreted, and its scope delimited, in light of the relevant rules of international law. Thus, the Aviation Directive could not apply as such to aircraft registered in third States that are flying over third States or the high seas. However, EU legislation may be applied to an aircraft operator when its aircraft is physically in the territory of one of the Member States. Thus, the ECJ held that the Aviation Directive does not infringe the principle of territoriality or sovereignty, since the aircraft are subject on that basis to the jurisdiction of the EU. As for the fact that the operator of an aircraft in such a situation is required to surrender emissions allowances calculated in light of the whole of the international flight, the EU legislature may in principle choose to permit a commercial activity to be carried out in the territory of the EU only on condition that operators comply with the criteria that have been established by the EU and are designed to fulfill the environmental protection objectives which it has set for itself, in particular where those objectives follow from an international agreement to which the EU is a signatory, such as the UNFCCC and the Kyoto Protocol. Finally, the ECJ held that EU law can apply to matters contributing to the pollution of the air, sea, or land territory of the Member States even where they originate in an event which occurs partly outside EU territory, citing its past decisions regarding the application of competition law and regarding hydrocarbons accidentally spilled beyond a Member State's territorial sea.[62]

[61] Appellate Body Report, *Argentina—Measures Affecting Imports of Footwear, Textiles, Apparel and other Items (Argentina—Textiles and Apparel)*, WT/DS56/AB/R, adopted April 22, 1998, para. 70.

[62] Case C-366/10 *Air Transportation* paras. 101–30.

One aspect of the ECJ ruling is troubling. Article 2(2) of the Kyoto Protocol also requires Annex I parties to pursue reduction of emissions of greenhouse gases to marine bunker fuels, working through the International Maritime Organization. The ECJ recognized the principle of customary international law that a vessel on the high seas is in principle governed only by the law of its flag, but held that this does not apply by analogy to aircraft flying over the high seas.[63] This reasoning is rather incongruent, since it means that air and marine transport would not be addressed in the same manner in a unilateral measure, even though they are addressed in the same manner in the Kyoto Protocol.

However, the ECJ ruling on the principle of territoriality and sovereignty is consistent with the approach of the Appellate Body to a similar argument in *US—Shrimp* and *US—Shrimp (Article 21.5—Malaysia)*.[64] The WTO Appellate Body found that, in certain circumstances, a Member can justify unilateral recourse to trade restrictions to induce changes in the environmental practices of other Members. The WTO Appellate Body interpreted GATT Article XX(g) to permit the United States to unilaterally impose trade barriers to pressure Malaysia to change its domestic environmental regime for the protection of sea turtles. Like the Aviation Directive case, the Shrimp case raises important questions regarding the relationship between trade law, environmental law, and the general principles of public international law. As such, these cases have important implications not only in the field of trade and environment, but more generally in the realm of public international law and global governance.

The United States banned shrimp imports from WTO members that did not comply with American legal requirements regarding the protection of sea turtles from incidental death in the shrimp harvesting process. The United States negotiated and concluded a regional international agreement on sea turtle protection and conservation with some countries in the Americas, but not other countries that were affected by the trade ban. The United States gave countries that were parties to the Inter-American Convention three years to introduce "turtle exclusion devices," while others were given only four months. This was not part of the original design of the measures, but resulted from a United States court order requiring the United States administration to apply the import ban to the entire world when it was only being applied in the Americas.[65] However, the Appellate Body held that the United States government was responsible for meeting its WTO obligations and it was no excuse under international law that the action was required by the courts.

All species of turtles involved were listed as being under threat of extinction under the Convention on International Trade in Endangered Species of Wild

[63] Case C-366/10 *Air Transportation* para. 106.

[64] Appellate Body Reports, *US—Shrimp*; Appellate Body Reports, *United States—Import Prohibition of Certain Shrimp and Shrimp Products, Recourse to Article 21.5 by Malaysia* (*US—Shrimp, Article 21.5—Malaysia*), WT/DS58/AB/RW, adopted November 21, 2001.

[65] *Earth Island Institute v. Christopher*, 20 CIT 1389, 948 F. Supp. 1062 (Court of International Trade 1996).

Fauna and Flora (CITES)[66] Appendix I, and occurred in American territorial waters as part of their migratory route. However, the American measures were not taken under CITES or any other multilateral environmental agreement. The Appellate Body rejected arguments that unilateral measures could not be included under Article XX(g), stating:

> It appears to us . . . that conditioning access to a Member's domestic market on whether exporting Members comply with, or adopt, a policy or policies unilaterally prescribed by the importing Member may, to some degree, be a common aspect of measures falling within the scope of one or another of the exceptions (a) to (j) of Article XX.[67]

The Appellate Body held that the measure met the requirements of Article XX(g), but not the chapeau. The Appellate Body emphasized a preference for multilateral solutions to international environmental problems, citing both WTO and other international instruments to that effect. It noted that the protection and conservation of migratory species demands concerted and cooperative efforts on the part of many countries. It cited the references in the Decision on Trade and Environment to Principle 12 of the Rio Declaration on Environment and Development and Agenda 21 as proof that the WTO has recognized both the need for such cooperative efforts and the inappropriateness of unilateral action in dealing with extraterritorial aspects of international environmental problems.

However, in *Shrimp 21.5*, the Appellate Body said:

> Requiring that a multilateral agreement be *concluded* by the United States in order to avoid "arbitrary and unjustifiable discrimination" in applying its measure would mean that any country party to the negotiations with the United States, whether a WTO Member or not, would have, in effect, a veto over whether the United States could fulfil its WTO obligations. Such a requirement would not be reasonable.
>
> . . .
>
> Principle 12 of the Rio Declaration . . . states, in part, that "[e]nvironmental measures addressing transboundary or global environmental problems should, as far as possible, be based on international consensus." Clearly, and "as far as possible," a multilateral approach is strongly preferred. Yet it is one thing to *prefer* a multilateral approach in the application of a measure that is provisionally justified under one of the subparagraphs of Article XX . . . ; it is another to require the *conclusion* of a multilateral agreement as a condition of avoiding 'arbitrary and unjustifiable discrimination' under the chapeau. . . . We see, in this case, no such requirement.[68]

Thus, while the Article XX chapeau required the United States to make good faith efforts to reach international agreements before imposing a unilateral import ban to further international environmental objectives, it could not be required to succeed. Certainly, if trade restrictions are to be permitted for the purpose of persuading

[66] Convention on International Trade in Endangered Species of Wild Fauna and Flora, Signed at Washington, D.C., March 3, 1973, Amended at Bonn, June 22, 1979, entered in force July 1, 1975 <http://www.cites.org> (accessed April 5, 2013).

[67] Appellate Body Report, *US—Shrimp* para. 121.

[68] Appellate Body Report, *US—Shrimp (Article 21.5—Malaysia)* paras. 123–4. Emphasis in the original.

other countries to participate in multilateral environmental protection efforts, this is a logical result. Moreover, the wording of the *Rio Declaration* ("as far as possible") clearly leaves open the possibility that unilateral measures may be needed in some circumstances.

The Appellate Body held that there was a sufficient jurisdictional nexus between the United States and the turtles for the purposes of Article XX(g). All of the species occur in waters over which the United States has jurisdiction, even though they migrate across national borders and international waters. In this regard, the Appellate Body's reasoning resembles that of the ECJ in the Aviation Directive case. At the same time, however, the Appellate Body expressly declined to decide whether there is an implied jurisdictional limitation in Article XX(g) and, if so, the nature or extent of that limitation. The Appellate Body recognized that a State has a legitimate interest in the protection of migratory species that occur within its territory.

There is an important distinction to be made regarding environmental concerns based on their geographic connection with the importing country. In international environmental law, both Agenda 21 and the Rio Declaration call upon countries to avoid "unilateral action to deal with environmental challenges *outside* the jurisdiction of the importing country" (emphasis added). It is only with respect to "environmental measures addressing transboundary or global environmental problems" that countries are urged, "as far as possible," to base their actions on international consensus. The phrase "as far as possible," leaves an opening for a country to take unilateral action where efforts at international negotiation fail, but only with respect to transboundary or global environmental problems.[69]

Under customary international law, a State acts in excess of its own jurisdiction when its measures purport to regulate acts which are done outside its territorial jurisdiction by persons who are not its own nationals, and which have no, or no substantial, effect within its territorial jurisdiction.[70] The territorial foundation of jurisdiction fails to resolve some modern jurisdictional conflicts (for example, the management of migratory species). Thus, the territorial basis for jurisdiction is subject to a developing principle of substantial and genuine connection between the subject matter of jurisdiction, on the one hand, and the territorial base and reasonable interests of the jurisdiction sought to be exercised, on the other.[71]

In *Shrimp*, the holding that the United States had a sufficient "jurisdictional nexus" for its measure to qualify for provisional justification under GATT Article XX(g) is consistent with customary international law insofar as harm to the turtles would have an effect within its territorial jurisdiction. This holding is also consistent with international environmental law in that the conservation of sea turtles is a transboundary environmental problem. However, the ruling might be viewed as

[69] Rio Declaration on Environment and Development refers to transboundary and global environmental problems. However, Agenda 21 only refers to "transborder problems."

[70] Ian Brownlie, *Principles of Public International Law* (2nd edn., Oxford University Press, Oxford 1973) 299–301.

[71] Ian Brownlie, *Principles of Public International Law* (6th edn., Oxford University Press, Oxford 2003) 297.

inconsistent with public international law to the extent that the American measure purports to regulate acts which are done outside its territorial jurisdiction by persons who are not its own nationals.

The geographic location of the resource and the legitimacy of the policy goal (which, in the *Shrimp* case, is reflected in the consensus that it is necessary to prevent the extinction of species) are factors that should be taken into account in determining whether the subject matter qualifies for provisional justification under Article XX(g). The question is whether a geographic connection between the resource and the enacting country is necessary or whether a less proximate connection should suffice. The governing principle of customary international law is that a State acts in excess of its own jurisdiction when its measures purport to regulate acts which are done outside its territorial jurisdiction by persons who are not its own nationals and which have no, or no substantial, effect within its territorial jurisdiction. GATT should be interpreted in a manner that is consistent with this principle, which means that some degree of territorial connection to the resource is necessary for a State to unilaterally regulate activities concerning that resource.

The American measures in the *Shrimp* cases were aimed at a transboundary environmental issue in which the United States had a territorial connection. Such cases should be relatively easy to identify. However, distinguishing between "global" and "extraterritorial" environmental problems is more difficult. Where the problem occurs entirely outside the territory of the importing country, it would appear to be extraterritorial. However, if one views the global ecosystem as interconnected, then problems that appear to be extraterritorial at first glance could be characterized as global. For example, does the United States or Europe have a jurisdictional nexus with respect to the Amazon because it contains a large percentage of global biodiversity and absorbs a significant percentage of global carbon dioxide emissions? This is not simply a legal question, but also a question of fact that requires scientific analysis. Where there exists a multilateral environmental agreement that indicates widespread international consensus regarding the global importance of a particular environmental problem, such as climate change, that may provide evidence that the issue is global rather than local and thus raise a rebuttable presumption regarding the categorization of the issue.

The existence of widespread consensus regarding the measures that are required to achieve the policy goal (as opposed to consensus regarding the legitimacy of the goal itself) is a factor to be considered in the Article XX chapeau analysis, since this is an implementation issue. The question is, how effective must the chosen measure be in order to pass the test? Where there is consensus, the measure is, in essence, deemed to be effective. However, the issue of the effectiveness of a measure in achieving the stated policy goal would have to be considered independently where such consensus is absent.

The Rio Declaration suggests that unilateral measures should only be used as a last resort. The urgency of the environmental problem is thus relevant to determining whether efforts to reach a negotiated resolution have been adequate. For example, where a resource is plentiful, it may not be possible to justify unilateral actions without devoting several years to negotiations. The role that urgency plays is thus central. The UNFCCC parties have identified climate change as an urgent problem, deciding "that Parties will urgently work towards the deep reduction in global greenhouse gas emissions required to hold the increase in global average temperature below 2°C above pre-industrial levels."[72]

A widely acknowledged general principle of international environmental law is that States are required to cooperate with each other in mitigating transboundary environmental risks. In the *Lac Lanoux* arbitration, the tribunal held that France had complied with its treaty and customary international law obligations to consult and negotiate in good faith before diverting a watercourse shared with Spain. However, the duty to negotiate did not require France to obtain Spain's consent.[73] Similarly, the Stockholm Declaration, Principle 24 provides that:

> co-operation through multilateral or bilateral arrangements or other appropriate means is essential to effectively control, prevent, reduce and eliminate adverse environmental effects resulting from activities conducted in all spheres, in such a way that due account is taken of the sovereignty and interests of all states.

The United Nations General Assembly endorsed this principle but noted that it should not be construed to enable other States to delay or impede the exploitation and development of natural resources within the territory of States.[74]

The panel in *US—Shrimp (Article 21.5—Malaysia)* noted that the good faith negotiations had to be ongoing. The panel emphasized that the right to take unilateral measures was provisional, not permanent, and subject to ongoing WTO supervision:

> [I]n a context such as this one where a multilateral agreement is clearly to be preferred and where measures such as that taken by the United States in this case may only be accepted under Article XX if they were allowed under an international agreement, or if they were taken further to the completion of serious good faith efforts to reach a multilateral agreement, the possibility to impose a unilateral measure to protect sea turtles under Section 609 is more to be seen, for the purposes of Article XX, as the possibility to adopt a *provisional* measure allowed for emergency reasons than as a definitive 'right' to take a permanent measure. The extent to which serious good faith efforts continue to be made may be reassessed at any time. For instance, steps which constituted good faith efforts at the beginning of a negotiation may fail to meet that test at a later stage.[75]

Making the right to take unilateral measures provisional is consistent with the manner in which the precautionary principle is incorporated into Article 5.7 of the SPS Agreement, which permits provisional measures where scientific evidence is insufficient.

[72] Draft decision –/CP.18, Agreed outcome pursuant to the Bali Action Plan para. 1.

[73] *Lac Lanoux (Spain v. France)* (1957) 24 ILR 101; *Nuclear Tests Cases (New Zealand v. France)* (1974) ICJ 457.

[74] UNGA Res. 2995 XXVII (1972); Patricia W. Birnie and Alan E. Boyle, *International Law and the Environment* (Oxford University Press, Oxford 1992) 102–9.

[75] Panel Report *United States—Import Prohibition of Certain Shrimp and Shrimp Products, Recourse to Article 21.5 by Malaysia* (*US—Shrimp, Article 21.5—Malaysia*), WT/DS58/RW, adopted November 21, 2001, para. 5.88. Emphasis in the original.

In the *Shrimp* case, the survival of highly migratory species depends on concerted and cooperative efforts on the part of many countries whose waters are traversed in the course of migration. The panel characterized the legal framework as follows: (1) the need to protect migratory species has been recognized by the WTO and numerous international instruments; (2) sustainable development is a WTO objective; (3) there was the common opinion of WTO membership expressed in the 1996 Report of the Committee on Trade and Environment endorsing "multilateral solutions based on international cooperation and consensus as the best and most effective way ... to tackle environmental problems of a transboundary or global nature"; and (4) the parties to the dispute have accepted almost all of the relevant multilateral environmental agreements.[76]

Climate change raises similar issues, and the jurisprudence in the *Shrimp* case will be highly relevant in the event of a WTO dispute over climate change mitigation or adaptation measures. This WTO jurisprudence also suggests some other factors that may be relevant to determining the WTO consistency of climate change measures. The key factors that need to be taken into account in assessing whether to permit unilateral trade measures to conserve transboundary or global resources can be summarized as follows. They must be preceded by good faith efforts to reach a negotiated solution, be applied flexibly to take into account different conditions among countries, and comply with transparency and procedural fairness. In some circumstances, they may require technical assistance. However, the degree of urgency may be a factor that requires much greater or lesser effort to reach a negotiated solution, particularly with respect to the time frame involved.

The US Safe Climate Act, which has not been implemented, is an example of the types of unilateral border taxes and subsidies that might be implemented to address competitiveness concerns. A stated purpose is to address the threat of carbon leakage, and it provides that an international agreement with GHG-emitting nations is a preferable alternative. It is a unilateral measure designed to pressure other countries to limit their GHG emissions. It would require an annual report to Congress regarding whether China and India have adopted GHG standards at least as strict as those adopted by the United States.[77] However, this type of regime requires careful design and implementation in order to avoid violating WTO rules.[78]

Like the EU Aviation Directive, the US Safe Climate Act shows that it is possible to make progress on climate change unilaterally, not just multilaterally. The same is true with respect to international economic law. In the case of the WTO, for example, countries can eliminate trade barriers unilaterally, as long as the most-favored nation (MFN) rule is observed, and bilaterally or plurilaterally, as long as GATT Article XXIV and the General Agreement on Trade in Services (GATS)

[76] Panel Report, *US—Shrimp (Article 21.5—Malaysia)* paras. 5.53–5.57.

[77] Morosini, "Trade and Climate Change" 734–5.

[78] Joost Pauwelyn, "U.S. Federal Climate Policy and Competitiveness Concerns: The Limits and Options of International Trade Law" (2007) Nicholas Institute for Environmental Policy Solutions Working Paper NI WP 07–02.

Article V are observed. In the case of climate change, however, the regime is less developed and there is greater uncertainty than in the WTO (and this in spite of some uncertainty in the WTO regarding the compliance of many free trade agreements with WTO rules). It might be useful to make explicit provision for principles such as MFN and MFN exceptions in the UNFCCC, to reduce the uncertainty regarding the consistency of unilateral, bilateral, and plurilateral approaches to climate change regulation. As with the WTO, unilateral, bilateral, and plurilateral approaches can complement multilateral approaches, by pushing countries to follow suit multilaterally. The UNFCCC already incorporates the language of GATT Article XX regarding arbitrary or unjustifiable discrimination between countries.

2.9 Conclusion

In general, the concepts and principles of international environmental law and those of WTO law are not inherently inconsistent with each other. Indeed, many of the same concepts and principles find expression in both areas of public international law, albeit in different ways: sustainable development, the precautionary principle, and common but differentiated responsibilities/special and differential treatment are key examples. The same principles find expression in the UNFCCC and the Kyoto Protocol. Moreover, the international jurisprudence has taken a consistent approach in key cases, in the WTO Appellate Body, the International Court of Justice and the ECJ. The boundaries of territoriality and sovereignty are still shifting, but recent jurisprudence has taken a consistent approach to unilateral measures in this regard, one that supports the limited use of unilateral measures to address climate change. In this chapter and in subsequent chapters, we analyze those limits.

The polluter-pays principle, which forms part of international environmental law, has encountered political resistance in international climate change law. In particular, the wide use of fossil fuel subsidies and the disincentives to subsidize clean energy technology reveal a need to change the multilateral regime, particularly the WTO SCM Agreement. This represents a weak point, where the multilateral consensus is lacking and unilateral measures might begin to fill the gap.

The science of climate change demonstrates that the need for mitigation and adaptation is becoming increasingly urgent. International regulatory capture has become a source of multilateral paralysis in both the UNFCCC and the WTO. Regulatory capture also creates risks that unilateral measures will serve as disguised restrictions on international trade rather than legitimate efforts to combat climate change. For this reason, unilateral measures should be designed and applied in accordance with GATT Article XX, to minimize the risk of unilateral measures that constitute arbitrary or unjustifiable discrimination or disguised restrictions on international trade. This same language has been incorporated into international environmental law and the UNFCCC. The political and economic context that has led to multilateral negotiation paralysis means that unilateralism may represent the future of climate change regulation, at least in the short to medium term. However, this does not mean that we cannot use the multilateral consensus that has been achieved so far to regulate the use of unilateral measures.

In subsequent chapters, we will analyze the limits of unilateralism. Chapter 3 analyzes the limits that WTO law places on measures that affect international trade. Chapter 4 will analyze the limits that foreign investment agreements place on climate change measures and consider the consequences of these measures for the attraction of foreign investment and the capital and technology that it can bring. Chapter 5 will analyze the implications of intellectual property rights for technology transfer to adapt farming to a changing climate and argue in favor of flexibility in multilateral and regional agreements to maintain a range of policy options. We favor an approach to intellectual property rights that maintains policy flexibility in response to an uncertain political and economic environment.

In the current political and economic context, multilateralism is not working in the UNFCCC and WTO systems. This requires unilateralism, but in a well-considered approach that creates economic incentives to engage the private sector and to push governments into effective multilateral agreements. For example, funding for adaptation and technology transfer should be made conditional on recipients implementing PPP-based mitigation measures, in order to channel funding and technology to combat climate change. This can be accompanied by unilateral trade measures on goods and services to create political will in developing countries; the private industry that opposes climate change action based on competitive concerns might change their stance if unilateral trade measures begin to affect market access and competitiveness. However, unilateral measures to combat climate change need to be taken in a manner that is consistent with existing obligations and principles of international environmental and economic law, in order to more effectively address this urgent global issue.

3

Unresolved Issues in WTO Law

3.1 Introduction

Measures aimed at addressing climate change raise legal issues regarding the relationship between WTO law and international environmental law and the relationship between various WTO agreements. This chapter analyzes several unresolved issues in WTO law that may affect the WTO-consistency of unilateral and multilateral measures that are likely to be taken to address climate change: (1) how to deal with environmental subsidies under the GATT 1994, the SCM Agreement,[1] and the Agreement on Agriculture;[2] (2) whether processing and production methods are relevant to determining the issue of "like products" in GATT Articles I and III, the SCM Agreement, the Agreement on Implementation of Article VI of the General Agreement on Tariffs and Trade 1994 (Antidumping Agreement),[3] and the Agreement on Technical Barriers to Trade (TBT Agreement);[4] (3) the relationship between GATT Article XX, other WTO Agreements, and multilateral environmental agreements; and (4) the interpretation and application of Article 2 of the TBT Agreement. The focus of this chapter is these unresolved issues rather than the WTO-consistency of specific climate change measures.

Recent WTO jurisprudence has shed some light on these issues. Recent jurisprudence on the TBT Agreement has further clarified the extent to which GATT jurisprudence can be used to interpret the similar provisions in other WTO agreements, in what might be called judicial cross-pollination. This jurisprudence is especially important regarding the interpretation of provisions that use the terms "like products" and "less favorable treatment" and provisions that incorporate language that is similar or identical to terms used in GATT Article XX. The interpretative approach of the WTO Appellate Body to these issues also should

[1] Agreement on Subsidies and Countervailing Measures GATT Secretariat, The Results of the Uruguay Round of Multilateral Trade Negotiations, the Legal Texts (Geneva, 1994) 231 (SCM Agreement).

[2] Agreement on Agriculture, GATT Secretariat, The Results of the Uruguay Round of Multilateral Trade Negotiations, the Legal Texts (Geneva, 1994), 33.

[3] Agreement on Implementation of Article VI of the General Agreement on Tariffs and Trade 1994, GATT Secretariat, The Results of the Uruguay Round of Multilateral Trade Negotiations, the Legal Texts (Geneva, 1994), 147 (Antidumping Agreement).

[4] Agreement on Technical Barriers to Trade, GATT Secretariat, The Results of the Uruguay Round of Multilateral Trade Negotiations, the Legal Texts (Geneva, 1994), 121 (TBT Agreement).

influence the approach taken by other international tribunals that interpret provisions in international agreements that incorporate language that is similar or identical to terms used in GATT Article XX.

The focus of this chapter is on unresolved issues. However, climate change measures also need to comply with other provisions, such as GATT Article XI (quantitative restrictions) and GATT Article II (maximum tariff levels). Moreover, competitiveness concerns due to differences in climate change regulation between countries (such as carbon taxes) cannot be addressed with antidumping duties (because the relevant comparison is not between regulatory differences, but rather prices in the domestic market and the export market, or a constructed price or third country market and the export market) or countervailing duties (because not taxing carbon, or taxing carbon at a lower level in one country than in another, does not qualify as a "subsidy").[5]

3.2 Emerging Climate Change Policies

The principal policy alternatives to address climate change fall under three categories: (1) the cap-and-trade approach; (2) standards-based policies, which require the adoption of specific measures or set source-specific emissions limits; and (3) carbon taxes.[6] Depending on the manner in which these policies are implemented, they may raise issues of WTO compatibility. If pollution permits are distributed or sold in a discriminatory manner, a cap-and-trade system could be inconsistent with the nondiscrimination obligations of GATT Articles I:1 and III:4. Similarly, if carbon taxes are applied in a discriminatory manner, there could be a violation of GATT Article III:2. If the revenue from carbon taxes is used to grant subsidies, those subsidies might be inconsistent with the SCM Agreement. Standards-based policies could also be implemented in a discriminatory manner, contrary to the GATT and the TBT Agreement. Environmental labels, such as those based on the carbon footprint of products, may be designed to favor domestic industry and may be inconsistent with these WTO Agreements.

A cap-and-trade system establishes a price on emissions through market forces. It places a limit on the combined emissions of a group of regulated pollution sources by creating a limited number of tradable pollution permits or emissions allowances for a given period and requiring firms to surrender a quantity of allowances equal to their emissions during that period, without imposing particular limits on emissions from any given firm or source. The government distributes pollution permits to firms, for free or sold at auction. The firms can then sell a portion of their permits to firms whose cost of reducing emissions is higher. The emissions cap can be imposed

[5] Joost Pauwelyn, "U.S. Federal Climate Policy and Competitiveness Concerns: The Limits and Options of International Trade Law" (2007) Nicholas Institute for Environmental Policy Solutions Working Paper NI WP 07-02.

[6] Robert N. Stavins, "A U.S. Cap-and-Trade System to Address Global Climate Change" <http://ssrn.com/abstract=1026353> (accessed March 15, 2013).

"upstream" (on fossil fuels at the point of extraction, processing, or distribution) or "downstream" (at the point of combustion). A cap-and-trade system can gradually reduce the caps over time to encourage investment in emission-reducing technology.[7] Cap-and-trade systems are easier to harmonize internationally, since pollution permits denominated in units of carbon content of fossil fuels or CO2 emissions create a natural unit of exchange for harmonization.[8]

Carbon taxes are a market-based alternative to a cap-and-trade system. Both policies put a price on CO2 emissions. However, whereas a carbon tax sets the price of CO2 emissions, a cap-and-trade system sets the amount of emissions and allows the price of the emissions to adjust to meet the emissions cap. A cap-and-trade system provides more certainty regarding whether emissions targets will be met, but less certainty regarding the cost of meeting those targets. Conversely, a carbon tax provides less certainty regarding whether emissions targets will be met, but more certainty regarding the cost of meeting the emissions targets. Under a cap-and-trade system, the firms that face higher adjustment costs can be granted pollution permits that can be sold in the market. Under a carbon tax system, those firms can be granted tax exemptions or benefit from the redistribution of the tax revenue. Tradable tax exemptions and redistribution of tax revenue could provide the same level of distributional flexibility as a cap-and-trade system, but political and practical considerations may limit what can be done in practice.[9] Moreover, the manner in which economic benefits are distributed may raise issues regarding the compliance of the system with the SCM Agreement. The principal advantage of a carbon tax is that it eliminates the potential price volatility of a cap-and-trade system, while the principal disadvantage of a carbon tax is political resistance to new taxes.[10]

Standards—a command-and-control system—could be used instead of or together with a market-based cap-and-trade system. However, standards may be less effective and less efficient than a cap-and-trade system, because: (1) standards are likely to apply to new equipment only, thereby creating a disincentive to replace old equipment; (2) standards are unlikely to address all sources of emissions, due to administrative limitations and pressure on legislators to grant exemptions; (3) the lack of flexibility inherent in uniform standards does not permit taking into account variations in the cost of compliance; and (4) it is more difficult to mitigate the distributional implications of standards, meaning that the costs of compliance can vary significantly.[11]

In addition to the foregoing policy alternatives, countries may choose to apply tariffs or border taxes that discriminate between different products based on differences in national climate change policies or differences in the carbon footprints of products, or may provide direct subsidies to domestic producers. Concerns

[7] Robert N. Stavins, "A U.S. Cap-and-Trade System to Address Global Climate Change" 7–8.
[8] Robert N. Stavins, "A U.S. Cap-and-Trade System to Address Global Climate Change" 53.
[9] Robert N. Stavins, "A U.S. Cap-and-Trade System to Address Global Climate Change" 49–51.
[10] Robert N. Stavins, "A U.S. Cap-and-Trade System to Address Global Climate Change" 52.
[11] Robert N. Stavins, "A U.S. Cap-and-Trade System to Address Global Climate Change" 48–9.

regarding the impact of carbon taxes on competitiveness can lead to the implementation of accompanying border tax adjustments or export subsidies to compensate for a lack of carbon taxes or equivalent measures in trade partners. As we noted in Chapter 2, the United States proposed to implement this type of system, in its Safe Climate Act.[12] The GATT consistency of such border tax adjustments is unclear.[13] Some companies have begun to measure the carbon footprint of their products and to display the results on their websites or on their packaging.[14] This trend coincides with a movement towards standards regarding the measurement of carbon footprints, and labeling and may facilitate the design and application of tariffs, border taxes, or other trade restrictions that are based on differences in the carbon footprints of products.

The EU enacted a biofuels directive to impose sustainability criteria and require certification of biofuels production according to their greenhouse gas emissions.[15] Malaysia and Indonesia account for the majority of biofuels made from palm oil, which would fall below the minimum EU requirement.[16] Brazil has complained about US and EU ethanol tariffs, which protect American corn-based ethanol and European ethanol made from sugar beets and wheat, and has sought their elimination in the Doha Round.[17] The United Nations Development Program has called for the elimination of these ethanol tariffs because Brazilian sugarcane-based ethanol reduces greenhouse gas emissions by up to 70 percent, compared to 13 percent for corn-based ethanol.[18] Brazil also has complained about US subsidies for corn-based ethanol and the EU has threatened to impose countervailing duties to address the same subsidies.[19]

[12] Fabio Morosini, "Trade and Climate Change: Unveiling the Principle of Common but Differentiated Responsibilities from the WTO Agreements" (2010) 42 George Washington International Law Review 713, 734–5.

[13] This issue has been analyzed elsewhere. In addition to Morosini, see Pauwelyn, "U.S. Federal Climate Policy and Competitiveness Concerns"; Richard G. Tarasofsky, "Heating Up International Trade Law: Challenges and Opportunities Posed by Efforts to Combat Climate Change" (2008) 2 Carbon & Climate Law Review 7, 11.

[14] Andrew Martin, "How Green Is My Orange?" New York Times, January 21, 2009 <http://www.nytimes.com/2009/01/22/business/22pepsi.html> (accessed March 8, 2009).

[15] "EU's Sustainable Biofuels Push Angers Malaysia, Brazil" mongabay.com (November 7, 2008) <http://news.mongabay.com/2008/1107-biofuels.html> (accessed March 10, 2009).

[16] "EU Biofuels Regulation" Directive 2009/28/EC of the European Parliament and of the Council of 23 April 2009 on the promotion of the use of energy from renewable sources and amending and subsequently repealing Directives 2001/77/EC and 2003/30/EC (Text with EEA relevance), Art. 17 <http://eur-lex.europa.eu/LexUriServ/LexUriServ.do?uri=OJ:L:2009:140:0016:0062:EN:PDF> (accessed April 5, 2013); Multa Fidrus, "Minister Lobbies EU over Palm Oil Restrictions" The Jakarta Post <http://www.orangutans.com.au/Manager/Item.aspx?id=Minister-lobbies-EU-over-palm-oil-restrictions&AspxAutoDetectCookieSupport=1> (accessed March 7, 2009).

[17] Bradley S. Klapper, "Brazil Weighs WTO Ethanol Case Against US" Associated Press (July 30, 2008).

[18] Jim Lane, "Brazilian sugarcane head said EU, US, Argentine biofuel trade policies about economic benefit, not global emission reductions" Biofuels Digest (January 17, 2008) <http://www.biofuelsdigest.com/blog2/2008/01/17/brazilian-sugarcane-head-said-eu-us-argentine-biofuel-trade-policies-about-economic-benefit-not-global-emission-reductions/> (accessed March 8, 2009).

[19] Inae Riveras, "Biofuels Protectionism Trumps Climate Concerns" Reuters (January 16, 2008) <http://www.reuters.com/article/GlobalAgricultureandBiofuels08/idUSN16611111120080116> (accessed March 8, 2009); *United States—Domestic Support and Export Credit Guarantees for Agricultural*

3.3 Environmental Subsidies

Subsidies will raise issues under GATT 1994, the Agreement on Agriculture and the SCM Agreement. The SCM Agreement applies cumulatively with GATT Articles VI and XVI. If the subsidies apply to agricultural goods, they may raise issues under the SCM Agreement or the Agreement on Agriculture, or both. In general, panels examine claims under the more specific agreement on trade in goods before examining claims under the GATT 1994, because a provision of the more specific agreement prevails over a GATT 1994 provision in the event of a conflict.[20] However, the SCM Agreement does not preclude action "under other relevant provisions of GATT 1994, where appropriate."[21] Moreover, while claims regarding agricultural subsidies are examined first under the Agreement on Agriculture,[22] some are also subject to the disciplines of the SCM Agreement.

An export subsidy that violates Articles 3.3 and 8 of the Agreement on Agriculture also violates SCM Agreement Article 3.1(a).[23] However, it seems less likely that an export subsidy that is *consistent* with the Agreement on Agriculture could be impugned under SCM Agreement Article 3.1(a). Since the former agreement permits certain export subsidies and the latter prohibits all export subsidies, there appears to be a conflict.[24] Given the conflict, the more specific provisions of the Agreement on Agriculture should prevail.[25] Not everyone agrees with this conclusion. However, the inconsistency between the two agreements is such that it would be difficult to reconcile through interpretation. If the application of the SCM Agreement means that the right to employ certain export subsidies in accordance with the Agreement on Agriculture is denied, the relevant provisions of the Agreement on Agriculture would not be effective. This result would run counter to the rule of effective treaty interpretation. Both agreements define export subsidies as being contingent upon export performance.[26] Nevertheless, this issue will become moot if WTO Members eliminate all agricultural export subsidies, as they

Products, DS365 (panel established December 17, 2007 but not composed) <http://www.wto.org/english/tratop_e/dispu_e/cases_e/ds365_e.htm> (accessed December 21, 2012).

20 General interpretative note to Annex 1A.

21 SCM Agreement note 56.

22 SCM Agreement art. 3.1.

23 Appellate Body Report, *United States—Subsidies on Upland Cotton* (*US—Upland Cotton*), WT/DS176/AB/R, adopted 21 March 2005, paras. 582–4.

24 In this situation, there is a conflict of norms in the sense that the exercise of rights under one norm constitutes a breach under the other norm. See Joost Pauwelyn, *Conflict of Norms in Public International Law: How WTO Law Relates to other Rules of International Law* (Cambridge University Press, Cambridge 2003) 275.

25 Agreement on Agriculture art. 1.

26 Agreement on Agriculture art. 1(e); SCM Agreement art. 3.1(a); Luís Yahir Acosta Pérez, "La Relación entre el Acuerdo sobre Agricultura y el Acuerdo sobre Subvenciones y Medidas Compensatorias" (August 2009), Documento de trabajo no. 9 <http://cdei.itam.mx/AcostaSMCAA.pdf> (accessed December 21, 2012); Bradly J. Condon, *El Derecho de la Organización Mundial de Comercio: Tratados, Jurisprudencia y Practica* (Cameron May, London 2007) 278.

agreed to do at the Ministerial conference in Hong Kong.[27] In contrast, even if WTO Members comply with their obligations regarding domestic support commitments in the Agreement on Agriculture, they could still violate Article 3.1(b) of the SCM Agreement if those domestic subsidies are made contingent on the use of domestic goods.[28]

Subsidies are also subject to Part III (actionable subsidies) and Part V (countervailing measures) of the SCM Agreement. Countervailing measures can be applied to imports to counter the effect of subsidized products where the subsidy causes injury to the domestic producers of like products. Part III can be used to attack subsidies that cause adverse effects in third country markets, for example due to lost sales and price suppression. Unlike countervailing measures, there is no obligation in the SCM Agreement to quantify the precise amount of the subsidy for purposes of an adverse effects claim.[29] Agricultural subsidies are also subject to these Parts of the SCM Agreement.[30] Thus, subsidies related to climate change policies could be subject to multilateral action under Part III or unilateral action under Part V.

In addition to more obvious subsidies, carbon taxes could be structured in a manner that violates provisions of the SCM Agreement (for example, differential taxation of "carbon-friendly" products). Since the SCM Agreement and GATT Article III:2 are not mutually exclusive, such measures could be subject to both sets of obligations. These two sets of provisions can apply cumulatively to different aspects of the same measure.[31] The same logic would apply to the relationship between the SCM Agreement and GATT Article I:1.

While some have argued in favor of imposing countervailing duties against products from countries that do not require emissions reductions, the definition of subsidy would likely preclude such actions unless, for example, a country applied a general carbon tax but then subsidized a specific industry by not collecting the tax.[32] The SCM Agreement only applies to a measure if it constitutes a subsidy within the meaning of SCM Agreement Article 1.1. A "financial contribution" and a "benefit" are two separate legal elements in Article 1.1, which together determine whether a subsidy exists.[33] The differential application of carbon taxes could constitute a "financial contribution by a government" within the meaning of Article 1.1(a)(1)(ii). The two principal cases on this point held that there was a subsidy within the meaning of SCM Agreement Article 1.1(a)(1)(ii) in the following

[27] Hong Kong Ministerial Declaration, WT/MIN(05)/DEC, adopted December 18, 2005, para. 6.

[28] Appellate Body Report, *US—Upland Cotton* para. 550.

[29] Appellate Body Report, *United States—Measures Affecting Trade in Large Civil Aircraft—Second Complaint (US—Large Civil Aircraft (2nd complaint))*, WT/DS353/AB/R, adopted 23 March 2012, para, 697.

[30] Condon, *El Derecho de la Organización Mundial de Comercio* 334.

[31] Panel Report, *Indonesia—Certain Measures Affecting the Automobile Industry* (*Indonesia—Autos*), WT/DS54/R, WT/DS55/R, WT/DS59/R, WT/DS64/R, adopted July 23, 1998, paras. 14.36, 14.97–14.99.

[32] Tarasofsky "Heating Up International Trade Law" 14.

[33] Appellate Body Report, *Brazil—Export Financing Programme for Aircraft* (*Brazil—Aircraft*), WT/DS46/AB/R, adopted August 20, 1999, para. 156.

situations: (1) different tax treatment for income from foreign and domestic sales (*US—FSC*) and (2) an exemption from payment of an MFN import duty that would otherwise apply to auto imports, conditional upon domestic production requirements (*Canada—Autos*).

The source of a subsidy affects the evidence required and the difficulty to prove that the subsidy is subject to the SCM Agreement. The SCM Agreement distinguishes between subsidies made by "a government or any public body" and those made by a "private body." Financial contributions from a private body are only subject to the SCM Agreement if there is an affirmative demonstration of the link between the government and the specific conduct of the private body, an element that does not need to be demonstrated in the case of a governmental entity. In the SCM Agreement, unlike Article 5 of the International Law Commission (ILC) Articles on State Responsibility, the question of attribution of conduct to a State requires an examination of both the particular conduct and the type of entity. A public body's conduct can be attributed directly to the State, whereas a private body's conduct can be attributed to the State only indirectly. A public body within the meaning of Article 1.1(a)(1) of the SCM Agreement must possess, exercise or be vested with governmental authority. The question of whether an entity is public or private depends on whether an entity is vested with authority to exercise governmental functions. However, apart from an express delegation of authority in a legal instrument, the existence of mere formal links between an entity and government is insufficient to establish the necessary possession of governmental authority. For example, the mere fact that a government is the majority shareholder of an entity does not demonstrate that the government exercises meaningful control or has bestowed it with governmental authority. However, where the evidence shows formal government control, and that such control has been exercised in a meaningful way, the evidence may permit an inference that the entity concerned is exercising governmental authority.[34]

According to the Appellate Body, "the mere fact that revenues are not 'due' from a fiscal perspective does not determine that the revenues are or are not 'otherwise due' within the meaning of Article 1.1(a)(1)(ii) of the SCM Agreement."[35] A "financial contribution" does not arise simply because a government does not raise revenue which it could have raised. The term "otherwise due" implies a comparison with a "defined normative benchmark," as established by the tax rules applied by the Member in question.[36] The determination of "whether revenue foregone is otherwise due must allow a comparison of the fiscal treatment of comparable income, in the hands of taxpayers in similar situations."[37]

[34] Appellate Body Report, *United States—Definitive Anti-Dumping and Countervailing Duties on Certain Products from China (US—Anti-Dumping and Countervailing Duties (China))*, WT/DS379/AB/R, adopted March 25, 2011, paras. 284–322.

[35] Appellate Body Report, *United States—Tax Treatment for "Foreign Sales Corporations" (Article 21.5—EC) (US—FSC (Article 21.5—EC))*, WT/DS108/AB/RW, adopted January 29, 2002, para. 88.

[36] Appellate Body Report, *United States—Tax Treatment for "Foreign Sales Corporations"* (*US—FSC*), WT/DS108/AB/R, adopted March 20, 2000, para. 90.

[37] Appellate Body Report, *US—FSC (Article 21.5—EC)* para. 98.

If a country taxes products according to their carbon footprint, the most probable result is that different products will be subject to different tax rates. One example is where fossil fuels are subject to a sales tax that is not applied to other products, as is already the case in some jurisdictions. For example, the Canadian province of British Columbia introduced a carbon tax on fossil fuels in 2008. A more elaborate scheme might apply different levels of sales taxes to different categories of products based on different ranges of carbon footprints, taking into account the production of carbon emissions during the lifecycle of the products. If such schemes are designed so that domestic products are subject to a lower tax than imported products, then the lower tax rate might constitute revenue foregone that is otherwise due. If both the domestic and imported products are substitutable inputs for domestic production (as is the case with fuels) and the foregone revenue confers a benefit, then there could be a violation of SCM Agreement Article 3.1(b). This could be the case if countries diverge in their regulation and reduction of carbon emissions, so that some countries engage in less carbon-intensive production than others. Moreover, such divergences in the carbon intensity of production would probably lead to differential treatment of imports, thereby raising issues regarding MFN treatment. The reference in footnote 1 to "the exemption of an exported product from duties or taxes borne by the like product" could indicate that Article 1.1(a)(1)(ii) is intended to apply to other cases where like products receive different consumption tax treatment. In this regard, the Appellate Body has stated: "The tax measures identified in footnote 1 as not constituting a 'subsidy' involve the exemption of exported products from product-based consumption taxes."[38] However, revenue is not otherwise due just because certain revenue is not taxed (or not taxed at as a high a level as it could be); a WTO Member is "free *not* to tax any particular categories of revenues."[39] Thus, it is not clear in which circumstances differential taxation of products based on their carbon footprints might constitute a "financial contribution by a government" within the meaning of the SCM Agreement.

In *Canada—Aircraft*, the Appellate Body interpreted the term "benefit" under Article 1.1(b) as follows: "a financial contribution will only confer a 'benefit', i.e., an advantage, if it is provided on terms that are more advantageous than those that would have been available to the recipient on the market."[40] "A 'benefit'... must be received and enjoyed by a beneficiary or a recipient" and "calls for an inquiry into what was conferred on the recipient"; the measurement of the benefit is not whether there was a cost to the government.[41] The person or entity receiving the benefit does not have to be the same as the one who received the financial contribution.[42] The assessment of benefit must examine the terms and conditions

[38] Appellate Body Report, *US—FSC* para. 93.

[39] Appellate Body Report, *US—FSC* para. 90.

[40] Appellate Body Report, *Canada—Measures Affecting the Export of Civilian Aircraft* (*Canada—Aircraft*), WT/DS70/AB/R, adopted August 20, 1999, para. 149.

[41] Appellate Body Report, *Canada—Aircraft* para. 154.

[42] Appellate Body Report, *United States—Countervailing Measures Concerning Certain Products from the European Communities*, WT/DS212/AB/R, adopted January 8, 2003, para. 110.

of the challenged transaction at the time it is made and compare them to the terms and conditions that would have been offered in the market at that time.[43] Thus, Article 1.1(b) could require an analysis of whether a benefit is obtained from differential carbon tax rates, by whom, whether such a benefit could have been otherwise obtainable in the marketplace, and what the relevant marketplace is.

The Panel Reports in *Canada—Renewable Energy* and *Canada—Feed-In Tariff Program* indicate that the issue of benefit might be used to exclude clean energy subsidies from the application of the SCM Agreement. The issue was which market provides the most appropriate benchmark in determining the existence and magnitude of a subsidy benefit for solar and wind power producers, in particular the extent to which the wholesale market for electricity in Ontario should be the appropriate focus of the benefit analysis.[44] In the absence of Ontario's feed-in-tariff (FIT) program, a competitive wholesale market for electricity in Ontario could not support commercially viable operations of solar and wind power producers.[45] The Panel rejected the complainants' argument that the analysis of benefit should compare the terms and conditions of participation in the FIT Program with those that would be available to generators participating in a wholesale electricity market where there is effective competition. The evidence indicated that competitive wholesale electricity markets will rarely operate to remunerate adequately the mix of generators needed to secure a *reliable* electricity system that pursues *human health and environmental* objectives through the inclusion of facilities using solar photovoltaic and wind technologies into the supply-mix. However, the Panel also rejected Canada's argument that the relevant market comparator must be the market for electricity produced from wind and solar power technologies. The majority held that none of the alternatives that had been advanced by the complainants or Canada could be used as appropriate benchmarks against which to measure whether the FIT Program conferred a benefit within the meaning of Article 1.1(b) of the SCM Agreement.[46] However, in a dissenting opinion, one panelist concluded that there was a benefit. By bringing these high cost and less efficient electricity producers into the wholesale electricity market, when they would otherwise not be present, Ontario's purchases of electricity from solar and wind power generators under the FIT Program conferred a benefit.[47] On appeal, the Appellate Body agreed that none of the benchmarks were appropriate, but for

[43] Appellate Body Report, *European Communities—Measures Affecting Trade in Large Civil Aircraft (EC and certain member States—Large Civil Aircraft)*, WT/DS316/AB/R, adopted June 1, 2011, para. 838; Appellate Body Report, *United States—Large Civil Aircraft (2nd complaint)*, WT/DS353/AB/R, adopted March 23, 2012, para. 636.

[44] Panel Report, *Canada—Certain Measures Affecting the Renewable Energy Generation Sector (Canada—Renewable Energy)*, WT/DS412/R and *Canada—Measures Relating to the Feed-in Tariff Program (Canada—Feed-In Tariff Program)*, WT/DS426/R, adopted May 24, 2013, para. 7.270. See also Appellate Body Report, WT/DS/426/AB/R, adopted May 24, 2013, paras. 5.79, 5.219.

[45] Panel Reports, *Canada—Renewable Energy and Canada—Feed-In Tariff Program* paras. 7.276–7.277.

[46] Panel Reports, *Canada—Renewable Energy and Canada—Feed-In Tariff Program* paras. 7.309–7.319.

[47] Panel Reports, *Canada—Renewable Energy and Canada—Feed-In Tariff Program* para. 9.23.

different reasons. The absence of a general environmental exception in the SCM Agreement makes the role of the benefit analysis important in saving clean energy subsidies from violating the SCM Agreement. The Appellate Body's analysis suggests that the appropriate comparison is between different renewable energy prices, not between renewable energy prices and fossil fuel prices. What are the implications for clean energy subsidies made under the Clean Development Mechanism?

In *Canada—Renewable Energy* and *Canada—Feed-In Tariff Program* the FIT Program imposed domestic input requirements on the recipients. The Appellate Body found this to be a violation of GATT Article III:4 and Article 2.1 of the Agreement on Trade-Related Investment Measures (TRIMS Agreement). Paragraph 1(a) of the Illustrative List in the Annex to the TRIMS Agreement provides that TRIMS are inconsistent with GATT Article III:4 when compliance is necessary to obtain an advantage and requires the purchase or use by an enterprise of products of domestic origin. GATT Article III:8(a) did not exclude the FIT Program from the application of Article III:4 because the generation equipment was not in a competitive relationship with the product purchased by the government (electricity).[48] Thus, while the benefit analysis saved the FIT Program from a finding of inconsistency with Article 3.1(b) of the SCM Agreement, it remains inconsistent with WTO law. However, a finding of inconsistency with GATT Article III:4 leaves open the possibility of justification under GATT Article XX, which is unlikely to be a possibility in the event of a violation of the SCM Agreement. Thus, the approach to the analysis of this issue in *Canada—Renewable Energy* and *Canada—Feed-In Tariff Program* leaves open the possibility that *non-discriminatory* clean energy subsidies could survive a WTO challenge.

Could environmental subsidies that are inconsistent with the SCM Agreement or the Agreement on Agriculture be justified under GATT Article XX? Marceau and Trachtman have suggested that it would require a "heroic approach to interpretation" to extend the application of GATT Article XX to justify a violation under another agreement of Annex 1A.[49] However, in *US—Shrimp (Thailand)* and *US—Customs Bond Directive*, the Appellate Body declined to express a view on whether a defense under GATT Article XX(d) was available to justify a measure found to constitute a "specific action against dumping" under Article 18.1 of the Antidumping Agreement.[50] Article 18.1 of the Anti-Dumping Agreement provides that "[n]o specific action against dumping of exports from another Member can be taken except in accordance with the provisions of the GATT 1994, as interpreted by this Agreement." Having found that the enhanced continuous bond requirement constituted "specific action against dumping" and that it was not a "reasonable security" under the Ad Note to Article VI of the GATT 1994, and thus was not "in accordance with the provisions of the GATT 1994, as interpreted by the

[48] Appellate Body Report, *Canada—Renewable Energy and Canada—Feed-In Tariff Program*, WT/DS/426/AB/R, adopted May 24, 2013, para. 5.79.

[49] Gabrielle Marceau and Joel Trachtman, "The Technical Barriers to Trade Agreement, the Sanitary and Phytosanitary Measures Agreement, and the General Agreement on Tariffs and Trade" (2002) 36 Journal of World Trade 811, 874.

[50] Appellate Body Report, *United States—Shrimp (Thailand)* and *United States—Customs Bond Directive*, WT/DS343/AB/R, WT/DS345/AB/R, adopted August 1, 2008, paras. 310, 319.

[Anti-Dumping] Agreement," the Panel in that case examined the United States' defense under Article XX(d) of the GATT 1994, but found that the measure could not be justified as necessary.

The chapeau of GATT Article XX indicates that the general exceptions apply to "this Agreement." This appears to exclude the application of Article XX beyond the provisions of the GATT itself. However, the provisions of the GATT serve as the starting point for the majority of the multilateral agreements on trade in goods.

In *China—Publications and Audiovisual Products*, the Appellate Body concluded that China could invoke GATT Article XX as a defense against a violation of section 5.1 of its Protocol of Accession.[51] In *China—Raw Materials*, both the Panel and the Appellate Body concluded that China could not invoke GATT Article XX as a defense against a violation of section 11.3 of its Protocol of Accession.[52] After observing that the WTO Agreement contains no general exception, the Panel concluded that the reference in the chapeau of Article XX to "this Agreement" indicates that its general exceptions apply only to the GATT, and not to other WTO Agreements. The Panel further noted that WTO Members had incorporated Article XX by reference, in the TRIMS Agreement, and that other WTO Agreements contained their own general exceptions, such as GATS Article XIV.[53] However, the Appellate Body limited its analysis to why GATT Article XX could not apply to section 11.3 of China's Protocol of Accession. The Appellate Body considered that Article XX could not be invoked to justify the violation of an obligation that was not regulated by the GATT. The obligation emanated exclusively from the Protocol. It also observed, as did the Panel, that section 11.3 made no reference to Article XX, even though it referred expressly to GATT Article VIII. Unlike sections 11.1 and 11.2, section 11.3 contained no obligation to ensure conformity with GATT. Moreover, unlike section 5.1 of the Protocol in *China—Publications and Audiovisual Products*, section 11.3 made no reference to the right of China to regulate trade in a manner compatible with the WTO Agreement.[54] The Appellate Body's reasoning in these cases indicates that the applicability of the general exceptions of GATT Article XX to other WTO Agreements depends on the specific provision and its context.

In the case of subsidies, GATT Articles VI (countervailing duties) and XVI (subsidies in general) apply together with the provisions of the SCM Agreement. Indeed, the principal object and purpose of the SCM Agreement is to augment and

[51] Appellate Body Report, *China—Measures Affecting Trading Rights and Distribution Services for Certain Publications and Audiovisual Entertainment Products (China—Publications and Audiovisual Products)*, WT/DS363/AB/R, adopted January 19, 2010; Bradly J. Condon, "Comentario sobre China—Publicaciones y productos audiovisuales, Informe del Órgano de Apelación" (2010) <http://cdei.itam.mx/medios_digitales/archivos/investigacion/ComentarioChinaPublicaciones.pdf> (accessed December 21, 2012).

[52] Appellate Body Report, *China—Measures Related to the Exportation of Various Raw Materials (China—Raw Materials)*, WT/DS394/AB/R, WT/DS395/AB/R, WT/DS398/AB/R, adopted February 22, 2012.

[53] Panel Report, *China—Measures Related to the Exportation of Various Raw Materials (China—Raw Materials)*, WT/DS394/R, WT/DS395/R, WT/DS398/R, adopted February 22, 2012, paras. 7.150–7.154.

[54] Appellate Body Report, *China—Raw Materials* paras. 290–303.

improve GATT disciplines regarding the use of subsidies and countervailing measures.[55] Also note that the name of the SCM Agreement (Agreement on Subsidies and Countervailing Measures), in contrast to that of the Antidumping Agreement (Agreement on Implementation of Article VI of the General Agreement on Tariffs and Trade 1994), does not indicate that it serves merely to interpret and apply GATT provisions. It would be odd if GATT Article XX could be applied to GATT Articles VI and XVI, but not to the SCM Agreement itself, absent evidence of a contrary intention (which SCM Agreement Article 8 might provide). This assumes that GATT Article XX could be applied to GATT Articles VI and XVI. However, it is not at all clear how this would work in the case of countervailing measures. Would environmental subsidies that meet the requirements of GATT Article XX be non-actionable and thus not subject to countervailing duties under Part V or multilateral action under Part III of the SCM Agreement? This was the case for a limited range of environmental subsidies before the expiry of SCM Agreement Article 8.[56] Since negotiators developed specific exceptions and language to address the issue of environmental subsidies, and did not incorporate the language of Article XX or incorporate Article XX by reference, it seems unlikely that GATT Article XX could be invoked to preclude action under Parts III and V. Moreover, in the case of actionable subsidies under Part III, negotiators specified that Articles 5 and 6 would not apply to subsidies maintained on agricultural products as provided in Article 13 of the Agreement on Agriculture.[57] While Article 13 has since expired, this indicates that negotiators turned their minds to the issue of whether to exclude certain types of subsidies from the application of Part III. Similarly, in Part V the non-actionability of certain types of subsidies, including environmental subsidies, was carefully circumscribed.[58]

Environmental subsidies could be non-actionable under Parts III and V of the SCM Agreement if differences in carbon footprints can be used to conclude that products are not "like" as that term is used in the SCM Agreement. The SCM Agreement uses the term "like products" for a variety of purposes. It is a pivotal issue in Part V regarding the initiation of countervailing duty investigations (Articles 11.2(i), 11.4, 16.1) and the determination of injury (Articles 15.1, 15.2, 15.3, 15.6) and, in Part III, regarding the determination of serious prejudice in paragraphs (a), (b), and (c) of Article 6.3.[59] In addition, environmental subsidies could be designed so as to not be specific to an enterprise or industry under SCM Agreement Article 2 and thereby be non-actionable under Parts III and V. However, if the importing Member produces environmental products that are like

[55] Appellate Body Report, *United States—Countervailing Duties on Certain Corrosion-Resistant Carbon Steel Flat Products from Germany*, WT/DS2133/AB/R, adopted December 19, 2002, para. 73; Appellate Body Report, *United States—Final Countervailing Duty Determination with Respect to Certain Softwood Lumber from Canada*, WT/DS257/AB/R, adopted February 17, 2004, para. 64; Appellate Body Report, *Brazil—Measures affecting Desiccated Coconut* (*Brazil—Desiccated Coconut*), WT/DS22/AB/R, adopted March 20, 1997, 15.

[56] SCM Agreement arts. 8.1, 8.2 (c), 8.3, 10, 31.

[57] SCM Agreement arts. 5, 6.9.

[58] SCM Agreement arts. 8, 10 fn. 35.

[59] Condon, *El Derecho de la Organización Mundial de Comercio* 331–2.

those that benefit from the subsidy in the exporting Member, the subsidies could be actionable under Parts III and V. Recent WTO disputes in this category relate to solar panel, wind power equipment and feed-in tariffs programs for the renewable energy sector.[60]

What about prohibited subsidies (export subsidies and subsidies contingent on the use of domestic goods)? Such subsidies are deemed to be specific under Article 2.3. SCM Agreement Article 32.1 provides that "[n]o specific action against a subsidy of another Member can be taken except in accordance with the provisions of GATT 1994, as interpreted by this Agreement." SCM Agreement note 56 provides that "[t]his paragraph is not intended to preclude action under other relevant provisions of GATT 1994, where appropriate." However, SCM Agreement Article 3.1 indicates that export subsidies and subsidies contingent on the use of domestic goods are prohibited "[e]xcept as provided in the Agreement on Agriculture." This could be interpreted as precluding the application of any other exceptions to SCM Agreement Article 3.1, including the general exceptions of GATT Article XX, which would make it difficult for export subsidies to be found consistent with WTO law, even those designed to address competitive disadvantages from domestic carbon taxes or other climate change measures with similar effects.

However, the preamble of the Agreement on Agriculture refers to "the need to protect the environment" and Article 14 of the Agreement on Agriculture indicates that the Agreement on the Application of Sanitary and Phytosanitary Measures[61] applies cumulatively to the Agreement on Agriculture. The preamble of the SPS Agreement indicates that it elaborates on GATT rules, in particular Article XX(b). Thus, the argument could be made that the Agreement on Agriculture opens the door to the application of GATT Article XX(b). However, the Agreement on Agriculture only applies to the "agricultural products" listed in Annex 1 of the Agreement on Agriculture.[62] Thus, even if one were to accept the foregoing argument, the application of GATT Article XX(b) to SCM Agreement Article 3.1 could be limited to measures affecting these agricultural goods. The more likely conclusion is that Article XX is not available to justify a violation of the Agreement on Agriculture. Rather, environmental subsidies that applied to agricultural products would have to comply with the commitments in the schedules of WTO Members. As with environmental subsidies for non-agricultural products,

60 *United States—Countervailing Duty Measures on Certain Products from China*, WT/DS437, Panel established September 28, 2012 <http://www.wto.org/english/tratop_e/dispu_e/cases_e/ds437_e.htm> (accessed December 21, 2012); *China—Measures Concerning Wind Power Equipment*, WT/DS419, in consultations December 22, 2010 <http://www.wto.org/english/tratop_e/dispu_e/cases_e/ds419_e.htm> (accessed December 21, 2012); Panel Reports, *Canada—Renewable Energy, Canada—Feed-In Tariff Program*; *European Union and certain Member States—Certain Measures Affecting the Renewable Energy Generation Sector*, in consultations November 5, 2012, WT/DS452, <http://www.wto.org/english/tratop_e/dispu_e/cases_e/ds452_e.htm> (accessed December 21, 2012).

61 Agreement on the Application of Sanitary and Phytosanitary Measures, GATT Secretariat, The Results of the Uruguay Round of Multilateral Trade Negotiations, the Legal Texts (Geneva, 1994) 59 (SPS Agreement).

62 Agreement on Agriculture art. 2.

environmental subsidies for agricultural products could be non-actionable under Parts III and V of the SCM Agreement either because they are not specific or because differences in carbon footprints are relevant in the like products analysis, as noted above.

A more general argument might be raised regarding the applicability of GATT Article XX to all Agreements in Annex 1A, including the SCM Agreement, based on the argument that all WTO Agreements are cumulative and apply simultaneously and that the effective interpretation principle requires that both rights (such as those in Article XX) and obligations are cumulative.[63] However, the foregoing analysis suggests that the applicability of GATT Article XX to the other agreements in Annex 1A would have to be considered one agreement at a time and one provision at a time. This approach is consistent with the view of the Appellate Body that the relationship between the GATT 1994 and the other agreements in Annex 1A must be considered on a case-by-case basis.[64] It is also consistent with *China—Raw Materials*, in which the Appellate Body found that China could not invoke Article XX as a defense under section 11.3 of its Protocol of Accession.[65] In *China—Publications*, the Appellate Body reasoned that Article XX could be invoked when the measure could be inconsistent with a GATT provision or a provision related to goods in another WTO agreement. Otherwise, a complainant could deprive a respondent of its rights under Article XX by avoiding claims under the GATT. In *China—Raw Materials*, the Appellate Body placed greater emphasis on the wording and the context of the specific provision. Taken together, these two cases suggest that, in order to invoke Article XX, a provision outside the GATT would have to contain a reference to Article XX, a right to regulate trade or other reference to the GATT. Nevertheless, where a provision contains no such reference, but incorporates language from Article XX, it might be interpreted in a manner that is consistent with Article XX, as the Appellate Body did in *US—Clove Cigarettes* with respect to TBT Agreement Article 2.1.[66] The foregoing series of Appellate Body reports, together with our analysis of the text of the SCM Agreement, indicate that Article XX could not be invoked as a defense under the SCM Agreement.

3.4 "Like Products" and Processing and Production Methods

The term "like products" is a key concept in the analysis of nondiscrimination obligations in GATT Articles I:1, III:2, and III:4, as well as in the TBT Agreement. The issue of whether processing and production methods (PPMs) are relevant to

[63] Marceau and Trachtman "The Technical Barriers to Trade Agreement" 874–5.

[64] Appellate Body Report, *Brazil—Desiccated Coconut* 13.

[65] Appellate Body Report, *China—Raw Materials* paras. 290–1.

[66] Danielle Spiegel Feld and Stephanie Switzer, "Whither Article XX? Regulatory Autonomy Under Non-GATT Agreements After *China—Raw Materials*" (2012) 38 Yale Journal of International Law 16, 29–30; Appellate Body Report, *United States—Measures Affecting the Production and Sale of Clove Cigarettes* (*US—Clove Cigarettes*), WT/DS406/AB/R, adopted April 24, 2012, para. 182.

determining the likeness of products also has implications for the SCM Agreement (and the Antidumping Agreement), since the term "like products" is relevant to a variety of key issues in those agreements.[67] In these two agreements this term is defined, whereas it is not in the GATT or the TBT Agreement. The determination of which environmental concerns can be taken into consideration to determine the likeness of products would have wide-ranging consequences in WTO law.

In *Japan—Alcoholic Beverages II*, the Appellate Body explained that:

> [the] concept of "likeness" is a relative one that evokes the image of an accordion. The width of the accordion must be determined by the particular provision in which the term "like" is encountered as well as by the context and the circumstances that prevail in any given case to which that provision may apply.[68]

The jurisprudence suggests that the same four criteria that apply to determine likeness in GATT Article III also apply in the context of GATT Article I.[69]

The Appellate Body has consistently applied the following four criteria to determine whether products are in a competitive relationship that would lead to the conclusion that they are like products under GATT Article III: (1) the physical properties, nature, and quality of the products; (2) the extent to which the products may serve the same or similar end uses in a given market; (3) the extent to which consumers perceive and treat the products as alternative means of satisfying a want or demand; and (4) tariff classification of the products. In *EC—Asbestos*, the Appellate Body noted that these criteria are "simply tools to assist in the task of sorting and examining the relevant evidence. They are neither treaty mandated nor a closed list of criteria that will determine the legal characterization of a product."[70] The source of the first three criteria is the Working Party Report on *Border Tax Adjustments*, while the fourth criterion was added by subsequent GATT panels. The purpose of the like products analysis is "to take account of evidence which indicates whether, and to what extent, the products involved are—or could be—in a competitive relationship in the marketplace."[71] The relevant market is the market in which the products compete, and the competitive relationship between the same products may be different in different markets.[72]

As the Appellate Body noted in *Japan—Alcoholic Beverages II*, the tariff classification of products is a problematic criterion where it is not sufficiently detailed. Since the Harmonized System only harmonizes the first six digits, each government has discretion to determine more detailed product classification. Moreover, the Harmonized System (HS) was not designed to resolve the issue of whether products

[67] Condon, *El Derecho de la Organización Mundial de Comercio* 331–2, 417.

[68] Appellate Body Report, *Japan—Taxes on Alcoholic Beverages* (*Japan—Alcoholic Beverages II*), WT/DS8/AB/R, WT/DS10/AB/R, WT/DS11/AB/R, adopted November 1, 1996, 21.

[69] Panel Report, *Indonesia—Autos* para. 14.141.

[70] Appellate Body Report, *European Communities—Measures Affecting Asbestos and Asbestos–Containing Products* (*EC—Asbestos*), WT/DS135/AB/R, adopted April 5, 2001, para. 102.

[71] Appellate Body Report, *EC—Asbestos* para. 103.

[72] Appellate Body Report, *EC—Asbestos* para. 103; Appellate Body Report, *Philippines—Taxes on Distilled Spirits (Philippines—Distilled Spirits)*, WT/DS403/AB/R, WT/DS396/AB/R, adopted January 20, 2012, paras. 168–9.

are similar according to the GATT.[73] However, even if HS classification cannot be applied directly to decide like products, it is useful to support the decision regarding likeness based on other criteria.

A key issue in the case of measures based on the carbon footprint of products is whether product PPMs can be used to determine likeness. Some might argue that, since the health effects of asbestos were relevant to determining likeness in *EC—Asbestos* (under the first and third criteria), the environmental effects of a production process are relevant to determine likeness. For example, Howse has argued that the Appellate Body ruling in *EC—Asbestos* supports the view that nondiscriminatory process-based measures are consistent with GATT Article III:4, based on their consideration of consumer preferences.[74] The same evidence can be considered under both Articles III and XX, for different purposes, according to the Appellate Body in *EC—Asbestos*:

> Under Article III:4, evidence relating to health risk may be relevant in assessing the competitive relationship in the marketplace between allegedly "like" products. The same, or similar, evidence serves a different purpose under Article XX(b), namely, that of assessing whether a Member has a sufficient basis for "adopting or enforcing" a WTO-inconsistent measure on the grounds of human health.[75]

However, the measures at issue in *EC—Asbestos* can be distinguished from measures based on the carbon footprint of a product's production process because the health effects of asbestos are related to the product as such, not its production method. Indeed, the principal argument against taking PPMs into account to determine likeness is that likeness should be based on the characteristics of the product as such. In *US—Tuna (Mexico)*, the GATT panel held that the term "like products" did not apply to production processes, but rather to products as such. It therefore did not permit differentiation between products based on production processes that had no effect on the quality of the product.[76] As an unadopted GATT report, *US—Tuna (Mexico)* has no normative value. However, a panel may find useful orientation in its reasoning.[77]

It would be more appropriate to determine whether non-product-related PPMs are relevant to determining likeness on a case-by-case basis, based on the evidence presented in each case. If a PPM determines the competitive relationship between products in a particular case, the PPM should be relevant to determining whether

[73] Appellate Body Report, *Japan—Alcoholic Beverages II* 25.

[74] Robert Howse, "The Appellate Body Rulings in the Shrimp/Turtle Case: A New Legal Baseline for the Trade and Environment Debate" (2002) 27 Columbia Journal of Environmental Law 491, 515.

[75] Appellate Body Report, *EC—Asbestos* para. 115.

[76] GATT Panel Report, *United States—Restrictions on Imports of Tuna* (*US—Tuna (Mexico)*), DS21/R, September 3, 1991, unadopted, BISD 39S/155, 41; Appellate Body Report, *United States—Safeguard Measures on Imports of Fresh, Chilled or Frozen Lamb Meat from New Zealand and Australia* (*US—Lamb Safeguard*), WT/DS177/AB/R, WT/DS178/AB/R, adopted May 16, 2001; Appellate Body Report, *United States—Transitional Safeguard Measure on Combed Cotton Yarn from Pakistan* (*US—Cotton Yarn*), WT/DS192/AB/R, November 5, 2001; Marceau and Trachtman "The Technical Barriers to Trade Agreement" 857–8.

[77] Appellate Body Report, *Japan—Alcoholic Beverages II* 14–15.

the products are like.[78] This approach would be consistent with the purpose of the like products analysis, "to take account of evidence which indicates whether, and to what extent, the products involved are—or could be—in a competitive relationship in the marketplace."[79] Moreover, the competitive relationship between products is not static, but rather evolves over time.[80] Thus, a factor that might not be considered relevant to this issue today could be relevant in the future.

In *Philippines—Distilled Spirits*, the Appellate Body held that differences in products' inputs that do not affect their competitive relationship are not an obstacle to finding that the products are similar with respect to their physical properties. Moreover, the production process was relevant to determining the likeness of the products in that case, since it was designed to ensure that the domestic products would be sufficiently similar to compete with the imported products. However, the Appellate Body also noted that GATT Article III:2 refers to "like products," not inputs, and that differences in inputs that do not affect the final products would not prevent a finding of likeness.[81] Thus, it is not clear from this case whether the approach in this case would permit a finding that products are not like where PPMs affect the competitive relationship but do not affect the final product as such. However, it makes it clear that the nature and extent of the competitive relationship between products is the central issue in determining likeness. Thus, for example, where the nature and extent of the competitive relationship between products depends on a particular PPM, that PPM should determine the likeness of products even where it does not affect the final product as such. Moreover, in *Philippines—Distilled Spirits*, the Appellate Body held that the products need not be in a competitive relationship in all segments of the market; their competitive relationship in one sector of the market was sufficient to establish likeness. This raises an interesting question. If consumer preferences in one segment of a market make the carbon footprint a determining factor in assessing the competitive relationship between products, would this be sufficient to sustain that the products are not like?

In *US—Tuna II (Mexico)*, the evidence showed that US consumers preferred dolphin-safe tuna products. Indeed, these consumer preferences determined the competitive relationship between tuna products, not their physical characteristics or their end use. The Panel considered that these preferences could be relevant in determining likeness of products under Article 2.1 of the TBT Agreement to the extent that they affected the competitive relationship between products. However, in this case, the evidence did not prove that the fishing methods for Mexican tuna

[78] For a different view, see Patrick Low, Gabrielle Marceau, and Julia Reinaud, "The Interface between the Trade and Climate Change Regimes: Scoping the Issues" (2011) WTO Staff Working Paper ERSD-2011-1, at 2 <http://www.wto.org/english/res_e/reser_e/ersd201101_e.pdf> (accessed October 20, 2012).

[79] Appellate Body Report, *EC—Asbestos* para. 103; Appellate Body Report, *Philippines—Distilled Spirits* para. 119.

[80] I thank Deborah Elms for this insight. One example is the competitive relationship between cellular phones and cameras, which has changed as technology has evolved.

[81] Appellate Body Report, *Philippines—Distilled Spirits* paras. 125–8.

were more harmful than those for US tuna or tuna from other countries.[82] Nevertheless, this case demonstrates that it is possible for PPMs to be the determining factor regarding the competitive relationship between products. The United States did not appeal the Panel's finding that Mexican tuna products are "like" US tuna products and tuna products from other countries within the meaning of Article 2.1 of the TBT Agreement, so the Appellate Body did not address this issue.[83] However, in *US—Clove Cigarettes*, the Appellate Body held that the analysis of like products in Article 2.1 of the TBT Agreement serves to determine the competitive relationship between products, just as it does in GATT Article III.[84] Moreover, the Appellate Body has stated that distinctions "based exclusively on such particular product characteristics or on particular processes and production methods" would not per se constitute less favorable treatment within the meaning of Article 2.1.[85]

Since the Appellate Body has signaled that the competitive relationship between products is the central issue in determining likeness, then it should be possible for differences in PPMs to determine likeness. The parallels between the dolphin-safe label in this case, and carbon footprint labels, mean that differences in carbon footprints might be sufficient to conclude that products are not "like" in markets where the carbon footprint determines the competitive relationship between products. Of course, there are other hurdles to overcome with respect to carbon labels, such as the accuracy of their measurement of GHGs produced in the life cycle of a particular product, the extent to which the carbon label discriminates between products from different countries based on factors other than emissions, and the extent to which consumers consider carbon footprints to be a determining factor in a given marketplace. In the context of the TBT Agreement, there is also an issue regarding whether the issuer of the carbon label qualifies as an international standardizing body, an issue we analyze below. The point here is that a properly designed and applied carbon label might not violate national treatment and MFN rules, in the GATT or the TBT Agreement, based on the issue of likeness. However, in *US—Clove Cigarettes* and in *Philippines—Distilled Spirits*, the Appellate Body held that the analysis of consumer preferences in the like products analysis (in Article 2.1 of the TBT Agreement and GATT Article III:2, respectively) need not show that products are substitutes for all consumers or that they compete in all segments of the market in order to conclude that they are similar.[86] This raises an interesting issue regarding carbon labels, since it implies that the carbon content of products would have to determine the competitive relationship between products

[82] Panel Report, *United States—Measures Concerning the Importation, Marketing and Sale of Tuna and Tuna Products (US—Tuna II (Mexico))*, WT/DS381/R, adopted June 13, 2012, para. 7.249.

[83] Appellate Body Report, *United States—Measures Concerning the Importation, Marketing and Sale of Tuna and Tuna Products (US—Tuna II (Mexico))*, WT/DS381/AB/R, adopted June 13, 2012, para. 230.

[84] Appellate Body Report, *US—Clove Cigarettes* paras. 108–13.

[85] Appellate Body Report, *United States—Certain Country of Origin Labelling Requirements (US—COOL)*, WT/DS386/AB/R, adopted July 23, 2012, para. 268; Appellate Body Report, *US—Tuna II (Mexico)* para. 211; Appellate Body Report, *US—Clove Cigarettes* para. 169.

[86] Appellate Body Report, *US—Clove Cigarettes* paras. 142–5.

in all segments of a particular market in order to conclude that the products are not similar. If the products competed in one segment irrespective of their carbon footprint, then they would be like products according to the logic of the Appellate Body in these cases.

Another possible approach is to consider the carbon "content" of a product to be analogous to alcohol content. In *US—Malt Beverages*, the GATT panel found that beer with a low alcohol content and beer with a high alcohol content were not like products under GATT Article III:4 because the differentiation in treatment of low alcohol beer and high alcohol beer did not afford protection to domestic production.[87] However, in *Chile—Alcoholic Beverages*, the Appellate Body found that the differentiation in taxation of alcoholic beverages based on alcohol content was inconsistent with the second sentence of GATT Article III:2 because the products were directly competitive or substitutable and because the design of the measure did afford protection to domestic production.[88] These cases might support the view that the likeness of low and high carbon content products under GATT Article III:4, or the issue of whether they are directly competitive or substitutable under the second sentence of GATT Article III:2, turns on the issue of whether the measure in question affords protection to domestic production. However, an important difference between alcohol content and carbon content is that the former is in fact part of the product itself, whereas carbon "content" refers to the production process and is not part of the product itself. Thus, if the likeness of products depends on factors that affect the product as such, these cases would be less relevant.

Yet another approach might be to apply a tax to carbon content for domestic production and apply a border tax adjustment to the carbon content of imported products. If the measure applied the same tax rate to the value of the carbon content, rather than the value of the products themselves, the measure could meet the national treatment requirement of Article III:2 first sentence.[89]

GATT Article I would be relevant if imports from different WTO Members were treated differently. GATT Article I requires that like products be granted unconditional market access, which may imply that nondiscriminatory access to the importing nation's market cannot be made conditional upon the exporting country's environmental policies. In *EC—Tariff Preferences*, in which access to tariff preferences were conditional upon countries suffering from an illicit drug problem, the Panel interpreted Article I:1 as prohibiting the placing of conditions on access to preferential treatment.[90] In *Indonesia—Autos* the panel held that an advantage

[87] GATT Panel Report, *United States—Measures Affecting Alcoholic and Malt Beverages* (*US—Malt Beverages*), DS23/R–39S/206, adopted June 19, 1992, paras. 5.71–5.73.

[88] Appellate Body Report, *Chile—Taxes on Alcoholic Beverages* (*Chile—Alcoholic Beverages*), WT/DS87/AB/R, WT/DS110/AB/R, adopted January 12, 2000.

[89] GATT Panel Report, *United States—Taxes on Petroleum and Certain Imported Substances*, L/6175–34S/136, adopted June 17, 1987, paras. 5.2.7–5.2.8.

[90] Panel Report, *European Communities—Conditions for the Granting of Tariff Preferences to Developing Countries* (*EC—Tariff Preferences*), WT/DS246/R, adopted April 20, 2004, para. 7.59.

could not be conditional upon criteria that were unrelated to the product itself.[91] However, it is not clear whether the issue of conditionality can be determined independently of the issue of whether products are like.[92] If the products under consideration are not like, would subjecting them to different treatment amount to a condition?

Finally, some have argued that, if regulatory categories are permissible in the determination of whether the treatment of imports was "no less favourable," then PPMs might be considered in this part of the analysis.[93] Others have argued that it is still better to address PPMs in Article XX.[94] The resolution of this issue could affect the analysis in GATT Article III and TBT Agreement Article 2.1.[95] In order to address climate change and other environmental issues in the SCM Agreement, WTO Members will have to consider whether to address environmental subsidies under the like products analysis, the extension of GATT Article XX to the SCM Agreement or both. In doing so, they will have to keep in mind the systemic implications of this decision.

3.5 GATT Article XX

Two paragraphs, (b) and (g), in GATT Article XX will play an important role in determining the kind of measures that may be used to combat climate change. In addition, the analysis under the chapeau of Article XX will determine how those measures should be applied.

Article XX(g) applies to measures "relating to the conservation of exhaustible natural resources if such measures are made effective in conjunction with restrictions on domestic production or consumption." This phrase raises four key issues. (1) Is the climate an "exhaustible natural resource"? (2) If a jurisdictional nexus is required between the Member enacting a measure and the natural resource, does a sufficient nexus exist between all WTO Members and the global climate? (3) How should a panel determine whether a specific measure relates to climate change? (4) Are the measures "made effective in conjunction with restrictions on domestic production or consumption"?

[91] Panel Report, *Indonesia—Autos* para. 14.145.

[92] Panel Report, *Canada—Certain Measures Affecting the Automotive Industry*, WT/DS139/R, WT/DS142/R, adopted June 19, 2000, para. 10.22; Mitsuo Matsushita, Thomas Schoenbaum, and Petros Mavroidis, *The World Trade Organization: Law, Practice and Policy* (2nd edn., Oxford University Press, Oxford 2006) 216.

[93] Robert Howse and Elizabeth Tuerk, "The WTO Impact on Internal Regulations—A Case Study of *Canada—EC Asbestos* Dispute" in Gráinne de Búrca and Joanne Scott (eds.), *The EU and the WTO: Legal and Constitutional Issues* (Hart Publishing, Oxford 2001).

[94] Michael Trebilcock and S. Giri, "The National Treatment Principle in International Trade Law," Law and Economics Workshop Series, Faculty of Law, University of Toronto, WS 2002–2003 (9), February 2003.

[95] Marceau and Trachtman "The Technical Barriers to Trade Agreement"; Andrew James Green, "Climate Change, Regulatory Policy and the WTO: How Constraining are Trade Rules?" (2005) 8 JIEL 143.

In *US—Shrimp*, the Appellate Body interpreted the term "exhaustible natural resources" to include both living and non-living natural resources.[96] The Appellate Body and GATT panels have found the following to be exhaustible natural resources: clean air;[97] migratory sea turtles;[98] salmon and herring;[99] tuna;[100] and dolphins.[101] In *US—Shrimp*, since the migratory sea turtles were listed under CITES as being in danger of extinction, the Appellate Body held that they were exhaustible natural resources. Preserving the global climate could be considered analogous to the preservation of clean air in *US—Gasoline*. Alternatively, the issue of the levels of carbon and other greenhouse gases in the atmosphere could be viewed as a clean air issue.[102] The US Environmental Protection Agency decision to address greenhouse gases under the Clean Air Act supports this view.[103] Multilateral environmental agreements on climate change might be taken into account to support the view that the global climate is an exhaustible natural resource. The following passage in *US—Shrimp* lends support to this view:

> The words of Art. XX(g), "exhaustible natural resources,"... must be read by a treaty interpreter in the light of contemporary concerns of the community of nations about the protection and conservation of the environment.... From the perspective embodied in the Preamble of the WTO Agreement [referring to the objective of sustainable development], the generic term of "natural resources" is not "static" in its content or reference but is rather, by definition, evolutionary.[104]

In *US—Shrimp*, the Appellate Body held there was a sufficient jurisdictional nexus between migratory sea turtles and the United States because they spent part of their migratory life cycle in American waters, without ruling on whether there was a jurisdictional limit implied in the language of Article XX(g). The effects of climate change are global. Therefore, there should be a sufficient jurisdictional nexus between all WTO Members and climate change.

The term "relating to" has been interpreted to mean "primarily aimed at," rather than "necessary or essential."[105] The term "relating to" requires an examination of "the relationship between the measure at stake and the legitimate policy of

[96] Appellate Body Report, *United States—Import Prohibition of Certain Shrimp and Shrimp Products* (*US—Shrimp*), WT/DS58/AB/R, adopted November 6, 1998, paras. 128–31.

[97] Appellate Body Report, *United States—Standards for Reformulated and Conventional Gasoline* (*US—Gasoline*), WT/DS2/AB/R, adopted May 20, 1996.

[98] Appellate Body Report, *US—Shrimp*.

[99] GATT Panel Report, *Canada—Measures Affecting Exports of Unprocessed Salmon and Herring* (*Canada—Salmon and Herring*), L/6268—35S/98, adopted March 22, 1988.

[100] GATT Panel Report, *United States—Prohibition of Imports of Tuna and Tuna Products from Canada* (*US—Tuna from Canada*), L/5198—29S/91, adopted February 22, 1982.

[101] GATT Panel Report, *US—Tuna (Mexico)*.

[102] I thank the moot team from the University of Melbourne in the 2009 ELSA moot court competition on WTO law in Taipei, Taiwan for this observation (Ms. Bellamy, Mr. Kruse, and Mr. Tran).

[103] See further *Massachusetts and others v. Environmental Protection Agency and others* (United States Supreme Court) 549 US (2007).

[104] Appellate Body Report, *US—Shrimp* paras. 129–30.

[105] GATT Panel Report, *Canada—Salmon and Herring*; Appellate Body Report, *US—Gasoline*; Appellate Body Report, *US—Shrimp*.

conserving exhaustible natural resources." This requires "a close and genuine relationship of ends and means" and an examination of "the relationship between the general structure and design of the measure ... and the policy goal it purports to serve."[106] Multilateral environmental agreements on climate change could serve as evidence that measures aimed at the reduction of greenhouse gas emissions relate to the conservation of the global climate. This could include measures such as differential tax treatment based on the different carbon emissions resulting from production processes, provided that there is a close and genuine relationship between the general structure and design of the measure and the policy goal of reducing carbon emissions to conserve the global climate. If the structure and design of the measure is based on specific obligations in a multilateral environmental agreement on climate change, it would be more likely to meet the requirements of paragraph (g). While this is probably not essential, such specific obligations would provide evidence that the measure does relate to climate change.

Article XX(g) also requires that conservation measures be "made effective in conjunction with restrictions on domestic production or consumption." In *US—Gasoline*, the Appellate Body interpreted "made effective" as referring to a governmental measure being "operative," as "in force," or as having "come into effect." The clause does not establish an empirical "effects test" for the availability of the Article XX(g) exception. Rather, this clause is a requirement of evenhandedness in the imposition of restrictions, in the name of conservation, upon the production or consumption of exhaustible natural resources, but does not require identical treatment of domestic and imported products.[107] In *China—Raw Materials*, the Appellate Body noted that the equivalent terms in Spanish and French of "made effective" ("se apliquen" and "sont appliqués") confirm this interpretation. There is no additional requirement that the trade measure be primarily aimed at making the domestic restrictions effective.[108] It is not clear whether differences in the treatment of products, based on their impact on climate change, could meet this requirement without the differences in treatment being justified by reference to the evidence regarding the reasons for the differential treatment, such as scientific evidence comparing the carbon footprints of different products.

Article XX(b) applies to measures "necessary to protect human, animal or plant life or health." This paragraph requires that the policy goal at issue falls within the range of policies designed to protect human, animal or plant life or health. In *Brazil—Retreaded Tyres*, the Panel accepted that measures aimed at protecting Brazil's environment fell within the range of policies covered by Article XX(b).

Once it is established that the policy goal fits the exception, the issue is whether the measure is "necessary" to achieve the policy goal. This analysis takes place in light of the level of risk that a Member sets for itself. To demonstrate that the measure is necessary involves weighing and balancing a series of factors. First, the greater the importance of the interests or values that the challenged measure is

[106] Appellate Body Report, *US—Shrimp*.
[107] Appellate Body Report, *US—Gasoline* 20–1.
[108] Appellate Body Report, *China–Raw Materials* para. 356.

intended to protect, the more likely it is that the measure is necessary. Second, the greater the extent to which the measure contributes to the end pursued, the more likely that the measure is necessary. Third, the less the trade impact of the challenged measure, the more likely that the measure is necessary. Fourth, whether a WTO-consistent alternative measure which the Member concerned could reasonably be expected to employ is available, or whether a less WTO-inconsistent measure is reasonably available. The weighing and balancing process of the first three factors also informs the determination of the fourth.[109]

There is no question that environmental protection would be considered an important interest or value in Article XX(b). In *Brazil—Retreaded Tyres* the Panel found that "few interests are more 'vital' and 'important' than protecting human beings from health risks, and that protecting the environment is no less important."[110] The Appellate Body agreed that protection of the environment is an important value.[111] However, the weight accorded to the objective of environmental protection could be less than that accorded to the objective of protecting human life or health, given the Appellate Body's characterization of the former as "important" (*Brazil—Retreaded Tyres*) and of the latter as "both vital and important in the highest degree" (*EC—Asbestos*).

The extent to which a climate change measure contributes to the end pursued would be difficult to measure. A measure must be "apt to produce a material contribution to the achievement of its objective."[112] A measure that only makes "a marginal or insignificant contribution" to the objective is not enough to be considered necessary.[113] Nevertheless, in *Brazil—Retreaded Tyres*, the Appellate Body emphasized the need to view the measure against the broader context of a comprehensive strategy to deal with a problem.[114] Moreover, the Appellate Body stated that the contribution of a trade-restrictive measure to address climate change, while not immediately observable, can be justified under Article XX(b):

> We recognize that certain complex public health or environmental problems may be tackled only with a comprehensive policy comprising a multiplicity of interacting measures. In the short-term, it may prove difficult to isolate the contribution to public health or environmental objectives of one specific measure from those attributable to the other measures that are part of the same comprehensive policy. Moreover, the results obtained from certain

[109] Appellate Body Report, *Korea—Measures Affecting Imports of Fresh, Chilled and Frozen Beef* (*Korea—Beef*), WT/DS161/AB/R, adopted January 10, 2001; Appellate Body Report, *EC—Asbestos*; Appellate Body Report, *United States—Measures Affecting the Cross-Border Supply of Gambling and Betting Services* (*US—Gambling*), WT/DS285/AB/R, adopted April 20, 2005; Appellate Body Report, *Dominican Republic—Measures Affecting the Importation and Internal Sale of Cigarettes* (*Dominican Republic—Cigarettes*), WT/DS302/AB/R, adopted May 19, 2005; Appellate Body Report, *Brazil—Measures Affecting Imports of Retreaded Tyres* (*Brazil—Retreaded Tyres*), WT/DS332/AB/R, adopted December 17, 2007.

[110] Panel Report, *Brazil—Measures Affecting Imports of Retreaded Tyres* (*Brazil—Retreaded Tyres*), WT/DS332/R, adopted December 17, 2007, para. 7.108.

[111] Appellate Body Report, *Brazil—Retreaded Tyres* para. 179.

[112] Appellate Body Report, *Brazil—Retreaded Tyres* para. 151.

[113] Appellate Body Report, *Brazil—Retreaded Tyres* para. 150.

[114] Appellate Body Report, *Brazil—Retreaded Tyres* para. 154.

actions—for instance, measures adopted in order to attenuate global warming and climate change . . . —can only be evaluated with the benefit of time.[115]

In *China—Raw Materials*, the Panel had to confront China's arguments based on this passage.[116] China argued that this passage indicates that the contribution should be evaluated in the present and in the future.[117] The Panel first focused on the contribution in the present. Taking into consideration how some regulatory policies cancelled the alleged environmental benefits of others, the panel found that the net effect of the export restrictions did not contribute significantly to environmental protection. The Panel then rejected China's argument that its policies would contribute to its economic development over the long-term, which would in turn contribute to its ability to protect the environment, in accordance with the Kuznet's curve, which shows a correlation between pollution and the level of economic development. The Panel concluded that, while economic growth makes environmental protection statistically more likely, this did not prove that China's export restrictions were necessary to obtain environmental benefits.[118]

Regarding the trade impact of the challenged measure, if a "comprehensive regulatory strategy" is relevant to the extent of the contribution, then it should also be examined in assessing the trade-restrictive impact of the measure. In that case, the cumulative impact of a series of climate change measures could together have much more significant restrictive effects than a measure considered in isolation.

The same issue arises regarding the issue of whether alternative measures would achieve the same objectives as the challenged measure. If the challenged measure is part of a comprehensive regulatory strategy and the effect of the measure might not be revealed in the near future, this will be a difficult point to argue. The Member defending the measure may point out why alternative measures would not achieve the same objectives as the challenged measure, but it is under no obligation to do so in order to establish, in the first instance, that its measure is "necessary." If the complainant raises a WTO-consistent alternative measure that, in its view, the respondent should have taken, the respondent will be required to demonstrate why its challenged measure nevertheless remains "necessary" in the light of that alternative or, in other words, why the proposed alternative is not, in fact, "reasonably available." If the respondent demonstrates that the alternative is not "reasonably available," in light of the interests or values being pursued and the party's desired level of protection, it follows that the challenged measure must be "necessary."[119] Should the alternatives be considered in light of relevant international norms, such as those set out in multilateral environmental agreements on climate change? Should any alternative measures be measures that the respondent can take alone, rather than measures that are beyond its control or that would require consultations

[115] Appellate Body Report, *Brazil—Retreaded Tyres* para. 151.
[116] Panel Report, *China—Raw Materials* paras. 7.470–7.471.
[117] Panel Report, *China—Raw Materials* para. 7.518.
[118] Panel Report, *China—Raw Materials* paras. 7.540–7.551.
[119] Appellate Body Report, *US—Gambling* paras. 310–11.

or negotiations with other countries?[120] What kind of scientific evidence will be required? The Appellate Body has stated the following in this regard: "In justifying a measure under Article XX(b) of the GATT 1994, a Member may also rely, in good faith, on scientific sources which, at that time, may represent a divergent, but qualified and respected, opinion."[121]

The purpose of the chapeau is to prevent the abuse of the exceptions in Article XX. The chapeau embodies the recognition on the part of WTO Members of the need to maintain a balance between the right of a Member to invoke an exception on the one hand, and the substantive rights of the other Members on the other hand.[122]

The chapeau requires that a measure that has been provisionally justified under one of the paragraphs of Article XX not be applied in a manner that constitutes: (1) arbitrary discrimination between countries where the same conditions prevail; (2) unjustifiable discrimination between countries where the same conditions prevail; or (3) a disguised restriction on international trade. The respondent has the burden of proof to show that the application of the measure meets the requirements of the chapeau. In order for the measure to pass the chapeau test, the respondent must prove that all three requirements have been met. In order for the measure to fail the chapeau test, the complainant only needs to show that one of these three requirements has not been met.

There are three elements in the chapeau analysis of whether a measure is applied in a manner that constitutes "arbitrary or unjustifiable discrimination between countries where the same conditions prevail": (1) the application of the measure results in discrimination; (2) the discrimination is arbitrary or unjustifiable; and (3) the discrimination occurs between countries where the same conditions prevail (between different exporting countries or between the exporting countries and the importing country). The chapeau also refers to disguised restrictions on international trade. The jurisprudence has tended to find that the evidence that supports a finding of arbitrary or unjustifiable discrimination also supports a finding of disguised restrictions on international trade.[123]

In *US—Gasoline* and *US—Shrimp*, the Appellate Body identified two main criteria to determine whether discrimination that has been shown to exist is arbitrary or unjustifiable: (1) a serious effort to negotiate with a view to achieving the policy goal of the measure at stake; and (2) flexibility of the measure (e.g. in taking into account the situation prevailing in other countries). With respect to the second criteria, in *US—Shrimp (Art. 21.5)*, the Appellate Body agreed with the Panel that conditioning market access on the adoption of a program *comparable in*

[120] Appellate Body Report, *US—Gambling* paras. 316–18.

[121] Appellate Body Report, *EC—Asbestos* para. 178, citing Appellate Body Report, *European Communities—Measures Concerning Meat and Meat Products (Hormones)* (*EC—Hormones*), WT/DS26/AB/R, WT/DS48/AB/R, adopted February 13, 1998, para. 194.

[122] Appellate Body Report, *US—Shrimp*.

[123] Appellate Body Report, *US—Gasoline* para. 23; Panel Report, *Brazil—Retreaded Tyres* para. 7.319; Appellate Body Report, *Brazil—Retreaded Tyres* para. 239.

effectiveness, allows for sufficient flexibility in the application of the measure, so as to avoid arbitrary or unjustifiable discrimination.[124]

In *US—Shrimp*, the Appellate Body found that the American regulations were arbitrary or unjustifiable because the US: (1) required WTO members to adopt "essentially the same policy" as that applied in the United States without taking into account other policies and measures a country may have adopted that would have a comparable effect on sea turtle conservation; (2) applied the same standard without taking into consideration whether it was appropriate for the conditions prevailing in other countries; (3) failed to engage in "serious, across-the-board negotiations with the objective of concluding bilateral or multilateral agreements for the protection and conservation of sea turtles, before enforcing the import prohibition"; and (4) failed to provide due process in the denial of certification compared to those who were granted certification.[125] However, the chapeau does not require that a Member succeed in its efforts to negotiate a multilateral solution to a transnational environmental problem.[126]

In *Brazil—Retreaded Tyres*, the Appellate Body held that:

> there is arbitrary or unjustifiable discrimination when a measure provisionally justified under a paragraph of Article XX is applied in a discriminatory manner "between countries where the same conditions prevail," and when the reasons given for this discrimination bear no rational connection to the objective falling within the purview of a paragraph of Article XX, or would go against that objective.[127]

This requirement that the reasons for the discrimination relate to the objective of the particular paragraph of Article XX might explain diverging WTO jurisprudence on whether the chapeau requires an effort to negotiate prior to employing trade restrictions to address environmental issues. In the two WTO cases involving paragraph (g), the Appellate Body found that a failure to negotiate led to a failure to comply with the nondiscrimination requirements of the chapeau.[128] In *US—Shrimp*, it was unclear whether the obligation to negotiate stemmed from multilateral environmental documents that expressed a preference for multilateral solutions to transboundary or global environmental problems, or whether it stemmed from the American failure to negotiate with Asian countries having done so with countries in the Americas. In the cases involving paragraph (b), the Appellate Body has not found any obligation to negotiate in order to comply with the nondiscrimination requirements of the chapeau.[129]

The divergence in the jurisprudence might be explained by arguing that paragraphs (b) and (g) apply to different matters.[130] This might explain why in some

[124] Appellate Body Report, *US—Shrimp (Art. 21.5)* para. 144.

[125] Appellate Body Report, *US—Shrimp* paras. 163–6, 181.

[126] Appellate Body Report, *US—Shrimp (Art. 21.5)*.

[127] Appellate Body Report, *Brazil—Retreaded Tyres* para. 227.

[128] Appellate Body Report, *US—Gasoline*; Appellate Body Report, *US—Shrimp*.

[129] Appellate Body Report, *EC—Asbestos*; Appellate Body Report, *Brazil—Retreaded Tyres*; Appellate Body Report, *US—Gambling* (considering a comparable provision in Article XIV of the GATS).

[130] Bradly J. Condon, "GATT Article XX and Proximity of Interest: Determining the Subject Matter of Paragraphs b and g" (2004) 9 UCLA Journal of International Law and Foreign Affairs 137.

cases the avoidance of arbitrary or unjustifiable discrimination requires an effort to negotiate. The rule of effective treaty interpretation requires that treaty terms be interpreted so as to avoid redundancy. This suggests that paragraphs (b) and (g) must apply to different matters. However, paragraph (g) has been applied to measures aimed at the conservation of animals (migratory turtles,[131] salmon,[132] herring,[133] tuna,[134] and dolphins[135]) and paragraph (b) has also been applied to a measure aimed at protecting animals (monkeys in Brazil[136]). The only obvious difference in these animals is that those considered under paragraph (g) are migratory, and hence a transboundary environmental issue, whereas the monkeys considered under paragraph (b) are not migratory, and hence a domestic environmental issue. In addition to migratory species, a clean air measure has been addressed under paragraph (g). While the clean air at issue was that of the United States, and hence domestic, clean air is a transboundary environmental issue. Air pollution does not respect national boundaries. Examples include forest fires in Mexico causing air pollution in the United States and air pollution in the United States causing acid rain in Canada. Thus, the case law supports the view that one difference between the two paragraphs might be that (b) addresses domestic issues and (g) addresses transboundary issues. An analysis under the Vienna Convention on the Law of Treaties, Articles 31, 32, and 33 also supports this view.[137]

The obvious objection to the notion that paragraph (g) addresses only transboundary issues is that this appears to exclude exhaustible natural resources that are contained within one country's borders, such as mineral resources. The answer to this objection is not obvious. One possibility is to consider that mineral resources are a finite global resource, even when they are contained within the borders of one country.[138] Another possibility is to address domestic resources under other exceptions, such as: GATT Article XX(i) (for restrictions on exports of domestic materials necessary to ensure essential quantities of such materials to a domestic processing industry); GATT Article XX(j) (for measures essential to the acquisition or distribution of products in general or local short supply); or GATT Article XXI (for measures necessary for the protection of a Member's essential security interests).[139] However, neither the Panel nor the Appellate Body considered the difference between paragraphs (b) and (g) in *China—Raw Materials*.

The different thresholds in paragraphs (b) and (g) also suggest that they apply to different matters. WTO jurisprudence has indicated that the term "necessary" sets a higher threshold than the term "relating to." At the same time, WTO jurisprudence has indicated that there is no interest or value more important than human life

[131] Appellate Body Report, *US—Shrimp*; Appellate Body Report, *US—Shrimp (Art. 21.5)*.
[132] GATT Panel Report, *Canada—Salmon and Herring*.
[133] GATT Panel Report, *Canada—Salmon and Herring*.
[134] GATT Panel Report, *US—Tuna from Canada*.
[135] GATT Panel Report, *US—Tuna (Mexico)*.
[136] Panel Report, *Brazil—Retreaded Tyres*.
[137] Vienna Convention on the Law of Treaties, opened for signature May 23, 1969, 1155 UNTS 331 (entered into force 27 January 1980); Condon "GATT Article XX and Proximity of Interest."
[138] I thank David Morgan for this idea.
[139] I thank David Morgan again for helping me on this point.

and health.[140] It would be an odd result to set a higher threshold for measures that aim to preserve human life and health than for the conservation of an exhaustible natural resource. Another possible reason for a stricter threshold in paragraph (b) is that the cause of protecting human, animal, or plant life or health can be more easily abused by Members because it is more subjective than the conservation of natural resources, which can be determined more objectively.[141] One solution to this conundrum is for WTO jurisprudence to evolve to a point where the threshold converges.[142] If one considers that the avoidance of arbitrary or unjustifiable discrimination under the chapeau requires WTO Members to seek multilateral solutions to address the conservation of transboundary resources, while no such requirement exists for measures that address the protection of domestic human, animal or plant life or health, the analysis under the chapeau would eliminate any difference in the thresholds in paragraphs (b) and (g). In other words, the term "relating to," in combination with a negotiation requirement, would set a higher threshold than the term "necessary" without a negotiation requirement. Indeed, a negotiation requirement could act as a barrier to litigation, not just a threshold issue in litigation. In the context of GATS Article XIV(a), the Appellate Body disagreed with the panel that the term "necessary" implied a negotiation requirement.[143] However, the circumstances in which there might be a negotiation requirement in the chapeau of GATT Article XX or GATS Article XIV has not been resolved in WTO jurisprudence. It is also inappropriate to require international cooperation or negotiations to address domestic health issues, since each WTO Member has the right to determine its appropriate level of health protection[144] and this issue is entirely within each Member's jurisdiction. Indeed, limiting the scope of paragraph (b) to domestic concerns resolves the question of whether there is an implicit jurisdictional limitation in paragraph (b) and paragraph (g).

Other paragraphs that use the term "necessary" are consistent with the idea that this threshold applies to domestic matters. GATT Article XX(a) applies this term to "public morals" and GATS[145] Article XIV(a) to "public morals" and "public order." Since the standards for public morals vary from one country to the next (and even among communities within the same country), it is reasonable to conclude that these paragraphs apply to domestic issues. Similarly, public order is a domestic issue. GATT Article XX(d) and GATS Article XIV(c) apply the term "necessary" to measures to secure compliance with laws or regulations. The Appellate Body has held that the term "laws or regulations" in GATT Article XX(d) refers to domestic laws or regulations.[146] GATS Article XIV(b) applies to the same subject matter as GATT Article XX(b).

[140] Appellate Body Report, *EC—Asbestos* para. 172.
[141] I thank an anonymous reviewer for making this point.
[142] I thank Professor Matsushita for this idea.
[143] Appellate Body Report, *US—Gambling* paras. 308, 317, 321.
[144] Appellate Body Report, *EC—Asbestos* paras. 167–8.
[145] GATT Secretariat, The Results of the Uruguay Round of Multilateral Trade Negotiations, the Legal Texts (Geneva, 1994) 284.
[146] Appellate Body Report, *Mexico—Tax Measures on Soft Drinks and Other Beverages* (*Mexico—Taxes on Soft Drinks*), WT/DS308/AB/R, adopted March 24, 2006, para. 69.

If Article XX(b) does not apply to transnational or global environmental concerns, measures aimed at addressing climate change would not fall within the range of policies covered by Article XX(b), unless it could be shown that the measures also addressed domestic environmental or health concerns. While climate change is a global issue, it can also affect domestic issues such as human health.[147] It is possible for more than one paragraph in Article XX to apply to different aspects of the same measure. The Appellate Body ruled that the GATT and the GATS could apply to different aspects of the same measure.[148] Thus, measures aimed at climate change could be characterized as addressing both transnational and domestic issues, allowing both paragraphs to apply. The issue of whether a measure addresses a transnational or domestic problem is a question of fact. The scope of paragraphs (b) and (g) is a question of law.

3.6 Technical Barriers to Trade

The TBT Agreement is likely to come into play with respect to some measures related to climate change, such as carbon labels for instance. When a Member alleges that a measure violates both the TBT Agreement and GATT 1994, the allegations under the former are examined first.[149] The Preamble of the TBT Agreement states that it is intended to "further the objectives of GATT 1994." While the Appellate Body has indicated that the TBT Agreement "does so through a specialized regime that applies solely to a limited class of measures,"[150] the jurisprudence also indicates that the two agreements apply cumulatively.[151] Marceau and Trachtman argue that compliance with the more stringent requirements of the TBT Agreement should give rise to a presumption of compliance with GATT. In particular, a technical regulation that complies with TBT Agreement Articles 2.1 and 2.2 is likely to be compatible with GATT Articles I, III, and XX. However, a measure could be justifiable under GATT Article XX and not meet the requirements of the TBT Agreement.[152]

[147] Bradly J. Condon and Tapen Sinha, *Global Lessons from the AIDS Pandemic. Economic, Financial, Legal and Political Implications* (Springer Verlag, Berlin 2008) 5–6; Bradly J. Condon and Tapen Sinha, "Chronicle of a Pandemic Foretold: Lessons from the 2009 Influenza Epidemic" (2010) 22 Florida Journal of International Law 1.

[148] Appellate Body Report, *European Communities—Regime for the Importation, Sale and Distribution of Bananas* (*EC—Bananas III*), WT/DS27/AB/R, adopted September 25, 1997, para. 221.

[149] Appellate Body Report, *European Communities—Trade Description of Sardines* (*EC—Sardines*), WT/DS231/AB/R, adopted October 23, 2002.

[150] Appellate Body Report, *EC—Asbestos* para. 80.

[151] Panel Report, *European Communities—Trade Description of Sardines* (*EC—Sardines*), WT/DS231/R, adopted October 23, 2002, paras. 7.15–7.16; Appellate Body Report, *EC—Sardines* para. 313; Appellate Body Report, *EC—Bananas III* para. 204.

[152] Marceau and Trachtman "The Technical Barriers to Trade Agreement" 873–4.

3.6.1 Technical regulations

For the key obligations in Article 2 of the TBT Agreement to apply to a regulation, it must meet the definition of "technical regulation" in Annex 1.1.[153] According to the Appellate Body, a document must meet three criteria to fall within this definition: (1) The document must apply to an identifiable product or group of products. The identifiable product or group of products need not, however, be expressly identified in the document. (2) The document must lay down one or more characteristics of the product. These product characteristics may be intrinsic, or they may be related to the product, for example to the means of identification, presentation, and appearance of the product.[154] They may be prescribed or imposed in either a positive or a negative form. (3) Compliance with the product characteristics must be mandatory.[155]

The mere fact that it is legally permissible to sell a product on the market without using a particular label is not determinative when examining whether a measure is a "technical regulation" within the meaning of Annex 1.1.[156] The subject matter of a particular measure does not determine whether it constitutes a technical regulation or a standard, since terminology, symbols, packaging, marking, and labeling requirements may be the subject matter of either technical regulations or standards. Certain features may be common to both technical regulations and standards, such as compulsory and enforceable conditions that must be met in order to use a label. Thus, it is necessary to consider additional characteristics of the measure, such as whether the measure consists of a law or a regulation enacted by a WTO Member, whether it prescribes or prohibits particular conduct, whether it sets out specific requirements that constitute the sole means of addressing a particular matter, and the nature of the matter addressed by the measure.[157] These factors would be relevant in determining whether carbon labels are subject to the obligations in Article 2 of the TBT Agreement, for example. Annex 1.1 permits the document to set out "related processes and production methods" or to "deal exclusively with ... labelling requirements as they apply to a product, process or production method." PPM labels are covered by the TBT Agreement.[158]

3.6.2 TBT Agreement Article 2.1

TBT Agreement Article 2.1 provides as follows: "Members shall ensure that in respect of technical regulations, products imported from the territory of any Member shall be accorded treatment no less favourable than that accorded to like

[153] Appellate Body Report, *EC—Asbestos* para. 59; Appellate Body Report, *EC—Sardines* para. 175; Appellate Body Report, *US—Tuna II (Mexico)*; Appellate Body Report, *US—COOL*; Appellate Body Report, *US—Clove Cigarettes.*

[154] Appellate Body Report, *EC—Asbestos.*

[155] Appellate Body Report, *EC—Sardines* para. 176.

[156] Appellate Body Report, *US—Tuna II (Mexico)* paras. 196–8.

[157] Appellate Body Report, *US—Tuna II (Mexico)* paras. 187–8.

[158] Panel and Appellate Body Reports, *US—Tuna II (Mexico).*

products of national origin and to like products originating in any other country." We analyzed the treatment of like products in *US—Tuna II (Mexico)* above.

Technical regulations establish distinctions between products according to their characteristics or their related processes and production methods. Thus, Article 2.1 does not mean that any distinctions would per se constitute "less favourable treatment" within the meaning of Article 2.1.[159] Technical regulations may pursue legitimate objectives but must not be applied in a manner that would constitute a means of arbitrary or unjustifiable discrimination.[160]

"Less favourable treatment" requires a determination of whether the contested measure modifies the conditions of competition to the detriment of imported products. However, the existence of such a detrimental effect is not sufficient to demonstrate less favorable treatment if the detrimental impact on imports stems exclusively from a legitimate regulatory distinction rather than reflecting discrimination against the group of imported products.[161] A measure may be *de facto* inconsistent with Article 2.1 even when it is origin-neutral on its face.[162] A Panel must scrutinize the design, architecture, revealing structure, operation, and application of the technical regulation and, in particular, whether that technical regulation is evenhanded in the manner in which it addresses the risks in a particular case.[163] The question of whether the measure has the effect of exerting pressure on another Member to modify its practices may be relevant, but would not be sufficient to establish a breach of Article 2.1.[164] The fact that the detrimental impact may involve some element of private choice does not relieve the Member of responsibility under the TBT Agreement, where the measure it adopts modifies the conditions of competition to the detriment of imported products.[165] A measure may create incentives for market participants to behave in certain ways, and thereby treat imported products less favorably.[166]

Technical regulations that have a *de facto* detrimental impact on imports may not be inconsistent with Article 2.1 when this impact stems exclusively from a legitimate regulatory distinction, provided that it is evenhanded.[167] Thus, a "legitimate regulatory distinction" serves as a defense, where risks are addressed in an evenhanded way, for example where distinctions in treatment are based on evidence that the risks are different in different situations, and therefore the different situations need to be addressed in different ways to achieve the ultimate policy goal. With respect to climate change, emissions from different fuels could be subject to different taxes where the different emissions pose different risks, for example due

[159] Appellate Body Report, *US—COOL* para. 228; Appellate Body Report, *US—Tuna II (Mexico)* para. 211; Appellate Body Report, *US—Clove Cigarettes* para. 169.

[160] Appellate Body Report, *US—Tuna II (Mexico)* para. 213.

[161] Appellate Body Report, *US—Tuna II (Mexico)* para. 215.

[162] Appellate Body Report, *US—Tuna II (Mexico)* para. 225.

[163] Appellate Body Report, *US—Tuna II (Mexico)* paras. 225, 232; Appellate Body Report, *US—Clove Cigarettes* para. 182.

[164] Appellate Body Report, *US—Tuna II (Mexico)* para. 226.

[165] Appellate Body Report, *US—Tuna II (Mexico)* para. 239.

[166] Appellate Body Report, *US—COOL* para. 270.

[167] Appellate Body Report, *US—COOL* para. 271.

to the nature and quantity of GHG emissions for each fuel or the GHG emissions from their productions processes. Different treatment of products, based on their PPMs, also might not constitute less favorable treatment, for example due to differences in their carbon footprint. The difficulty is that carbon footprints would be difficult to measure and the design of carbon labeling programs runs the risk of being distorted to benefit domestic industry lobbies, in countries where the policy process seeks to address their needs in this type of regulation. *De facto* discrimination, which creates incentives for private actors to choose domestic inputs over imported ones, could be incorporated into some element of the design of a regime of carbon taxes and border tax adjustments, for example where the taxes themselves do not discriminate but the reporting or filing requirements are more burdensome for the imported products. For example, in *US—COOL*, the least costly way of complying with the country-of-origin labeling program was to rely exclusively on domestic livestock. This type of "direct practical effect" in the market denies competitive opportunities to imports as a result of governmental intervention that affects the conditions of competition. Thus, when the operation of a measure in the market creates incentives for private actors systematically to make choices that benefit domestic products to the detriment of like imported products, then the measure may be found to treat imported products less favorably.[168]

The Appellate Body Report in *US—COOL* indicates that it will be challenging to design and apply carbon labeling schemes in a manner that is consistent with Article 2.1 of the TBT Agreement. For example, labeling requirements may have to provide consumers with information commensurate with the type of information that upstream producers and processors are required to maintain and transmit. It may require the labels to list not just the total carbon footprint of a product, but to do so by production steps as well. For example, if the relevant production steps took place in more than one country, the label might need to identify the emissions involved with the production step that took place in each of those countries, indicate which production step took place in which country and in which order. Any exemptions from the labeling requirements would have to be carefully designed and applied as well. The burden of the recordkeeping and verification requirements would also have to be considered.[169] To avoid creating incentives that benefit domestic products to the detriment of like imported products, the labeling scheme may end up being too costly and burdensome for many firms and governments to handle, particularly small firms and governments in least-developed countries.

The interpretation of "no less favourable treatment than" in the TBT context is consistent with that of GATT Article III, where the Appellate Body has indicated that whether or not products are treated less favorably should be assessed by examining whether a measure modifies the conditions of competition in the relevant market to the detriment of imported products.[170] However, if a detrimental effect

[168] Appellate Body Report, *US—COOL* paras. 287–8.
[169] Appellate Body Report, *US—COOL* paras 343–50.
[170] Appellate Body Report, *Korea—Beef* para. 137; Appellate Body Report, *Philippines—Distilled Spirits* para. 119.

on an imported product is explained by factors or circumstances unrelated to the foreign origin of the product, this does not necessarily imply that the challenged measure accords less favorable treatment to imports within the meaning of GATT Article III:4.[171] If the categorization of products is in fact based on their origin, this should relieve the complainant of the obligation to prove that the products are similar, in accordance with WTO jurisprudence under GATT Article I:1, III:2, and III:4 that, where differential treatment is based on the origin of products, that distinction is sufficient to find a violation.[172]

3.6.3 TBT Agreement Article 2.2

TBT Agreement Article 2.2 provides as follows:

> Members shall ensure that technical regulations are not prepared, adopted or applied with a view to or with the effect of creating unnecessary obstacles to international trade. For this purpose, technical regulations shall not be more trade-restrictive than necessary to fulfil a legitimate objective, taking account of the risks non-fulfilment would create. Such legitimate objectives are, inter alia: national security requirements; the prevention of deceptive practices; protection of human health or safety, animal or plant life or health, or the environment. In assessing such risks, relevant elements of consideration are, *inter alia*: available scientific and technical information, related processing technology or intended end-uses of products.

The first issue under Article 2.2 is whether the regulation in question pursues a legitimate objective. The assessment of what a Member seeks to achieve by means of a technical regulation may take into account the texts of statutes, legislative history, and other evidence regarding the structure and operation of the measure. The next question is whether a particular objective is legitimate. The objectives listed in Article 2.2 provide a reference point for which other objectives may be considered to be legitimate, as do those of the preamble of the TBT Agreement and the objectives recognized in the provisions of other covered agreements.[173] Thus, for example, the objectives of sustainable development, which is recognized in the preamble of the WTO Agreement, would be a legitimate objective. Environmental protection would also be legitimate, since it is expressly listed. Pursuing these objectives in accordance with common but differentiated responsibilities would also be consistent with the WTO Agreements, as we noted in Chapter 2. The objective of the climate change regulations is likely to be presented as the "protection

[171] Appellate Body Report, *Dominican Republic—Cigarettes* para. 96.

[172] III:2: Panel Report, *Argentina—Measures Affecting the Export of Bovine Hides and the Import of Finished Leather* (*Argentina—Hides and Leather*), WT/DS155/R, adopted February 16, 2001, paras. 11.168–11.170; Panel Report, *Indonesia—Autos* para. 14.113; Panel Report, *India—Measures Affecting the Automotive Sector* (*India—Autos*), adopted April 5, 2002, para. 7.174; III:4: Panel Report, *Korea—Measures Affecting Imports of Fresh, Chilled and Frozen Beef* (*Korea—Beef*), WT/DS161/R, WT/DS169/R, adopted January 10, 2001, para. 627; I:1: Panel Report, *United States—Import Measures on Certain Products from the European Communities* (*US—Certain EC products*), WT/DS165/R, adopted January 10, 2001, para. 6.54.

[173] Appellate Body Report, *US—Tuna II (Mexico)* paras. 313–14.

of . . . animal or plant life or health, or the environment." *Brazil—Retreaded Tyres* supports the argument that the protection of the animal and plant life or health through the reduction of carbon emissions, like the protection of animal and plant life or health in Brazil through the reduction of air, water, and soil pollution, falls within the range of policies covered by the "protection of . . . animal or plant life or health, or the environment."

The second issue is whether a technical regulation "fulfils" the legitimate objective. This requires a determination of the degree of contribution that the technical regulation makes toward the achievement of the legitimate objective. It also requires a determination of the level at which a WTO Member seeks to pursue the particular legitimate objective, something the WTO Member articulates either implicitly or explicitly. The degree of achievement of a particular objective may be discerned from the design, structure, and operation of the technical regulation, as well as from evidence relating to the application of the measure. As in the context of Article XX of the GATT 1994, here a panel must assess the contribution to the legitimate objective actually achieved by the measure at issue.[174] This is an important point with respect to climate change, since there will be a degree of scientific uncertainty regarding the impact of measures taken to mitigate and adapt to climate change. As we noted above, the Appellate Body recognized, in *Brazil—Retreaded Tyres*, that some uncertainty would have to be allowed in circumstances such as climate change, where the contribution of climate change mitigation measures might not be immediately discernible.[175] Thus, as in the context of Article XX, the assessment of the contribution in the Article 2.2 analysis could be quantitative or qualitative.

The third issue under Article 2.2 is whether the regulation in question creates unnecessary obstacles to international trade. The assessment of necessity requires an analysis of three criteria: (1) the trade-restrictiveness of the technical regulation; (2) the degree of contribution that it makes to the achievement of a legitimate objective; and (3) the risks non-fulfillment would create.[176] Some degree of trade-restrictiveness is permitted. Article 2.2 is concerned only with restrictions on international trade that exceed what is necessary to achieve the degree of contribution that a technical regulation makes to the achievement of a legitimate objective. The risks non-fulfillment of the objective would create are assessed in light of available scientific and technical information, related processing technology or intended end-uses of products. Moreover, the comparison of the challenged measure with a possible alternative measure should be made in light of the nature of the risks at issue and the gravity of the consequences that would arise from non-fulfillment of the legitimate objective.[177] The contribution that the challenged measure makes to the achievement of its objective must be determined objectively,

[174] Appellate Body Report, *US—Tuna II (Mexico)* paras. 315–17.
[175] Appellate Body Report, *Brazil—Retreaded Tyres* para. 151.
[176] Appellate Body Report, *US—Tuna II (Mexico)* paras. 318–19; Appellate Body Report, *US—COOL* para. 374.
[177] Appellate Body Report, *US—Tuna II (Mexico)* para. 321.

and then evaluated along with the other factors mentioned in Article 2.2.[178] The measure does not have to fulfill the objective completely or satisfy some minimum level of fulfillment of that objective.[179]

With respect to climate change, the availability of scientific information is a key issue, given the scientific uncertainty regarding many aspects of climate change. Related processing technology is relevant to assessing the consistency of PPM-based measures with Article 2.2, since the capacity of a country to reduce GHG emissions is connected to the PPM technology it has available. The gravity of the risks associated with climate change should provide greater leeway to reject proposed alternatives. One difficult issue would be proximate cause, however, since there are many connecting points between burning coal to heat a factory and restricting imports from that factory to address the risk that they will contribute to climate change and hence to extreme weather, droughts, famine, and spread of diseases. In civil litigation, it might be difficult to meet the burden of proof as a plaintiff to prove causation between burning coal in China and a case of mosquito borne disease in Italy, for example. What standard of proof would be sufficient under Article 2.2? It is important to recall that the Preamble of the TBT Agreement allows each Member to determine the level of protection it considers appropriate. Moreover, any alternative measures should be measures that the respondent can take, rather than measures that are beyond its control and that would require consultations or negotiations with other WTO Members.[180] The alternatives, in addition to being less trade restrictive and making an equivalent contribution to the achievement of the relevant legitimate objective, also should be reasonably available to the relevant country, which would make economic and technological capacity to adopt alternative measures a relevant consideration.[181]

In the case of climate change measures that form part of a comprehensive scheme, the Appellate Body has pointed out that evidence of the effect of the measure is not essential in the context of GATT Article XX(b).[182] The same should be true in the context of TBT Agreement Article 2.2. Moreover, since the TBT Agreement does not explicitly regulate risk assessment or require scientific bases for regulations, the implicit requirement for some scientific basis should be significantly less rigorous than the explicit requirements of the SPS Agreement.[183] Article 2.2 only requires a consideration of "*available* scientific and technical information" (emphasis added). This does not require *conclusive* scientific evidence. According to the Appellate Body, "[i]n justifying a measure under Article XX(b) of the GATT 1994, a Member may also rely, in good faith, on scientific sources which, at that time, may represent a divergent, but qualified and respected, opinion."[184] The same reasoning should apply in the context of TBT Agreement Article 2.2.

178 Appellate Body Report, *US—COOL* para. 461.

179 Appellate Body Report, *US—COOL* para. 46.

180 Appellate Body Report, *US—Gambling* paras. 316–18.

181 Appellate Body Report, *US—Tuna II (Mexico)* para. 323; Appellate Body Report, *Brazil—Retreaded Tyres*.

182 Appellate Body Report, *Brazil—Retreaded Tyres* para. 151.

183 Marceau and Trachtman "The Technical Barriers to Trade Agreement" 836.

184 Appellate Body Report, *EC—Asbestos* para. 178.

The assessment of the degree of contribution that a measure makes to the achievement of a legitimate objective requires a panel to identify the measure's objective, by examining evidence and arguments presented by the parties relating to the measure's text, design, architecture, structure, and legislative history, as well as its operation.[185] The legitimacy of the specific objectives pursued also must be determined.[186] Environmental labeling (and thus carbon labeling) is likely to be considered a legitimate objective, since the practice is widespread and there is ample evidence that consumers are interested in it. Of course, since environmental protection is already listed in Article 2.2 as a legitimate objective, this enquiry need not take place for carbon labels, unless complainants allege that this is not the real objective. However, that line of argument not likely to bear fruit in most cases, since it would imply bad faith on the part of the WTO Member applying the labeling program, a conclusion that panels and the Appellate Body would be reluctant to reach.

3.6.4 International standards

What constitutes an "international standard" for the purposes of the TBT Agreement is an important issue for climate change measures. Article 2.4 requires WTO Members to use a "relevant international standard" as a basis for their technical regulations, except when such standard would be an ineffective or inappropriate means for the fulfillment of the legitimate objectives pursued by the Member in question. Article 2.5 creates a rebuttable presumption of compliance with Article 2.2 where a technical regulation for one of the explicitly mentioned legitimate objectives is in accordance with relevant international standards. Each of these provisions privilege multilaterally agreed technical regulations over unilateral ones. Unilateral measures are not prohibited, but they must pass a stricter legal test than multilateral measures.

Contrary to the SPS Agreement, which defines "international standards, guidelines and recommendations" by reference to specific organizations, the TBT Agreement does not contain a list of international standardizing organizations. International standardizing bodies must ensure representative participation and transparency in the development of international standards for them to qualify as such under the TBT Agreement.[187] Thus, an international organization, in which membership requires that the parties take the decision to issue an invitation by consensus, is not an "international" body and its standards do not qualify as "international standards" for the purposes of the TBT Agreement.[188] The characteristics of the entity approving a standard make it "international," not its subject matter. The Appellate Body has yet to decide whether the TBT Agreement also

[185] Appellate Body Report, *US—COOL* para. 396.
[186] Appellate Body Report, *US—COOL* paras. 438–50.
[187] Appellate Body Report, *US—Tuna II (Mexico)* para. 379.
[188] Appellate Body Report, *US—Tuna II (Mexico)* paras. 398–9.

requires an "international standard" to be based on consensus or whether it has to be made available to the public.[189]

In *EC—Hormones*, the Appellate Body refused to read into SPS Agreement Article 3.2 a reverse presumption that noncompliance with an international standard leads to an inference of noncompliance, so deviations from international standards are not prohibited.[190] A multilateral environmental agreement on climate change might qualify as a relevant international standard if membership is open to all WTO Members. Unlike in the SPS Agreement, where the standard setting bodies are clearly and exhaustibly identified,[191] the organizations or bodies that could develop "standards" within the definition of TBT Annex 1 are not. On this basis, it could be argued that an environmental agreement with quasi-universal membership could develop international standards that may fall within the scope of the TBT Agreement.[192]

TBT Agreement Article 2.4 provides as follows:

> Where technical regulations are required and relevant international standards exist or their completion is imminent, Members shall use them, or the relevant parts of them, as a basis for their technical regulations except when such international standards or relevant parts would be an ineffective or inappropriate means for the fulfilment of the legitimate objectives pursued, for instance because of fundamental climatic or geographical factors or fundamental technological problems.

The complainant bears the burden of proof in seeking a ruling of inconsistency with Article 2.4.[193] Members have to use relevant international standards that currently exist or whose completion is imminent with respect to the technical regulations that are already in existence.[194]

In *US—COOL*, the Appellate Body's approach to whether an international standard is ineffective or inappropriate to fulfill a legitimate objective suggests that the respondent's characterization of the objective can influence the outcome of the analysis. In that case, the international standard established a labeling scheme that provided consumers with information on where livestock were slaughtered, but not where the livestock were born and raised. The COOL measure's objective was to provide consumer information on origin, and the United States defined the origin of meat based on the place where an animal from which meat is derived was born, raised, and slaughtered. The Panel found that CODEX STAN 1 1985 would not provide information to consumers about the countries in which an animal was born, raised and slaughtered. Thus, it was "ineffective and inappropriate for the

[189] Appellate Body Report, *US—Tuna II (Mexico)* para. 353.

[190] Marceau and Trachtman "The Technical Barriers to Trade Agreement" 842.

[191] CODEX, OIE, IPPC—see SPS Agreement, Annex A.

[192] On this question, the TBT Committee Decision on Principles for the Development of International Standards, Guides and Recommendations with Relation to Articles 2, 5 and Annex 3 of the Agreement, in WTO document G/TBT/1/Rev.10, Decisions and Recommendations adopted by the WTO Committee on Technical Barriers to Trade since January 1, 1995 (June 9, 2011), can be informative, though it remains a Committee Decision (see section B of Annex to Part I).

[193] Appellate Body Report, *EC—Sardines* paras. 274–5.

[194] Panel Report, *EC—Sardines* para. 7.74; Appellate Body Report, *EC—Sardines* para. 205.

fulfillment of the specific objective as defined by the United States." In its analysis under Article 2.2 of the TBT Agreement, the Appellate Body noted that a mandatory labeling regime based on CODEX STAN 1 1985 "would, at best, contribute only partially to the objective of providing information to consumers on where livestock from which meat is derived were born, raised, and slaughtered." However, there were insufficient Panel findings to enable the Appellate Body to determine the degree of contribution made by the COOL measure to the United States' objective or to compare the degree of the COOL measure's contribution with that of an alternative measure proposed by the complainants.[195]

Even if not adopted by consensus, an international standard can constitute a relevant international standard.[196] To be a "relevant international standard," the standard at issue would have to "bear upon, relate to, or be pertinent to" the regulation.[197] There must be a very strong and very close relationship between two things in order to be able to say that one is "the basis for" the other. A standard is used as a basis for a technical regulation when it is used as the principal constituent or fundamental principle for the purpose of enacting the technical regulation.[198] A multilateral environmental agreement on climate change that sets standards might qualify as a relevant international standard if membership is open to all WTO Members, in light of the definition in Annex 1.4 of "international body or system" as a "Body or system whose membership is open to the relevant bodies of at least all Members." Article 12.4 provides that "developing country Members should not be expected to use international standards as a basis for their technical regulations ... which are not appropriate to their development, financial and trade needs." This might entitle them to more leeway, but this provision is not mandatory.

In order to constitute an international standard for the purposes of the TBT Agreement, a standard has to be adopted by an "international standardizing body" that has recognized activities in standardization.[199] Thus, relevant evidence includes recognition by WTO Members and by national standardizing bodies. The recognition of those who participate in the development of a standard may not be sufficient, since TBT Agreement Articles 2.4 and 2.5 apply with respect to all WTO Members, not just to those who participated in the development of the respective standard. However, a larger number of countries participating in the development of a standard makes it more likely that the body's activities in standardization are "recognized."[200] Nevertheless, a body's standards need not be widely used for it to have recognized activities in standardization for the purposes of the TBT Agreement.[201] Finally, an international organization that develops a single

[195] Appellate Body Report, *US—COOL* paras. 485–6.
[196] Appellate Body Report, *EC—Sardines* para. 227.
[197] Panel Report, *EC—Sardines* para. 7.68.
[198] Appellate Body Report, *EC—Sardines* para. 245.
[199] Appellate Body Report, *US—Tuna II (Mexico)* para. 356.
[200] Appellate Body Report, *US—Tuna II (Mexico)* para. 390.
[201] Appellate Body Report, *US—Tuna II (Mexico)* para. 392.

standard could have recognized activities in standardization, for example, if a large number of WTO Members participate in the development of the standard, acknowledge the validity and legality of the standard, or the body follows the principles contained in the TBT Committee Decision on Principles for the Development of International Standards, Guides and Recommendations with Relation to Articles 2, 5, and Annex 3 to the Agreement.[202]

With respect to climate change standards, the UNFCCC Conference of Parties could qualify as an "international standardizing body." However, the openness requirement, the uncertainty regarding the type of voting procedures that the TBT Agreement requires of international standards bodies, and the difficulty of determining the type of organization and process that qualifies, together with the phenomenon of multilateral negotiation paralysis we discussed in Chapter 2, may encourage unilateral approaches. In turn, unilateral measures will face a greater hurdle than multilateral measures that qualify as international standards under the TBT Agreement, by virtue of Articles 2.4 and 2.5.

Since the TBT Agreement is relevant interpretative context for the GATT, and vice versa, GATT XX also needs to be interpreted to take a stricter view of unilateral measures than multilateral measures. This approach is consistent with the Appellate Body Report in *US—Shrimp* and the Rio Declaration, both of which express a preference for multilateral approaches over unilateral ones.[203] Of course, the TBT Agreement applies not just to transboundary environmental measures, but also to purely domestic measures regarding environmental protection, as well as other legitimate objectives, including health and safety. Thus, both the GATT and the TBT Agreement discourage Baptist-and-Bootlegger approaches to the design of technical regulations, but do not prohibit unilateral approaches, in order to leave countries free to adopt their own level of protection.

At the end of the day, multilateral climate change regulation will likely prove insufficient to tackle climate change effectively. This gives WTO Members an argument under Article 2.4 to adopt unilateral technical regulations, since the international standards, if any exist, will be ineffective, in light of the growing scientific evidence of the urgency of addressing climate change. Multilateral negotiation paralysis, and the dramatic changes in the economic growth, technological capacity, and GHG emissions of developing countries since 1992, have made the UNFCCC approach outdated and ineffective to address climate change adaptation and mitigation. Moreover, new evidence indicates that the climate is changing faster than expected.

[202] Appellate Body Report, *US—Tuna II (Mexico)* para. 394; TBT Committee Decision on Principles for the Development of International Standards, Guides and Recommendations with Relation to Articles 2, 5, and Annex 3 to the Agreement, 46–8.

[203] For a detailed analysis of the treatment of unilateral and multilateral environmental measures in the GATT, see Bradly J. Condon, *Environmental Sovereignty and the WTO: Trade Sanctions and International Law* (Transnational Publishers, Ardsley, NY 2006).

3.7 Conclusion

The ongoing implementation of climate change policies could raise several unresolved issues in WTO law. GATT Article XX will play an important part in determining the WTO consistency of climate change measures. The scope of paragraphs (b) and (g) in GATT Article XX still needs to be defined in many aspects, as does the relationship between these two paragraphs. Multilateral environmental agreements on climate change will probably be relevant to determining the consistency of climate change measures with GATT Article XX and the provisions of the TBT Agreement. However, it is unlikely that GATT Article XX will be applied to the SCM Agreement, the Agreement on Agriculture or the TBT Agreement. Its application to provisions in other agreements in Annex 1A will have to be analyzed on a case-by-case basis.

If processing and production methods are relevant to determining the issue of "like products" in GATT Articles I and III, the SCM Agreement, the Antidumping Agreement and the TBT Agreement, then this may provide an alternative analytical approach to determine the WTO consistency of climate change measures. Again, this will have to be analyzed on a case-by-case basis in light of specific climate change measures. However, if environmental subsidies are designed so that they are not specific to certain enterprises, they will be not be subject to multilateral action under Part III or unilateral action under Part V. If the subsidies apply to agricultural products, they will have to comply with the commitments of Members under the Agreement on Agriculture. In the case of export subsidies, compliance with the Agreement on Agriculture may shield subsidies on agricultural products from action under SCM Agreement Article 3.1(a). However, opinion differs on this issue. In the case of subsidies contingent on the use of domestic products, it will be necessary to comply with both the SCM Agreement and the Agreement on Agriculture.

4

International Investment Agreements

4.1 Introduction

International investment agreements (IIAs) generally take two forms: chapters in Free Trade Agreements (FTAs), such as Chapter 11 of the North American Free Trade Agreement (NAFTA), and Bilateral Investment Treaties (BITs). There are thousands of BITs in existence. Both types of treaty set out State obligations regarding foreign investors. When States violate their obligations regarding foreign investors, foreign investors can seek compensation through investor–State arbitration.

In an OECD working paper, Gordon and Pohl have surveyed environmental references in FTAs with investment chapters and in BITs. They found that references to environmental concerns are common in FTAs with investment chapters (100 percent of 30 FTAs surveyed) while they are rare in BITs (6.5 percent of 1,593 BITs surveyed).[1] These references are either general references to environmental concerns or specific references to sanitary and phytosanitary objectives and conservation objectives.[2] The latter are often expressed as human, animal, and plant life or health and conservation of living or non-living exhaustible natural resources,[3] using the language of paragraphs (b) and (g) of GATT Article XX (see Chapter 3). Gordon and Pohl suggest that this use of specific references may prove less versatile when it comes to adapting this language to regulation favoring biodiversity or attenuating climate change, but acknowledge that analysis on the effect of including any kind of environmental language in international investment agreements has yet to be done.[4]

The Gordon and Pohl study raises some important legal issues, including: (1) whether the inclusion of references to environmental concerns in international investment agreements facilitates reconciling potential conflicts between foreign investment protection and environmental protection, and (2) which approach provides treaty parties with the most appropriate balance between predictability and flexibility with respect to the relationship between environmental and

[1] Kathryn Gordon and Joachim Pohl, "Environmental Concerns in International Investment Agreements: A Survey" (2011) OECD Working Papers on International Investment, No. 2011/1, OECD Investment Division <www.oecd.org/daf/investment> (accessed November 24, 2012).

[2] Gordon and Pohl, "Environmental Concerns in International Investment Agreements" 25–6.

[3] Gordon and Pohl, "Environmental Concerns in International Investment Agreements" 27.

[4] Gordon and Pohl, "Environmental Concerns in International Investment Agreements" 27.

investment norms. We have argued in Chapter 3 that WTO jurisprudence indicates that the language of GATT Article XX is sufficiently flexible to accommodate a variety of climate change regulations. In this chapter, we will argue that the language of international investment agreements is sufficiently flexible to accommodate climate change regulation as well. However, many issues remain unresolved in both areas of international economic law.

This chapter will analyze the extent to which States may be obliged to pay compensation to foreign investors for climate change regulation. Our analysis will be based on specific provisions in NAFTA Chapter 11 and other IIAs and the relevant jurisprudence from international investment tribunals. While international investment tribunals do not create precedent that is binding upon other tribunals, this jurisprudence does influence other tribunals. However, the approach of different international investment arbitrators to similar issues can vary considerably, which creates a degree of uncertainty regarding the outcome of international investment litigation.

Governments must strike a balance between regulation that discourages foreign investment and foreign investment protection that discourages regulation. If countries implement international agreements or domestic climate change policies in a way that violates the rights of foreign investors, they may have to pay compensation to the foreign investors. This risk can create disincentives to regulation, particularly in countries where the responsible government officials are unsure of the scope of their obligations to foreign investors. It is thus very important to understand the scope of these obligations when designing and implementing climate change regulation.

At the same time, foreign investment is an important source of knowledge and technology diffusion, together with trade in goods and services.[5] Thus, it is important to create adequate incentives for foreign investors to transfer best practices and technologies that can address climate change adaptation and mitigation. This means that governments must provide adequate protection to foreign investors, through IIAs and intellectual property rights. If these rights of foreign investors are watered down, in an effort to enhance access to technology, the effect could be to create disincentives to transfer technology and best practices through foreign investment (although intellectual property policies do need to vary with the technology, as we shall see in Chapter 5). In addition, IIAs can lower regulatory and political risks for foreign investors, and thus lower the cost of and create incentives for foreign investment in clean energy or in carbon mitigation technologies.[6]

[5] Bernard Hoekman and Beata Smarzynska Javorcik (eds.), *Global Integration and Technology Transfer* (Palgrave Macmillan and the World Bank, Washington 2006); Keith E. Maskus and Jerome H. Reichman (eds), *International Public Goods and Transfer of Technology under a Globalized Intellectual Property Regime* (Cambridge University Press, Cambridge 2005); Gill Wilkins, *Technology Transfer for Renewable Energy* (The Royal Institute of International Affairs and Earthscan Publications, Oxford 2002); Stephen O. Andersen, K. Madhava Sarma, and Kristen N. Taddonio, *Technology Transfer for the Ozone Layer: Lessons for Climate Change* (Global Environment Facility and Earthscan Publications, Oxford 2007).

[6] Anatole Boute, "Combating Climate Change through Investment Arbitration" (2012) 35 Fordham International Law Journal 613; UNCTAD, *World Investment Report 2010: Investing in a Low-*

Traditionally, debates regarding climate change and financial and technology flows have been framed as North–South. However, this has changed, with the emergence of new low-carbon technology companies in developing countries, such as China and India, which could diffuse clean technology nationally and internationally. In addition, the financial crisis has limited the financial capacity of major developed countries, notably those in the European Union, the United States, and Japan. Moreover, investment banks can help to mobilize capital for low-carbon technology, making their own investments and structuring investments for classes of investors with different risk–reward profiles and return expectations. This paradigm shift in financial and technology flows should inform debates regarding both.[7]

The legal analysis in this chapter begins with an examination of the scope of IIAs and their relationship to WTO law. We will consider the extent to which WTO jurisprudence can be applied to interpret IIAs. This analysis is important in order to determine the extent of these obligations and how to design environmental regulations to avoid inconsistency with obligations in both international trade agreements and international investment agreements.

IIAs impose three principal types of obligations on governments with respect to their treatment of foreign investors: (1) nondiscrimination between domestic and foreign investors, and between foreign investors from different countries; (2) a minimum standard of fair and equitable treatment for foreign investors; and (3) an obligation to pay compensation for expropriation. However, not all government regulations are subject to these obligations. This chapter will consider how environmental measures can either escape the application of the foregoing obligations or be justified under exceptions. First, we analyze the extent to which environmental regulation that affects foreign investors can be considered as measures that "relate to" foreign investment and foreign investors. This analysis determines whether the obligations in the IIA apply in a specific case. If they do not, the host government will not have to compensate foreign investors for economic loss caused by environmental regulation. This analysis is important to ensure that environmental regulations are designed so as to avoid triggering obligations to compensate foreign investors.[8] Next, this chapter will consider how the limited incorporation of

Carbon Economy (United Nations, Geneva 2010) <http://unctad.org/en/Docs/wir2010_en.pdf> (accessed November 2, 2012).

[7] Abyd Karmali, "Observations from the Carbon Emission Markets: Implications for Carbon Finance" in Cary Krosinsky and Nick Robins (eds.), *Sustainable Investing: The Art of Long-Term Performance* (Earthscan Publications, London 2008) 59, 71.

[8] Boute, "Combatting Climate Change"; Marie-Claire Cordonier Segger, Markus W. Gehring, and Andrew Newcombe (eds.), *Sustainable Development in World Investment Law* (Kluwer Law International, The Hague 2010); Marie-Claire Cordonier Segger and Markus Gehring, "Trade and Investment Implication of Carbon Trading for Sustainable Development", The Center for International Sustainable Development Law and The International Development Law Organization, 2010 <http://es.scribd.com/doc/51575616/Trade-and-Investment-Implication-of-Carbon-Trading-for-Sustainable-Development> (accessed April 9, 2013); Lise Johnson, "International Investment Agreements and Climate Change: The Potential for Investor–State Conflicts and Possible Strategies for Minimizing It" (2009) 39 Environmental Law Reporter 11147; Stephan W. Schill, "Do Investment Treaties Chill Unilateral State Regulation to Mitigate Climate Change?" (2007) 24 Journal of International Arbitration 469; Jacob Werksman, Kevin A. Baumert, and Navroz K. Dubash, *Will*

environmental exceptions into IIAs affects their interpretation and application in cases involving environmental regulation. NAFTA Articles 1106 and 1114 are the relevant NAFTA provisions in this regard. This analysis will also inform the design of climate change measures.

Next, we will analyze the nondiscrimination obligations in light of the relevant jurisprudence and with climate change regulation in mind. We then consider whether the interpretation and application of the minimum standard of treatment for foreign investors can or should vary in cases involving environmental regulation. This standard comes from public international law. To what extent does this area of public international law need to be interpreted to avoid conflicts with international legal obligations related to climate change? NAFTA Article 1105 is the relevant NAFTA provision in this regard. This analysis will also inform the design of environmental measures to address climate change. Finally, we will analyze the obligations regarding compensation for expropriation and measures that are tantamount to expropriation. This analysis will also inform the design of environmental measures to address climate change.

4.2 WTO Law and International Investment Agreements

The law on the relationship between environmental law and international economic law is most developed in WTO agreements regarding trade in goods, particularly GATT, the TBT Agreement, and the SPS Agreement. Even so, as we noted in Chapter 3, many legal issues remain unresolved in WTO law on trade in goods. The relationship between international intellectual property law and environmental law is far less developed. Foreign investment law also remains underdeveloped in this regard. However, there is more jurisprudence on this issue in foreign investment law than in international intellectual property law. Nevertheless, the WTO jurisprudence remains more developed and is more sophisticated than the jurisprudence in foreign investment law. All three areas of international economic law employ similar concepts and similar terminology, for example in national treatment and most-favored-nation obligations. However, the nature and context of these three fields differ in many respects. Thus, it is not easy to determine the extent to which developments in one area can be incorporated into another.

There is a growing interest in the relationship between WTO law and foreign investment law. This is due to several factors. First, the two areas use similar concepts and terminology.[9] Second, the WTO jurisprudence is of a higher quality,

international investment rules obstruct climate protection policies? (World Resources Institute, Washington 2001); Kate Miles, "International Investment Law and Climate Change: Issues in the Transition to a Low Carbon World" Society of International Economic Law (SIEL) Inaugural Conference, 2008 <http://ssrn.com/abstract=1154588> (accessed October 15, 2012).

[9] Nicholas DiMascio and Joost Pauwelyn, "Nondiscrimination in Trade and Investment Treaties: Worlds Apart or Two Sides of the Same Coin?" (2008) 102 AJIL 48.

more developed and is often taken into account by international investment arbitration panels. Indeed, the legal reasoning of international investment tribunals can be deficient.[10] Third, it is now clear that some measures can violate obligations in both areas. However, the consequences are very different. A violation of WTO law merely produces an obligation to bring the offending measure into compliance, with no liability for damages. A violation of international investment law can lead to a significant award of damages against the host country.

In *Archer Daniels Midland v. Mexico* and *Corn Products International v. Mexico*, foreign investors succeeded in their damage claims under NAFTA Chapter 11 for a Mexican tax on the use of high fructose corn syrup that discriminated in favor of Mexican sugar producers.[11] A WTO panel had found that the tax violated Mexico's national treatment obligations in GATT Article III:2.[12] The NAFTA Chapter 11 panels considered the WTO panel report in their determination of the violation of the national treatment obligation in Chapter 11, although they emphasized it was not determinative due to the differences in the terms and contexts of the national treatment obligations in GATT and in Chapter 11. Mexico invoked the public international law defense of countermeasures, arguing that its tax was justifiable as a countermeasure to induce US compliance with its NAFTA obligations regarding Mexican sugar imports. One panel held that public international law did not permit the use of countermeasures against foreign investors; they could only be taken against a foreign government. The other panel held that, while Mexico could invoke the public international law defense of countermeasures, it had not met the relevant criteria. This divergence of opinion between the two Chapter 11 panels raises the issue of whether foreign investors might succeed in a claim for damages against a WTO Member that applies countermeasures authorized by the WTO Dispute Settlement Body. The answer should be no, since the presumption against conflicts in international law would require investment arbitrators to interpret investment law in a manner that does not conflict with WTO law. Nevertheless, these cases demonstrate the potential for WTO law to influence foreign investment law and vice versa. There are tobacco cases regarding Australia's plain packaging law that raise issues regarding the relationship between WTO law and foreign investment law.[13] In addition, there are cases regarding clean energy subsidies in both areas of international economic law, which we discuss below.

[10] Federico Ortino, "Legal Reasoning of International Investment Tribunals: A Typology of Egregious Failures" (2012) 3 Journal of International Dispute Settlement 31.

[11] *Archer Daniels Midland Co. and Tate & Lyle Ingredients Americas, Inc. v. United Mexican States*, ICSID Case No. ARB(AF)/04/5, Award (November 21, 2007); *Corn Products International, Inc. v. United Mexican States*, ICSID Case No. ARB(AF)/04/1, Decision on Responsibility (January 15, 2008); *Cargill, Incorporated v. United Mexican States*, ICSID Case No. ARB(AF)/05/2, Award (September 18, 2009); *Mexico v. Cargill, Incorporated* 2010 ONSC 4656 (Ontario Superior Court of Justice); *Mexico v. Cargill, Incorporated* 2011 ONCA 622 (Ontario Court of Appeal); *Mexico v. Cargill, Incorporated* (2012), leave to appeal to Supreme Court of Canada denied.

[12] Appellate Body Report, *Tax Measures on Soft Drinks and Other Beverages (Mexico—Taxes on Soft Drinks)*, WT/DS308/AB/R, adopted March 24, 2006.

[13] *Australia—Certain Measures Concerning Trademarks and Other Plain Packaging Requirements Applicable to Tobacco Products and Packaging*, DS434 (complaint by Ukraine), DS435 (complaint by Honduras), DS441 (complaint by the Dominican Republic) <www.wto.org/english/tratop_e/dispu_e/

NAFTA Article 1131(1) requires the Tribunal to "decide the issues in dispute in accordance with this Agreement and applicable rules of international law." Article 38(1) of the Statute of the International Court of Justice sets out the sources of international law: international conventions, international custom, general principles of law, judicial decisions and the teachings of the most highly qualified publicists. The Vienna Convention on the Law of Treaties, which codifies the customary rules of treaty interpretation, applies to the interpretation of international investment treaties just as it applies to the interpretation of WTO law. In *Methanex v. United States*, the Tribunal and the parties agreed that GATT and WTO jurisprudence fall within the scope of "judicial decisions" that may be employed as "subsidiary means for the determination of rules of law" pursuant to Article 38 of the ICJ Statute. However, the Tribunal also recognized that the relevance of GATT and WTO jurisprudence depended on differences between the relevant texts, contexts and the objects and purposes of the different treaties, in accordance with Article 31 of the Vienna Convention. In this regard, the *Methanex* Tribunal cited the International Tribunal for the Law of the Sea in the *MOX Plant* case (as also applied in the *OSPAR* case): "the application of international law rules on interpretation of treaties to identical or similar provisions of different treaties may not yield the same results, having regard to, *inter alia*, differences in the respective contexts, objects and purposes, subsequent practice of parties and *travaux préparatoires*."[14]

In relation to environmental regulation, there are some important differences between international trade agreements and international investment agreements. Here, we illustrate these differences with respect to the WTO and NAFTA Chapter 11. First, only governments have access to the dispute settlement system of the WTO (and comparable dispute settlement systems in free trade agreements, including NAFTA Chapter 20). This filters out challenges to some measures. In contrast, private investors can file claims directly against host governments under international investment agreements, such as NAFTA Chapter 11. Private investors may be less likely to question the wisdom of challenging governments' right to regulate than government themselves.[15] Second, GATT contains general exceptions for environmental measures. Other WTO agreements on trade in goods (not

cases_e/ds434_e.htm> (accessed November 20, 2012), raises issues under the TRIPS Agreement, TBT Agreement, and GATT. *Philip Morris Asia Ltd. v. The Commonwealth of Australia*, UNCITRAL, PCA Case No. 2012–12. Regarding tobacco investment cases, see Tania Voon and Andrew Mitchell, "Time to Quit? Assessing International Investment Claims against Plain Tobacco Packaging in Australia" (2011) 14 JIEL 515.

[14] *Methanex Corporation v. United States*, NAFTA/UNCITRAL, Final Award of the Tribunal on Jurisdiction and Merits (3 August 2005) para. 16. *The MOX Plant case (Ireland v. United Kingdom)*, Order on Provisional Measures, 3 December 2001, 41 ILM 405, 413, para. 51, quoted in *Access to Information under Article 9 of the OSPAR Convention (Ireland v. United Kingdom)*, 42 ILM 1118, 1144, para. 141.

[15] Of course, this depends on the government. For example, Ukraine has filed a complaint against Australia at the WTO regarding its plain packaging tobacco regulations. See *Australia—Certain Measures Concerning Trademarks and Other Plain Packaging Requirements Applicable to Tobacco Products and Packaging*, DS434 <http://www.wto.org/english/tratop_e/dispu_e/cases_e/ds434_e.htm> (accessed December 20, 2012).

all) incorporate the language of the GATT exceptions, but framed as obligations. In both cases, provision is made to balance trade obligations against the right to regulate environmental protection. In contrast, NAFTA Chapter 11 does not contain a general exception for environmental measures, even though environmental protection and sustainable development are recognized as legitimate goals in the NAFTA Preamble, and environmental goals are reconfirmed in the subsequent North American Agreement on Environmental Cooperation (NAAEC). This means that the right to regulate should be addressed differently in the context of Chapter 11 than it is in the WTO context.[16]

The presumption against conflict in international law, as well as the rules that resolve conflicts, should come into play to avoid or to resolve conflicts between WTO law and IIAs. The presumption against conflicts requires tribunals to avoid conflicting interpretations of different international agreements. Where there is insufficient ambiguity in the treaty terms to avoid conflicts through interpretation, the general rules are that the specific prevails over the general (*lex specialis*) and the later prevails over the earlier treaty (*lex posterior*) in the event of a conflict. However, as we noted above, international investment tribunals vary in their approach to issues. This uncertainty is combined with need to consider specific provisions of specific treaties, which together make it difficult to generalize regarding how conflicts might be avoided or resolved.[17]

4.3 International Investment and Climate Change Measures

A variety of climate change measures may be subject to IIAs, provided that they relate to foreign investments or foreign investors and that the substance or procedural aspects of the measures violate the obligations owed to foreign investors according to the terms of the specific IIA. These measures can be for mitigation (carbon taxes, emissions trading schemes, border tax adjustments, carbon labels, standards, clean energy subsidies, and other infrastructure projects) or for adaptation (zoning bylaws and infrastructure projects such as flood barriers).

The majority of IIAs do not contain references to environmental concerns. This highlights the importance of determining the extent to which environmental regulation is subject to the disciplines contained in these agreements. NAFTA Article 1101 establishes that Chapter 11 only applies to measures "relating to" foreign "investors" and "investments." This in turn raised three issues. What categories of climate change "measures" might be the subject of claims under IIAs? What types of "investments" are covered by this term? To what extent can environmental regulation that affects foreign investors be considered a measure

[16] *Methanex Corporation v. United States*, Amicus Curiae Submission (submitted by Howard Mann and Don McRae, International Institute for Sustainable Development) (March 9, 2004) para. 23 <http://www.iisd.org/publications/pub.aspx?id=608> (accessed August 10, 2012).

[17] See generally Joost Pauwelyn, *Conflict of Norms in Public International Law: How WTO Law Relates to Other Rules of International Law* (Cambridge University Press, Cambridge 2003).

"relating to" foreign investments or foreign investors? This analysis determines whether the obligations in an IIA apply in a specific case. If they do not, the host government will not have to compensate foreign investors for economic loss caused by climate change regulation. This analysis is important to ensure that climate change regulations are designed and applied so as to avoid triggering obligations to compensate foreign investors.

Changes to land-use zoning have served as a basis for an investment claim. When rezoning affects an international investor, it may provide the basis for a claim of compensation, depending on the facts of each case. For example, in the *Metalclad* case, the area in which a hazardous waste disposal facility was located was changed to an ecological zone, which in turn prevented the hazardous waste plant from operating. The tribunal concluded that this rezoning constituted a measure "tantamount to expropriation" under NAFTA Article 1110 and ordered Mexico to pay compensation. Since the Mexican government had assured Metalclad that it had complied with all the relevant environmental laws prior to making its investment, Mexico could not rely on the exception in Article 1114, which permits host States to ensure that investment activity in its territory is undertaken in a manner sensitive to environmental concerns. While the state government made the change in zoning, the federal government was liable under NAFTA. Hurricane Sandy prompted proposed changes to zoning bylaws in New York, to take into account the expanded flood zone. This type of rezoning is likely to become more frequent as climate change causes more flooding. While it is not possible to anticipate all of the factors that might engender a claim for compensation under IIAs, there is a risk that such claims will occur.

Infrastructure projects—whether related to mitigation or adaptation—may also be the subject of foreign investor claims. The nature of the claims, and their chances of success, will depend on the facts of each case. There are three general categories of climate change mitigation measures that might be subject to IIAs: (1) subsidies that induce investment in clean energy projects or carbon mitigation projects; (2) carbon taxes; and (3) regulations that set specific standards or prohibit specific activities. These categories are not mutually exclusive. For example, there are cases proceeding under NAFTA Chapter 11 that involve clean energy infrastructure projects in Ontario, which combine elements of infrastructure, subsidies, and standards.

4.3.1 Subsidies

The Clean Development Mechanism, defined in Article 12 of the Kyoto Protocol, promotes the transfer of low-carbon technologies from industrialized countries to developing countries via projects that reduce GHG emissions and generate Certified Emission Reductions (CERs) that can be used by Annex I parties to meet their emission reduction targets.[18] National support for renewable energy,

[18] Maria Netto and Kai-Uwe Barani Schmidt, "CDM Project Cycle and the Role of the UNFCCC Secretariat" in David Freestone and Charlotte Streck (eds.), *Legal Aspects of Implementing the Kyoto Protocol Mechanisms: Making Kyoto Work* (Oxford University Press, Oxford 2005) 175.

energy efficiency, or other low-carbon investments may be implemented via "green certificates," "feed-in tariffs" (FITs), "premiums," or direct subsidies to electricity producers that use renewable energy sources.[19] Feed-in tariff schemes set a minimum price for the electricity produced from renewable energy installations and establish purchase obligations for electricity suppliers. In contrast, premium schemes set a premium on the electricity market price, so that the amount of support paid to the producers fluctuates with the market price of electricity. The investment issue is whether investors can claim compensation when governments induce foreign investment via these types of subsidy schemes, but then reduce the amount or duration of the subsidy after the investment has been made.

Government subsidies for renewable energy and other carbon mitigation investments may be inconsistent with the WTO Agreement on Subsidies and Countervailing Measures. As we noted in Chapter 3, there are several disputes before the WTO on such subsidies, including feed-in tariff schemes. If national subsidies are found to be inconsistent with the SCM Agreement, the government granting the subsidy is obliged to comply with its WTO obligations, normally by eliminating the subsidy. If a subsidy is withdrawn under these circumstances, can an IIA be interpreted to make the compliance with WTO law a compensable violation of the IIA? If the withdrawal of a subsidy is required under WTO and prohibited in the IIA, there would be a conflict between the two treaties. The presumption against conflict in international law would require an interpretation of the IIA to avoid a conflict with WTO law. One way to avoid this conflict would be to find that the elimination of a subsidy to comply with WTO law does not constitute a "legally significant connection" between a measure and an investor or an investment, and thus does not "relate to" the investment, as we discuss below. IIAs do not eliminate a State's right to regulate in the public interest.

With subsidies that are granted via the CDM, the subsidy is granted in accordance with international legal provisions in the Kyoto Protocol. Since the presumption against conflicts cannot override explicit treaty language, such a subsidy might still be incompatible with the SCM Agreement. However, such multilaterally based subsidies might not be challenged at the WTO. To date, no measures implemented in accordance with a multilateral environmental agreement have been challenged at the WTO. We analyze this issue in Chapter 8.

On 6 July 2011, Mesa Power Group LLC, a US-based company, served the Government of Canada with a Notice of Intent to Submit a Claim to Arbitration under NAFTA Chapter 11. Mesa Power Group's complaint concerns measures taken by the Government of Ontario, as they relate to the FIT program enabled by the Green Energy and Green Economy Act.[20] Mesa Power applied to participate in

[19] Commission of the European Communities, The Renewable Energy Progress Report: Commission Report in Accordance with Article 3 of Directive 2001/77/EC, Article 4(2) of Directive 2003/30/EC and on the Implementation of the EU Biomass Action Plan, COM (2009) 192 Final, at 6–7 (April 2009); Commission of the European Communities, Communication on the Support of Electricity from Renewable Energy Sources, COM (2005) 627 Final, at 4–5 (December 2005).

[20] *Mesa Power Group LLC v. Canada*, UNCITRAL <http://www.italaw.com/cases/1619> (accessed March 15, 2013).

Ontario's FIT program through four wind-farm projects. Renewable energy producers in the FIT program receive a premium price and a guaranteed market for the renewable energy that they produce, through long-term fixed price contracts with the Ontario Power Authority. Initially, participation in the wind power projects required 25 percent domestic content. This was raised to 50 percent for projects that became operational after 1 January 2012. Wind power projects over 10kW were required to obtain a minimum of this domestic content from Ontario sources. Projects were given a priority ranking for awarding contracts. Mesa Power claims that it lost its priority ranking when the rules were changed and was not offered contracts. Mesa Power claims that the ranking and contract-awarding process was arbitrary and non-transparent, in violation of NAFTA Article 1105. It also claims that the local content rules constitute performance requirements that violate Article 1106. It claims further that Samsung was given more favorable treatment, in violation of the MFN requirement of Article 1103, and that a Canadian company was given more favorable treatment, in violation of the national treatment requirement of Article 1102. Mesa Power claims CAD 775 million in damages.[21]

The nature of Mesa Power's claims is not unique to the renewable energy sector. Rather, its claims are related to alleged procedural irregularities in the awarding of renewable energy subsidies and contracts. The same FIT program is the subject of WTO disputes alleging that the subsidies violate the WTO Agreement on Subsidies and Countervailing Measures. The latter cases are more likely to succeed than the Mesa Power claims, since the SCM Agreement prohibits subsidies that are contingent on the use of domestic inputs. In *Metalclad v. Mexico*, the British Columbia Supreme Court (BCSC) held that a lack of transparency does not constitute of violation of Article 1105, because transparency does not form part of customary international law. The other claims are more properly characterized as government procurement claims, but the NAFTA Chapter on government procurement excludes obligations regarding government procurement by provincial governments.

The rule of effective treaty interpretation requires Chapter 11 to be interpreted harmoniously with the other provisions of NAFTA. Allowing an investor to succeed on claims that are excluded from the application of NAFTA would not be in accordance with this rule. Moreover, there is a presumption against conflict in international law. In effect, Mesa Power claims that it is entitled to subsidies that are prohibited by the SCM Agreement. Finally, Mesa Claims that NAFTA Chapter 11 requires the Ontario government to enter into contractual arrangements, a claim that would seem inconsistent with the principle of freedom of contract in contract law. Moreover, as the tribunal held in *Waste Management Inc. v. Mexico*, NAFTA Chapter 11 does not provide a forum for the settlement of contractual disputes. In that case, there was a contract between the Municipality of Acapulco and the claimant. If a breach of contract claim cannot succeed in Chapter 11, it seems that a claim based on no contract cannot.

[21] *Mesa Power Group LLC v. Canada*, UNCITRAL, Notice of Intent to Arbitrate (July 6, 2011).

Nevertheless, the FIT program cases demonstrate the importance of understanding the way that different areas of international economic law may interact in renewable energy regulation. In this instance, it appears that the design of the Ontario program failed to take into account the litigation risks stemming from NAFTA Chapter 11 and the SCM Agreement. The result is that the Canadian federal government must dedicate human and financial resources to international litigation on several fronts, since it is responsible in international law for the action of sub-national governments. Thus, the faulty design and implementation of Ontario's program increases the cost of its program and the risks that it will have to be redesigned to comply with Canada's international legal obligations.

On November 8, 2012, the US company Lone Pine Resources Inc. submitted a Notice of Intent to Submit a Claim to Arbitration against the Government of Canada under the dispute settlement provisions of NAFTA Chapter 11. The case concerns Quebec's decision to cancel oil and gas exploration permits for deposits under the St. Lawrence River as part of a wider moratorium on the controversial extraction technique of fracking. It is claiming CAD 250 million in damages. Lone Pine Resources claims a lack of due process, compensation, and public purpose in the revocation of its permits, in violation of NAFTA Articles 1105 and 1110. The Government of Quebec established a strategic environmental assessment committee to evaluate shale gas development in Quebec, but established the moratorium on June 13, 2011 with the Act to Limit Gas and Oil Activities before the environmental assessment was completed.[22]

Does the granting of oil and gas exploration permits imply that the ensuing activities, including exploratory drilling and eventual extraction activities should the resources be found, are not subject to environmental law or environmental impact assessment? Does the answer depend on the procedural and scientific basis for imposing the moratorium? In *Methanex v. United States*, a NAFTA Chapter 11 tribunal held that a ban on the gasoline additive methanol was not a measure relating to investment, but rather an environmental measure based on scientific evidence. As such, it was not subject to NAFTA Chapter 11. However, in *EC—Hormones*, the WTO Appellate Body held that the status of the precautionary principle in customary international law has not been established and found a ban on imports of hormone-treated beef to be inconsistent with the WTO Agreement on Sanitary and Phytosanitary Measures. If Quebec's moratorium on the extraction technique of fracking is based on the precautionary principle, and there is insufficient scientific evidence regarding its environmental impact, would this be a sufficient basis for a tribunal to conclude that the moratorium is not a measure relating to investment, but rather an environmental measure?

As we noted in Chapter 2, the precautionary principle has been recognized in international environmental instruments. Is that recognition sufficient to sustain such a finding under NAFTA Chapter 11? What degree of latitude does the precautionary principle allow for governments to regulate foreign investments

[22] *Lone Pine Resources Inc. v. The Government of Canada*, UNCITRAL, Notice of Intent to Submit a Claim to Arbitration (November 8, 2012) <http://www.italaw.com/cases/1606> (accessed March 13, 2013).

that may pose environmental risks? In *Metalclad v. Mexico*, the Tribunal found that NAFTA Article 1114 allows governments to require that foreign investors comply with environmental regulations. However, because it had assured Metalclad that it had complied with all relevant environmental regulations when in advance of its decision to invest in a hazardous waste facility, Mexico was found liable to pay compensation for a measure tantamount to expropriation when the area in which the plant was located was later declared an ecological zone by the state government. Does this mean that Quebec had to deny the exploration permits *ex ante* or lose its right to regulate fracking *ex post* based on environmental concerns?

On October 17, 2012, Windstream Energy LLC filed a claim against Canada relating to Ontario's FIT program, alleging violations of NAFTA Articles 1102, 1103, 1105, and 1110. Windstream claims compensation for a moratorium introduced on offshore wind energy projects after it was awarded a contract under the FIT program. The Ontario Government justified the moratorium on the grounds that further scientific research had to be completed before offshore wind development could proceed, but Windstream alleges that other factors in the decision were political opposition to offshore wind development on the Great Lakes and the higher cost of energy from offshore wind projects, as compared to onshore wind projects. Windstream was included in the moratorium even though it already had a contract. Windstream invoked the *force majeure* clause of the contract on the basis of the moratorium. The contract entitles the Ontario Power Authority to unilaterally terminate the contract if the *force majeure* delays the commercial operation for more than 24 months. Windstream claims expropriation of its investment in the project, a violation of the principle of fair and equitable treatment, as well as less favorable treatment than Samsung and less favorable treatment than Canadian companies, neither of which were subject to the moratorium (because their projects were land-based). Windstream claims damages of CAD 475 million.[23]

Like the *Mesa Power* case, the *Windstream* case involves Ontario's FIT program. In contrast to Mesa Power, Windstream had a contract. However, Windstream's invocation of the *force majeure* clause may entitle the Ontario Power Authority to unilaterally terminate the contract, in which case there would be no contract either. Were that not the case, the dispute might still be considered a contractual dispute that is not within the scope of Chapter 11, as in the Waste Management case. Like the *Lone Pine Resources* case, the *Windstream* case involves a moratorium that affects the rights of a foreign investor, though the rights are contractual rather than based on the issuance of permits. In both cases, the justification for the moratorium appears to be political opposition and a desire for further scientific research.

4.3.2 Carbon taxes and border tax adjustments

In *Archer Daniels Midland v. Mexico*, *Corn Products International v. Mexico*, and *Cargill v. Mexico*, foreign investors succeeded in their damage claims under NAFTA

[23] *Windstream Energy LLC v. Canada*, UNCITRAL, Notice of Intent (October 17, 2012) <http://www.italaw.com/cases/1788> (accessed March 15, 2013).

Chapter 11 for a Mexican tax on the use of high fructose corn syrup that discriminated in favor of Mexican sugar producers. In *Cargill v. Mexico*, the Tribunal compensated Cargill not only for the lost sales of the Mexican subsidiary, but for the lost sales of the American company to the Mexican subsidiary. Mexico sought judicial review at the Ontario Superior Court of Justice, unsuccessfully, on the grounds that the American company's losses were not compensable because it was not a foreign investor. Mexico lost a further appeal to the Ontario Court of Appeal and was denied leave to appeal to the Supreme Court of Canada. These cases show that trade-related losses from discriminatory taxes may be compensable under IIAs as a violation of national treatment, where the taxes discriminate between a foreign investor and a domestic competitor that is in "like circumstances." We analyze the national treatment obligation below.

It might be possible to argue that the taxes are countermeasures, if they are used to induce compliance with international climate change obligations. In *Archer Daniels Midland v. Mexico*, the Tribunal concluded that a countermeasures defense could constitute a defense to a claim under NAFTA Chapter 11, but that Mexico did not prove the elements of the defense. In contrast, in *Corn Products International v. Mexico*, the Tribunal found that this defense is not available, because the measures were applied to the private sector, not to a government. Where national climate change regulation is based on a multilateral agreement, the agreement provides evidence of the legitimacy of the regulation, and would buttress the countermeasures defense. However, where discriminatory taxes are used to induce a successful conclusion to climate change negotiations, this defense would not be available.

If the taxes are nondiscriminatory, legitimate environmental measures, they might not be subject to IIA obligations, since they would not qualify as measures "relating to" investments. The term "relating to" can be interpreted to exclude legitimate environmental regulation from the application of IIAs, a point we discuss below.

4.3.3 Regulations, labels, and standards

What if access to carbon labels is discriminatory? As with discriminatory taxes in *Archer Daniels Midland v. Mexico*, *Corn Products International v. Mexico*, and *Cargill v. Mexico*, foreign investors could succeed in claiming a violation of national treatment. A key issue is whether the foreign investor and a competing domestic company are in "like circumstances." As we saw in Chapter 3, the WTO Appellate Body Report in *US—Tuna II (Mexico)* suggests that processing and production methods could be relevant to a determination of the likeness of products in the TBT Agreement. If consumer preferences for products labeled low carbon in the relevant market determine the competitive relationship between domestic and imported products, and there is evidence that demonstrates that the products in fact have different carbon footprints, the products might not be considered "like" in the TBT Agreement and there would not be a violation of national treatment. A similar situation could lead to the conclusion that a foreign investor and a domestic competitor are not in "like circumstances" with respect to the national treatment obligation in an IIA. However, carbon labels are a recent development and consumer awareness of carbon footprints is not sufficiently developed to determine the competitive relationship between products. If governments introduce carbon labeling schemes it will be difficult to design the measurement of carbon footprints in a manner that avoids discrimination between otherwise competing products. If governments introduce carbon labeling schemes, this could constitute not just a violation of GATT or the TBT Agreement, but form the basis of a claim under an IIA. The same would be true for technical regulations regarding the carbon intensity of processing and production methods.

4.4 Applying Investment Agreements to Climate Measures

Boute analyzes whether international and national government support schemes for renewable energy qualify as investments.[24] The *Salini* test, developed on the basis of the Convention on the Settlement of Investment Disputes Between States and Nationals of Other States (the ICSID Convention), defines an "investment" according to four criteria: contributions by the investor; certain duration of performance; the existence of operational risks; and the contribution to the economic development of the host State.[25] Boute argues that some regulatory and contractual rights relating to national subsidies for low-carbon investments could qualify as "investments," and thus qualify for protection under international investment agreements. State withdrawal of financial support for low-carbon investment is unlikely to constitute a measure tantamount to expropriation, since it would not destroy the economic value of the overall investment or deprive investors of full ownership and control of their assets. However, while contractual rights may not be compensable on their own, where rights can be sold independently on the international carbon market and thus have intrinsic economic value, Boute argues that State interference may constitute partial expropriation.[26] However, Boute does not

[24] Boute, "Combatting Climate Change" 628–36.

[25] Boute, "Combatting Climate Change" 625–7; Convention on the Settlement of Investment Disputes Between States and Nationals of Other States, opened for signature 18 March 1965, 575 UNTS 159 (entered into force 14 October 1966) art. 1; *Salini Costrutorri S.p.A. and Italstrade S.p.A. v. Kingdom of Morocco*, ICSID Case No. ARB/00/4, Decision on Jurisdiction (July 23, 2001), (2004) 6 ICSID Rep. 400; *Fedax N.V. v. Republic of Venezuela*, ICSID Case No. ARB/96/3, Decision on Objections to Jurisdiction (July 11, 1997), (2002) 5 ICSID Rep. 186; Anna Turinov, "'Investment' and 'Investor' in Energy Charter Treaty Arbitration: Uncertain Jurisdiction" (2009) 26 Journal of International Arbitration 1. The Energy Charter Treaty definition of "investment" refers to "any investment associated with an Economic Activity in the Energy Sector and to investments or classes of investments designated by a Contracting Party in its Area as 'Charter efficiency projects' and so notified to the Secretariat." Energy Charter Treaty (1994) art. 1(6). <http://www.encharter.org/fileadmin/user_upload/document/EN.pdf> (accessed March 13, 2013).

[26] Boute, "Combatting Climate Change" 628–36; *Waste Management, Inc. v. United Mexican States*, ICSID Case No. ARB(AF)/00/3, Award (April 30, 2004), (2004) 43 ILM 967; Ursula Kriebaum, "Partial Expropriation" (2007) 8 Journal of World Investment and Trade 69.

analyze whether withdrawal of subsidies would constitute measures "relating to" investment.

The issue of whether contractual rights constitute investments does not resolve the issue of whether climate change regulations relate to an investment. While there may be an issue of what constitutes an investment, this is not the real issue in the case of regulation. With genuine climate change measures that are based on available scientific evidence and are designed and applied in a nondiscriminatory fashion, the more important issue in determining the applicability of IIAs is whether the measures relate to climate change mitigation or adaptation and are thus not investment measures. If climate change regulations are not investment measures, it does not matter whether there is an investment that is affected by those measures. In this Section, we briefly address the circumstances in which contractual rights may be compensable, before addressing the issue of whether a measure relates to investment.

4.4.1 Contractual rights

In *Waste Management v. Mexico*, the NAFTA Chapter 11 Tribunal declined to find a violation of Article 1105 for a breach of contract. The Tribunal noted that NAFTA Chapter 11 is not a forum for the resolution of contractual disputes, and that, "Investment Treaties are not insurance policies against bad business judgments."[27] The Tribunal also declined to find that the breach of contract was tantamount to expropriation. The Tribunal noted that any serious breach of contract could result in the loss of benefits or expectations, but this is not a sufficient criterion for concluding that there is an expropriation; NAFTA Article 1110 is not intended to compensate for failed business ventures, in the absence of arbitrary intervention by the State amounting to a virtual taking of the enterprise.[28]

The *Waste Management* Tribunal addressed the question when a persistent and serious breach of a contract by a State organ can constitute expropriation of the right in question, or at least conduct tantamount to expropriation of that right, for the purposes of Article 1110.[29] There are three categories of cases: (1) cases where a whole enterprise is terminated or frustrated because its functioning is simply halted by decree or executive act, usually accompanied by other conduct; (2) cases where there has been a taking of property and associated contractual rights are affected in consequence, in which cases the bundle of rights to be compensated includes all the associated contractual and other incorporeal rights, unless these are severable and retain their value in the hands of the claimant notwithstanding the seizure of the

[27] *Waste Management, Inc.*, Award, para. 114; *Emilio Agustín Maffezini v. Kingdom of Spain*, ICSID Case No. ARB/97/7, Award (November 13, 2000), (2002) 5 ICSID Reports 419, para. 64; *CMS Gas Transmission Company v. Argentine Republic*, ICSID Case No. ARB/01/8, Decision of the Tribunal on Objections to Jurisdiction (July 17, 2003), (2003) 42 ILM 788, para. 29; *Eudoro A. Olguín v. Republic of Paraguay*, ICSID Case No. ARB/98/5, Award (July 26, 2001), (2004) 6 ICSID Reports 164, paras. 72–5.

[28] *Waste Management, Inc.*, Award, paras. 159–60.

[29] *Waste Management, Inc.*, Award, para. 165.

related property; and (3) cases where the only right affected is incorporeal, which come closest to a claim of contractual nonperformance and in which a simple assertion that "property rights are created under and by virtue of a contract" is not sufficient. Whereas any private party can fail to perform its contracts, nationalization and expropriation are inherently governmental acts. An investor faced with a breach of contract by its governmental counter-party, which does not take the form of an exercise of governmental prerogative, should sue in the appropriate court to remedy the breach. Claims for expropriation are appropriate only where such access is legally or practically foreclosed, so that the breach could amount to a definitive denial of the right that amounts to an effective taking of the investor's property. Thus, the Tribunal concluded that foreign investment law distinguishes between the expropriation of a right under a contract and failure to comply with the contract. Noncompliance by a government with contractual obligations is not equivalent or tantamount to an expropriation. The test for an expropriatory taking of contractual rights requires an investor to prove an effective repudiation of the right, unredressed by any available remedies, which has the effect of preventing its exercise entirely or to a substantial extent.[30]

4.4.2 Connection between measure and investment

NAFTA Chapter 11 tribunals have considered the extent to which WTO jurisprudence on the same term in GATT Article XX(g) is relevant to interpret NAFTA Article 1101. The general exception in GATT Article XX(g) permits measures "relating to" the conservation of exhaustible natural resources. GATT and WTO jurisprudence has interpreted "relating to" to mean "primarily aimed at" in this context. In *Pope & Talbot v. Canada*, Canada contended that a measure could only relate to an investment if it was "primarily directed" at that investment. The tribunal did not reject Canada's argument that it was insufficient that a measure affects an investor, but did reject the contention that the measure must be primarily directed at the investment.[31] In *Methanex v. United States*, the Tribunal noted that the WTO interpretation of the term "relating to" in GATT Article XX(g) was quite different from the interpretation in the *Pope & Talbot* case, which confirms the need to interpret a term in accordance with the particular context, object and purpose.[32]

In *Methanex v. United States*, California banned the use of methanol as a gasoline additive, for environmental reasons. The United States succeeded in arguing that, as a nondiscriminatory environmental measure, the methanol ban was not a measure "relating to" foreign investment or foreign investors under NAFTA Article 1101. As such, it was not subject to NAFTA Chapter 11 and the tribunal had no

[30] *Waste Management, Inc.*, Award, paras. 171–6.

[31] *Pope & Talbot Inc. v. The Government of Canada*, UNCITRAL (2000) (Award in relation to Preliminary Motion by Government of Canada to dismiss the claim because it falls outside the scope and coverage of NAFTA Chapter 11 "measures relating to investment" motion) 33–4 <http://italaw.com/sites/default/files/case-documents/ita0672.pdf> (accessed April 9, 2013).

[32] *Methanex Corporation v. United States*, NAFTA/UNCITRAL, Partial Award (August 7, 2002).

jurisdiction to hear the claim. The *Methanex* Tribunal found that Article 1101(1) requires "something more than the mere effect of a measure on an investor or an investment" and that the term "relating to" requires a "legally significant connection" between a measure and an investor or an investment.[33] Methanex had argued that the California ban was, in fact, intended to harm producers and marketers of methanol. Thus, the question was whether the scientific conclusions which were presented to the Governor were so faulty that the Tribunal could reasonably infer that the science merely provided a convenient excuse for the hidden regulation of methanol producers. However, the scientific and administrative record established clearly that Governor Davis and the California agencies acted with a view to protecting the environmental interests of the citizens of California, and not with the intent to harm foreign methanol producers. Faced with widespread and potentially serious methyl tertiary-butyl ether (MTBE) contamination of its water resources, California ordered a careful assessment of the problem and thereafter responded reasonably to independent findings that large volumes of the state's ground and surface water had become polluted by MTBE and that preventive measures were called for. Thus, on the facts of this case, there was no legally significant connection between the US measures, Methanex, and its investments. As such, the US measures did not "relate to" Methanex or its investments as required by Article 1101(1).[34]

In contrast to the *Methanex* case, in *S.D. Myers v. Canada*, in which Canada introduced a ban on the export of polychlorinated biphenyls (PCBs) to protect the Canadian PCB disposal industry from US competition, the Tribunal concluded that there was no legitimate environmental reason for introducing the ban. While the Basel Convention on the Control of Transboundary Movements of Hazardous Wastes and Their Disposal discourages transboundary movements of hazardous wastes and encourages the creation of domestic capacity for their disposal, in this case, it was environmentally preferable to ship hazardous wastes from Central Canada to Ohio than to Canada's only PCB disposal facility in Alberta, due to Ohio's much closer proximity. The tribunal held that the requirement in Article 1101 that the import ban relate to S.D. Myers and its investment was "easily satisfied," because the "specific inspiration for the export ban" was the prospect that S.D. Myers would carry through with its plans to expand its Canadian operations.[35] The S.D. Myers Tribunal also concluded that the applicability of other provisions of the NAFTA, regarding trade in goods or services, was not an obstacle to pursuing an investment claim. The Tribunal considered that, as in WTO law, different NAFTA chapters and provisions could be applied in a cumulative and complementary fashion, as long as there was no conflict in the sense that adherence to one provision would cause a violation of another. In this

[33] *Methanex v. United States*, Partial Award, para. 4.

[34] *Methanex v. United States*, Final Award, paras. 19–20, 22.

[35] *S.D. Myers v. The Government of Canada*, UNCITRAL, Partial Award (November 13, 2000) para. 234.

case, there was no reason why a measure which concerns trade in goods cannot be a measure relating to an investor or an investment.[36]

The *S.D. Myers* and *Methanex* cases highlight the importance of scientific evidence regarding the contribution of a particular measure to environmental protection. Gordon and Pohl found 16 treaties (FTAs and BITs) that contain provisions related to the recourse to environmental experts by arbitration tribunals, including NAFTA Article 1133.[37] The clauses in BITs provide:

> Without prejudice to the appointment of other kinds of experts where authorized by the applicable arbitration rules, a tribunal, at the request of a disputing party or, unless the disputing parties disapprove, on its own initiative, may appoint one or more experts to report to it in writing on any factual issue concerning environmental, health, safety, or other scientific matters raised by a disputing party in a proceeding, subject to such terms and conditions as the disputing parties may agree.[38]

This is a useful provision to have, given the important role of scientific evidence in environmental cases. However, it is not essential, since the parties have the right to present scientific evidence and tribunals have the right to consider this evidence in their determinations, without the presence of this type of clause.

Unlike GATT, NAFTA Chapter 11 has no general exception for environmental measures that is comparable to GATT Article XX. NAFTA Chapter 3 incorporates GATT Article XX by reference. However, Article XX cannot be invoked to justify a violation of Chapter 11. Instead, Chapter 11 contains two environmental exceptions that are of limited use. Article 1106 contains environmental exceptions that only apply to certain provisions in that article. Article 1114 contains a more general exception for environmental measures, but its scope is limited. We will consider each of these environmental exceptions below. However, the host government can argue that Chapter 11 does not apply to an environmental measure, because the measure does not relate to investors or investments according to Article 1101. Thus, the interpretation and application of Article 1101 is an important part of the analysis of the relationship between environmental law and investment law.

To what extent can WTO jurisprudence regarding the term "relating to" in GATT Article XX(g) be applied to interpret NAFTA Article 1101? On the one hand, the ordinary meaning of the term "relating to" should be the same, since the wording is identical. Moreover, the purpose of the term "relating to" is to determine the nature of the measure, in both GATT Article XX(g) and NAFTA Article 1101. These two factors suggest that the interpretation of this term should not be significantly different and that WTO jurisprudence should be relevant to the interpretation of NAFTA Article 1101. On the other hand, there are significant differences in the contexts of the two provisions. Unlike GATT Article XX(g), NAFTA Article 1101 is not an exception. In addition, to justify a measure in

[36] *S.D. Myers v. Canada*, Partial Award, paras. 292–4.

[37] Gordon and Pohl, "Environmental Concerns in International Investment Agreements" 22.

[38] Canada–Jordan BIT (2009); Canada–Peru BIT (2006); Mexico–United Kingdom BIT (2006); United States–Rwanda BIT (2008); United States–Uruguay BIT (2005); US Model BIT (2004) art. 32; Canada Model BIT (2004) art. 42.

GATT Article XX(g), the measure must relate to the conservation of exhaustible natural resources, meet a requirement of evenhandedness through the application of restrictions on domestic production or consumption, and the application of the measure must comply with the chapeau of Article XX. There are no comparable requirements in Article 1101. Moreover, while NAFTA incorporated GATT Article XX by reference in Chapter 3, it did not do so in Chapter 11. It would be inappropriate to incorporate this exception through the back door of treaty interpretation, particularly since it would not be subject to the additional requirements of Article XX(g) and the chapeau, whose terms cannot be read into Article 1101.

Nevertheless, a measure that meets the requirements of Article XX(g) is not likely to meet the requirements of Article 1101. That is, if a measure qualifies as a measure relating to conservation of exhaustible natural resources, it is not likely to qualify as a measure relating to investors or investments. For example, if a WTO Member implements its obligations under a multilateral environmental agreement (MEA) on climate change in a nondiscriminatory manner, the measure is likely to comply with GATT Article XX(g).[39] The same measure is unlikely to meet the requirements of Article 1101, as it was interpreted and applied in *Methanex v. United States.* The MEA would provide evidence that the measure is an environmental measure, not an investment measure. Of course, a measure could be both an environmental measure and an investment measure. However, the presumption against conflicts in international law would favor an interpretation of Article 1101 that avoids a conflict between the treaty obligations in Chapter 11 and the treaty obligations in the MEA. If the MEA requires the measure and the provisions of Chapter 11 prohibit the measure, there would be a conflict. Similarly, if GATT Article XX(g) permits the measure, then Chapter 11 should too. It is important to note that, in *S.D. Myers v. Canada*, there was no conflict with the Basel Convention and there was evidence that the government banned the export of PCBs to protect domestic industry, not the environment. Thus, the ruling in that case is consistent with our argument.

A key issue is the legitimacy of the disputed environmental measure. One way to define legitimacy is by asking whether the measure serves the public interest or a private interest. In *Methanex*, the measure served the former. In *S.D. Myers*, the measure served the latter. Of course, a measure can simultaneously serve both public and private interests. The real question here is whether the evidence demonstrates bad faith, protectionist intent, or intent to harm foreign investors on the part of the legislator or the judiciary. NAFTA Chapter 11 has been criticized as lacking legitimacy, due to the pressure it places on governments to ensure that laws are consistent with its provisions.[40] However, NAFTA Chapter 11 is far from unique, in a universe of thousands of BITs and dozens of investment chapters in

[39] Bradly J. Condon, *Environmental Sovereignty and the WTO: Trade Sanctions and International Law* (Transnational Publishers, Ardsley NY 2006) ch. 7.

[40] José E. Alvarez, "Critical Theory and The North American Free Trade Agreement's Chapter Eleven" (1997) 28 University of Miami Inter-American Law Review 303; Charles H. Brower, II, "Structure, Legitimacy, and NAFTA's Investment Chapter" (2003) 36 Vanderbilt Journal of Transnational Law 37; Jeffery Atik, "Repenser NAFTA Chapter 11: A Catalogue of Legitimacy Critiques" (2003) 3 Asper Review of International Business and Trade Law 215.

FTAs. Moreover, as far as environmental regulation is concerned, the legitimacy of environmental measures is the real issue. With respect to climate change regulation, as with other areas of environmental regulation, where national regulation is based on a multilateral agreement, the agreement provides evidence of the legitimacy of the regulation. However, a multilateral agreement is not required to prove legitimacy.

Article 1101 also should be interpreted in a manner that is consistent with customary international law. In *Maffezini v. Spain*, Spain's application of environmental impact assessment requirements was held to be consistent with the bilateral investment treaty between Argentina and Spain. The Tribunal noted that "the environmental impact assessment procedure is basic for the adequate protection of the environment and the application of appropriate preventive measures . . . not only under Spanish and EEC law, but also increasingly so under international law."[41] The Claimant had sought compensation for the additional costs resulting from the environmental impact assessment, claiming that it had been pressured to go ahead with the investment before that process was finalized. The Tribunal found that both Mr. Maffezini and his employees were aware that the project required an environmental impact assessment. Spain had done no more than insist on the strict observance of the European Economic Community (EEC) and Spanish law applicable to the industry in question. Therefore, the Tribunal held that Spain could not be held responsible for the decisions taken by the Claimant with regard to the environmental impact assessment. Furthermore, Spain's action was fully consistent with Article 2(1) of the Argentina–Spain Bilateral Investment Treaty, which calls for the promotion of investment in compliance with national legislation.[42] While the provision of the investment treaty in this case was different from NAFTA Article 1101(1), the same logic applies under the latter, since an environmental requirement that applies in general to industries, regardless of whether they are national or foreign investment, is not a measure relating to investment. In addition, while the environmental law in this case was national and regional, the same logic should apply where the source of the environmental law is international. In *Pulp Mills on the River Uruguay (Argentina v. Uruguay)*, the International Court of Justice held that environmental impact assessment is required under customary international law.[43] Were customary international law the source of the environmental impact assessment requirement, the conclusion should be the same.

The term "relating to" in Article 1101 should be subject to evolutionary interpretation and be interpreted to exclude legitimate climate change regulation from the scope of application of international investment agreements. In *US—Shrimp*, the WTO Appellate Body held that the definition of measures relating to "exhaustible natural resources" in GATT Article XX(g) is subject to evolutionary interpretation.[44] As noted above, one must be cautious about importing interpretations from trade law to investment law. However, when the issue is whether a

[41] *Maffezini v. Spain*, Award para. 67. [42] *Maffezini v. Spain*, Award paras. 65–71.

[43] *Pulp Mills on the River Uruguay (Argentina v. Uruguay)* (2010) (Judgment) ICJ Reports 2010 paras. 203–19.

[44] Appellate Body Report, *US—Shrimp*, WT/DS58/AB/R, adopted November 6, 1998.

measure relates to environmental protection or to investment, NAFTA Article 1101 and GATT Article XX(g) are addressing the same kind of question. It is the evolution of international environmental law that requires evolutionary interpretation of GATT Article XX(g). The same should be true for NAFTA Article 1101.

One could argue that Article 1101 is not intended to avoid or to resolve conflicts between investment law and environmental law. Rather, such conflicts should be resolved according to the specific environmental exceptions in Articles 1106 and 1114. However, the limited scope of those environmental exceptions mean that many legitimate environmental measures would not be subject to those exceptions. In such a case, one way to avoid the conflict would be through the interpretation and application of Article 1101. Moreover, Article 1114 supports the view that legitimate environmental regulation is beyond the scope of Chapter 11, by confirming the right of parties to protect the environment and recognizing that, far from discouraging environmental regulation, Chapter 11 discourages the relaxation of environmental regulation in order to attract investment.

Article 1114 provides as follows:

> Article 1114. Environmental Measures
>
> 1. Nothing in this Chapter shall be construed to prevent a Party from adopting, maintaining or enforcing any measure otherwise consistent with this Chapter that it considers appropriate to ensure that investment activity in its territory is undertaken in a manner sensitive to environmental concerns.
>
> 2. The Parties recognize that it is inappropriate to encourage investment by relaxing domestic health, safety or environmental measures. Accordingly, a Party should not waive or otherwise derogate from, or offer to waive or otherwise derogate from, such measures as an encouragement for the establishment, acquisition, expansion or retention in its territory of an investment of an investor. If a Party considers that another Party has offered such an encouragement, it may request consultations with the other Party and the two Parties shall consult with a view to avoiding any such encouragement.

In *Metalclad v. Mexico*, Metalclad invested in a hazardous waste plant. The Mexican state of San Luis Potosi declared the area where the plant was located to be an ecological zone, which prevented Metalclad from operating the plant. Both the NAFTA tribunal and the British Columbia Supreme Court held that this ecological decree was a measure equivalent to expropriation that required Mexico to pay compensation. Mexico raised Article 1114(1) as a defense. The Tribunal rejected this argument because Mexico had taken steps to satisfy itself that Metalclad's investment would be undertaken in a manner consistent with and sensitive to environmental concerns, through an agreement with Metalclad and federal permits.[45] On judicial review, the BCSC held that this conclusion was not unreasonable.[46]

[45] *Metalclad Corporation v. United Mexican States*, ICSID Case No. ARB(AF)/97/1, Award (August 30, 2000), (2002) 5 ICSID Rep. 212; a similar case, for which a decision had not been issued at the time of writing, is *Abengoa, S.A. y COFIDES, S.A. v. United Mexican States*, ICSID Case No. ARB (AF)/09/2, Hearing on Merits (June 2012); Katia Fach Gómez, "ICSID Claim by Spanish Companies against Mexico over the Center for the Integral Management of Industrial Resources" (2010) <http://ssrn.com/abstract=1631835> (accessed March 13, 2013).

[46] *Mexico v. Metalclad Corporation* 2001 BCSC 664 para. 104.

The ruling in the Metalclad case suggests that Article 1114(1) can only justify environmental regulations that are in force prior to an investment being made. Many existing investments could be affected by future climate change regulation. If one accepts the reasoning in *Metalclad* and in *Methanex*, such post-investment regulations would not be subject to NAFTA Chapter 11 because they do not relate to investment and thus would not have to be justified under Article 1114(1) in any event.

The type of language found in NAFTA Article 1114(1) is the most common category of environmental provision language, appearing in 82 of the 1623 treaties surveyed by Gordon and Pohl. For example, the Netherlands–Costa Rica BIT (1999) provides:

> The provisions of this Agreement shall, from the date of entry into force thereof, apply to all investments made, whether before or after its entry into force, by investors of one Contracting Party in the territory of the other Contracting Party in accordance with the laws and regulations of the latter Contracting Party, including its laws and regulations on labour and environment.

In 49 of the treaties surveyed, provisions discourage lowering environmental regulation for the purpose of attracting investment, in the same manner as NAFTA Article 1114(2). This type of provision was inserted to address concerns that FTAs would lead to the creation of "pollution havens"—countries that would loosen environmental regulations in order to attract investment. These types of provisions generally are not legally enforceable. For example, NAFTA Article 1114(2) merely provides that the countries "should not" relax environmental standards (as opposed to the mandatory "shall not"), and the only consequence of doing so is an obligation to discuss the matter should another country request consultations. This leaves countries free to change environmental regulations.

Another argument regarding the right of States to regulate climate change is found in Article 1112(1), which states: "In the event of any inconsistency between this Chapter and another Chapter, the other Chapter shall prevail to the extent of the inconsistency." NAFTA Article 2101 incorporates the general exceptions of GATT Article XX for the purposes of provision regarding trade in goods and technical barriers to trade. WTO jurisprudence has held that WTO Members have the right, under GATT Article XX(b), to determine the level of environmental and health protection that they consider appropriate.[47] Similarly, NAFTA Article 712 confirms the same right with respect to sanitary and phytosanitary measures, as does the WTO Agreement on Sanitary and Phytosanitary Measures. It is possible for the same measure to violate both WTO provisions and NAFTA Chapter 11 provisions. In the case of the former, the right to regulate is confirmed in the treaty texts and the jurisprudence. It would be inconsistent to restrict the same right to regulate by making it subject to the obligations in Chapter 11 and the rights of investors to receive compensation.

[47] Appellate Body Reports, *European Communities—Measures Affecting Asbestos and Asbestos-Containing Products* (*EC—Asbestos*), WT/DS135/AB/R, adopted April 5, 2001; *Brazil—Measures Affecting Imports of Retreaded Tyres* (*Brazil—Retreaded Tyres*), WT/DS332/AB/R, adopted December 17, 2007.

However, this argument might be countered by arguing that the NAFTA parties incorporated a modified version of the relevant GATT Article XX exceptions in Article 1106 and that this indicates their intention to limit the right to regulate environmental issues to the matters addressed in this article. However, one could argue that Article 1106 merely clarifies that its specific obligations are not intended to impair the right of States to regulate environmental protection and that this clarification was not necessary in the case of the preexisting customary international investment law reflected in Article 1105 (which can be resolved in accordance with customary international law) or the nondiscrimination obligations in Articles 1102, 1103 and 1104 (which prohibit discrimination rather than environmental regulation per se). Below, we discuss Article 1106, followed by Articles 1102, 1103, 1104, and 1105.

Gordon and Pohl found several BITs that incorporate general provisions, based on the language of GATT Article XX or based on security interests and sanitary and phytosanitary concerns.[48] Their paper sets out several examples, but we will only mention one of each category here:

> Provided that such measures are not applied in a discriminatory or arbitrary manner or do not constitute a disguised restriction on foreign investment, nothing in this Agreement shall be construed to prevent a Contracting Party from adopting measures to maintain public order, or to protect public health and safety, including environmental measures necessary to protect human, animal or plant life.[49]

> Nothing in this Agreement precludes the host Contracting Party from taking, in accordance with its laws applied reasonably and on a non-discriminatory basis, measures necessary for the protection of its own essential security interests or for the prevention of diseases or pests.[50]

The first example incorporates language from the chapeau and paragraph (b) of GATT Article XX, *inter alia*. However, the language is adapted to the investment context. Thus, while WTO jurisprudence would be relevant to its interpretation, the differences in terms and context would have to be taken into account. The second example does not borrow as much from GATT Article XX, but does incorporate the a necessity test that also is used in GATT Article XX(b). While WTO jurisprudence would be relevant to the interpretation of the term "necessary," the differences in the context, object, and purpose would have to be taken into account. In both examples, the use of GATT language is helpful to avoid conflicts or inconsistencies between WTO law and international investment law.

The majority of State practice is consistent with the view that international investment agreements do not negate the right to regulate climate change, since the overwhelming majority of IIAs do not incorporate explicit exceptions to preserve the right to regulate in the public interest. The term "relating to" (and similar

[48] Gordon and Pohl, "Environmental Concerns in International Investment Agreements" 17–18.

[49] Canada–Egypt BIT (1996); Canada–El Salvador BIT (1999); Canada–Lebanon BIT (1997); Canada–Panama BIT (1996); Canada–Philippines BIT (1995); Canada–South Africa BIT (1995); Canada–Thailand BIT (1997); Canada–Trinidad and Tobago BIT (1995).

[50] Australia–India BIT (1999).

terminology) can be interpreted and applied to exclude *bona fide* environmental regulation from the applications of such agreements. The fact that some recent agreements have incorporated environmental exceptions, some using the language of GATT Article XX, should not be viewed as an indication that other international investment agreements restrict governments' right to regulate in the public interest. Rather, the incorporation of such exceptions should be viewed as a clarification or a codification of existing customary international law, possibly in response to the unpredictability of international investment tribunals, or to public concern that these agreements were not addressing environmental concerns explicitly or possibly to avoid conflicts or inconsistencies between WTO law and international investment law. Thus, there are many reasons for incorporating this type of provision that do not suggest that such provisions are essential to preserve the right to regulate climate change and other matters of public interest.

4.5 Performance Requirements

NAFTA Article 1106 prohibits the imposition of performance requirements on investments of an investor of a party or of a non-party. Thus, unlike Articles 1102–1105, Article 1106 creates obligations regarding investors of non-parties, not just investors of NAFTA parties. The prohibited performance requirements include: minimum levels of domestic content; preference for domestic goods or services; and requirements to transfer technology, a production process, or other proprietary knowledge. Regarding technology transfer, there is an exception for measures that require an investment to use a technology to meet generally applicable health, safety or environmental requirements, as long as the measure complies with the non-discrimination obligations in Articles 1102 and 1103.

In *Mobil Investments Canada Inc. & Murphy Oil Corporation v. Canada*, the investors alleged that Guidelines for Research and Development Expenditures adopted in 2004 by the Canadian Newfoundland and Labrador Offshore Petroleum Board (the 2004 Guidelines) were more restrictive and onerous than the provisions of existing agreements concerning two offshore petroleum development projects. The 2004 Guidelines required the investors in offshore petroleum projects to pay millions of dollars per year for research and development in the Province of Newfoundland and Labrador. The Tribunal held that the 2004 Guidelines violated Article 1106.[51]

The *Mobil* Tribunal found that the 2004 Guidelines purported to impose conditions on the Claimants with respect to their management, conduct, or operation in the Province. The question before the Tribunal was whether the 2004 Guidelines imposed requirements "to purchase, use or accord a preference to goods produced or services provided in its territory, or to purchase goods or services from persons in its territory" within the meaning of Article 1106(1)(c), and whether Research and Development (R&D) and Education and Training (E&T) constitute "services"

[51] *Mobil Investments Canada Inc. & Murphy Oil Corporation v Government of Canada*, ICSID Case No. ARB(AF)/07/4, Decision on Liability and on Principles of Quantum (May 22, 2012).

within the meaning of Article 1106(1)(c).[52] The Tribunal found that, while the policy purposes may differ between different types of performance requirements, the requirement to use domestic sources of R&D and E&T was a performance requirement imposed on an investor. An interpretation that would exclude R&D and E&T from a definition of "services," based on whether the transmission was cross-border or the policy purpose not just strictly economic, is not supported by the NAFTA text, the NAFTA classification system for services, or the ordinary meaning of the term "services."[53] Moreover, the Tribunal gave a broad interpretation to the term "requirement," finding that the requirement to spend millions of dollars on R&D and E&T in the Province would in practice require the investor to purchase, use, or accord a preference to, domestic goods or services.[54]

The Tribunal noted that in most of the other cases that have involved NAFTA performance requirements, the requirements were incidental effects. The central purpose of the 2004 Guidelines was to require expenditures in the Province.[55] In contrast, in *S.D. Myers*, the Claimant argued that a Canadian export ban was a violation of Article 1106 because it required carrying out the physical disposal of PCB waste in Canada, requiring it to consume goods and services in Canada. However, the majority of the Tribunal noted that the ban "was not cast in the form of an express condition attached to a regulatory approval" and found that no requirements were imposed.[56] In *Pope & Talbot Inc.*, Canada introduced a regime to comply with the US–Canada Softwood Lumber Agreement, which permitted a certain level of softwood lumber to be exported from Canada without any fees, but imposed fees above that level. The *Pope & Talbot Inc.* Tribunal found that, while these fees deterred exports, there was no requirement to export at any particular level in return for the right to operate in Canada and thus the regime did not breach Article 1106(1)(a).[57] In *Merrill & Ring*, the Claimant argued that the implementation of Canada's Log Export Regime imposed performance requirements in breach of Article 1106, notably in connection with the obligation to cut and sort timber, to scale timber rafts metrically, and to follow certain other rules for properties located in remote areas, all of which impacted the way it managed its investments. The Tribunal held that a requirement related to the advertisement of goods as a step in the process of obtaining an export permit cannot be seen as a restriction on exports themselves. It further held that the cutting requirements are not a performance requirement designed to restrict or enhance exports, and that metric scaling is a measurement system used throughout Canada. Thus, while these requirements may have an incidentally adverse effect, they do not amount to the kind of prohibited performance requirement banned by Article 1106.[58] In *ADF Group Inc. v. United States of America*, a Canadian company

52 *Mobil v. Canada*, Liability and Quantum, paras. 211–12.
53 *Mobil v. Canada*, Liability and Quantum, para. 222.
54 *Mobil v. Canada*, Liability and Quantum, para. 238.
55 *Mobil v. Canada*, Liability and Quantum, paras. 240–2.
56 *S.D. Myers v. Canada*, Partial Award, para. 273.
57 *Pope & Talbot*, Award in relation to Preliminary Motion paras. 74–5.
58 *Merrill & Ring Forestry v. The Government of Canada*, UNCITRAL, Award (March 31, 2010) paras. 114–16, 118.

challenged the Buy America clause that applied to the fabrication of steel. The Tribunal ruled that the US measures were local content requirements contrary to Article 1106(1)(b), as well as a requirement to accord a preference to goods or services produced or provided in the US under Article 1106(1)(a), but came within Article 1108 reservations.[59] In the *Mobil* case, the 2004 Guidelines were designed to ensure that expenditures for R&D and E&T services occur in the Province, and thereby implied a legal requirement for the purposes of Article 1106. Such spending on R&D and E&T in the Province was a central feature of the 2004 Guidelines, and not an ancillary objective or consequence.[60]

The foregoing cases provide examples of the types of performance requirements that may be inconsistent with performance requirement obligations in IIAs.

Regarding domestic content requirements and preferences for domestic goods or services, Article 1106(6) establishes an exception that incorporates language from GATT Article XX:

> Provided that such measures are not applied in an arbitrary or unjustifiable manner, or do not constitute a disguised restriction on international trade or investment, nothing in paragraph 1(b) or (c) or 3(a) or (b) shall be construed to prevent any Party from adopting or maintaining measures, including environmental measures:
>
> (a) necessary to secure compliance with laws and regulations that are not inconsistent with the provisions of this Agreement;
>
> (b) necessary to protect human, animal or plant life or health; or
>
> (c) necessary for the conservation of living or non-living exhaustible natural resources.

The introductory paragraph of Article 1106(6) differs from the language of the GATT Article XX chapeau in three important respects: (1) it eliminates the reference to discrimination between countries where the same conditions prevail; (2) it adds a reference to investment, in addition to the reference to international trade; and (3) it clarifies that the term "measures" includes environmental measures. The second and third differences are relatively minor. The reference to investment is an understandably necessary adaptation to incorporate the exception into an investment chapter. Subsequent WTO jurisprudence has applied GATT Article XX to environmental measures, which means that the Article 1106(6) clarification is consistent with WTO jurisprudence. However, the first difference is more significant. In GATT Article XX, the reference to discrimination between countries where the same conditions prevail has been interpreted to require flexibility in the application of environmental measures to take into account differences in prevailing conditions in different countries.[61] The absence of this requirement in Article 1106 suggests that the conditions prevailing in the home State of a foreign investor

[59] *ADF Group Inc. v. United States*, ICSID Case No. ARB(AF)/00/1, Award (January 9, 2003) paras. 159–60, 170–4.

[60] *Mobil v. Canada*, Liability and Quantum, para. 242.

[61] Appellate Body Report, *US—Shrimp*.

need not be taken into consideration in the design of environmental measures that may affect the foreign investor in the host State. This makes sense if one assumes that the relevant environmental measures are aimed at protecting the environment of the host State.

Article 1106(6)(c) modifies the language of GATT Article XX(g) in three ways: (1) it replaces the term "relating to" with the stricter necessity requirement; (2) it clarifies that exhaustible natural resources may be living or non-living; and (3) it eliminates the evenhandedness requirement of GATT Article XX(g), which requires that conservation measures include restrictions on domestic production or consumption. Unlike the first and third modifications, the second modification is consistent with the interpretation of the term "exhaustible natural resources" in subsequent WTO jurisprudence regarding GATT Article XX(g). The first modification places a stricter test on environmental measures than GATT Article XX(g), whereas the third modification suggests that performance requirements can be applied to foreign investors without being applied to domestic investors. The absence of the GATT Article XX language regarding discrimination in the introductory paragraph of Article 1106(6) confirms this interpretation.

What types of climate change measures might be justifiable under Article 1106(6)? It is unlikely that the local content requirements of Ontario's FIT program could be justified under Article 1106(6). Measures aimed at promoting investment in clean energy sources could qualify as measures aimed at protecting human, animal or plant life or health from the effects of climate change or the conservation of the climate as an exhaustible natural resource. However, the FIT program's local content requirements are unlikely to qualify as necessary to meet these objectives, since they create a barrier to technology transfer from more efficient suppliers. It is necessary to reduce the cost of clean energy technology in order to transition to a green economy and thereby combat climate change. By requiring sourcing from less efficient suppliers, the Ontario requirement would achieve the opposite effect. Moreover, it likely would constitute a disguised restriction on international trade. The WTO panel ruling that the FIT program's local content requirement violates the national treatment obligations of TRIMS and GATT provides evidence of a restriction on international trade, although it does not prove that the restriction is a disguised one. What about a hypothetical requirement to use low-carbon inputs, which happens to favor local suppliers because they are the most efficient suppliers? That might pass these tests, as long as there was no restriction on the source of the low-carbon inputs and the method of measuring the carbon footprint of the inputs favors the lowest carbon footprint and not the local suppliers.

4.6 Nondiscrimination Obligations

NAFTA Chapter 11 contains the nondiscrimination obligations of national treatment (Article 1102) and most-favored-nation treatment (Article 1103), as well as an obligation to provide the better treatment of the two (Article 1104). These provisions prohibit discrimination between investors and investments in "like circumstances." Article 1102 prohibits less favorable treatment of foreign investors than domestic investors, while Article 1103 prohibits discrimination between foreign investors from other parties and any non-party. As we noted above, differences in the impact that investors or investments have on climate change could be relevant to determine whether they are in "like circumstances," to the extent that WTO jurisprudence can be applied to interpret these obligations as focusing on the competitive relationship between investors and investments in a manner that takes into account the impact of processing and production methods. However, the different context and terminology in Articles 1102 and 1103 leave open the question of which criteria should be used to make this determination. Moreover, those criteria are likely to vary according to the nature of the investor or the investment. For example, it is unlikely that goods manufacturers and service providers could be subject to the same criteria to determine likeness. Indeed, in the context of the WTO, goods and services are subject to different likeness criteria in the GATT and the GATS nondiscrimination provisions, respectively.[62]

Recent WTO jurisprudence on the meaning of "less favorable treatment" in the TBT Agreement also could be relevant to the interpretation of nondiscrimination obligations in IIAs. The Appellate Body held that "less favorable treatment" requires a determination of whether the contested measure modifies the conditions of competition to the detriment of imported products. However, the existence of such a detrimental effect is not sufficient to demonstrate less favorable treatment if the detrimental impact on imports stems exclusively from a legitimate regulatory distinction rather than reflecting discrimination against the group of imported products.[63] In the context of IIAs, this test could be applied with respect to the differential treatment of domestic and foreign investors or investments. Indeed, this test dovetails nicely with the concept of the right of governments to regulate, which has been discussed in the context of IIAs.

In *S.D. Myers v. Canada*, the Tribunal interpreted Article 1102 in light of the context of the NAFTA as a whole, the NAAEC and the principles that are affirmed by the NAAEC (including those of the Rio Declaration), including its concern for the environment and the need to avoid trade distortions that are not justified by environmental concerns. The Tribunal also took into account the part of the OECD Declaration on International and Multinational Enterprises of June 21, 1976 regarding national treatment, as well as OECD commentary on the "like situation" test from 1993. Finally, the Tribunal considered Supreme Court of Canada jurisprudence regarding discrimination against individuals. The Tribunal

[62] Mirreille Cossy, "Determining 'likeness' under the GATS: Squaring the circle?" (2006) WTO Working Paper ERSD-2006–08 <http://www.wto.org/english/res_e/reser_e/ersd200608_e.pdf> (accessed March 15, 2013); Nadja Dorothea Ruiz Euler, "El Trato Nacional y la Nación Más Favorecida en el Acuerdo General sobre el Comercio de Servicios de la OMC" (2012) 2 Revista de Derecho Económico Internacional 5 <http://dei.itam.mx/archivos/articulo3/primer_articulo_marzo.pdf> (accessed March 15, 2013).

[63] Appellate Body Report, *United States—Measures Concerning the Importation, Marketing and Sale of Tuna and Tuna Products (US—Tuna II (Mexico))*, WT/DS381/AB/R, adopted June 13, 2012, para. 215.

concluded that the assessment of "like circumstances" must take into account whether the foreign and national investors are in the same economic or business sector and circumstances that would justify governmental regulations that treat them differently in order to protect the public interest. The Tribunal compared S.D. Myers with Canadian competitors who also provided PCB waste remediation services.[64] Thus, in this case, the determination was in fact based on the comparators providing the same services and being competitors. This approach is consistent with the WTO focus on the competitive relationship between products in the determination of whether they are "like."

Regarding the issue of whether there was less favorable treatment, the *S.D. Myers* Tribunal based its decision on whether the practical effect of the measure is to create a disproportionate benefit for nationals over non-nationals and whether the measure, on its face, appears to favor nationals over non-nationals. In this regard, the Tribunal stated that protectionist intent would only be relevant if the measure produced an adverse effect on the foreign complainant. It held that it was a legitimate goal to want to maintain the ability to process PCBs within Canada, and consistent with the policy objectives of the Basel Convention. However, there were alternative measures that Canada could have taken to achieve this objective that would have been consistent with the NAFTA.[65] Thus, the Tribunal appears to have first interpreted Article 1102 to including *de facto* as well as *de jure* discrimination, and then introduced a test that resembles the least-trade-restrictive test of GATT Article XX.

In *Methanex v. United States*, a Canadian company sued the US government for damages under NAFTA Chapter 11 for a California ban on methanol. Methanex argued that the obligation to provide national treatment to foreign investors in "like circumstances" should be interpreted in the light of the GATT/WTO "like products" test, which focuses on the competitive relationship between products. It argued that if two or more investors or their investments compete for the same business, they are in "like circumstances" for the purposes of Article 1102. Methanex also argued that the "like circumstances" test was met because ethanol and methanol are "like products" according to the GATT/WTO test.[66] However, the Tribunal found that the California methanol ban had the same effect on the American investors and investments in methanol as it had on the Canadian investor, Methanex. Given the existence of domestic methanol producers, they were the appropriate point of comparison, not ethanol producers. Methanex did not receive less favorable treatment than the identical domestic comparators, producing methanol. As a result, the Tribunal did not need to determine how to interpret the term "like circumstances."[67] The interesting point of this aspect of the case is the question of the appropriate comparator in a given case.

[64] *S.D. Myers v. Canada*, Partial Award, paras 247–51.
[65] *S.D. Myers v. Canada*, Partial Award, paras. 254–5.
[66] *Methanex v. United States*, Final Award, Part IV(B) paras. 5–7.
[67] *Methanex v. United States*, Final Award, Part IV(B) paras. 18–22.

Regarding the applicability of the "like products" test to the "like circumstances" test in Article 1102, the *Methanex* Tribunal noted that "like goods" is never used with respect to the investment regime of NAFTA Chapter 11 and "like circumstances" is not used in relation to goods. The text shows that trade provisions were not to be transported to investment provisions. Accordingly, Article 1102 should be read on its own terms and not as if the words "any like, directly competitive or substitutable goods" appeared in it.[68] However, this sheds no light on how to interpret the term "like circumstances."

The term "like circumstances" requires tribunals to take into account the nature of investments on a case-by-case basis in the comparison of likeness. Relevant evidence could include the type of enterprise, whether it produces goods or sells services, its location, its technology and its environmental impact. Thus, "circumstances" cannot be limited to the physical characteristics of a product produced by an investment. The term "circumstances" is much broader than the term "products."[69]

NAFTA Article 1202 on services also uses the term "like circumstances." In the US–Mexico *Trucking Services* case, the Panel accepted that differential treatment for legitimate regulatory objectives related to safety was a valid consideration. It further stated that "such differential treatment should be no greater than necessary for legitimate regulatory reasons such as safety, and that such treatment be equivalent to the treatment accorded to domestic service providers."[70] The Panel interpreted Article 1202 in light of Article 2101 of NAFTA, which allows for exceptions for environmental and human health reasons.[71] Since Chapter 11 has no comparable exception for the nondiscrimination provisions, one could argue that legitimate regulatory objectives (like the legitimate regulatory distinctions test in the TBT Agreement) should be relevant circumstances, to enable a State to establish distinctions between investors on the basis of the actual impacts and effects of their investments.[72] This view is consistent with the statement of the *Feldman* Tribunal that "the concept of discrimination has been defined to imply unreasonable distinctions between foreign and domestic investors in like circumstances."[73] With respect to climate change regulation, would distinguishing between investments based on their carbon footprint be reasonable where the investors are otherwise in like circumstances? In other words, are differences in greenhouse gas emissions sufficient to negate a finding of like circumstances and thus reach a conclusion that the nondiscrimination obligations are not violated? Alternatively, would differential treatment based on differences in GHG emissions meet the

[68] *Methanex v. United States*, Final Award, Part IV(B) paras. 33–8.

[69] *Methanex Corporation v. United States*, Amicus Curiae Submission para. 36–9.

[70] *In the Matter of Cross-Border Trucking Services*, Final Report of the Panel, February 6, 2001 (Secretariat File No. USA-MEX-98-2008-01), para. 258, <http://www.worldtradelaw.net/nafta20/truckingservices.pdf> (accessed April 9, 2013).

[71] Final report of the Panel, *In The Matter of Cross-Border Trucking Services* (2001) USA-Mex-98-2008–01, paras. 257–258.

[72] *Methanex Corporation v. United States*, Amicus Curiae Submission paras. 40–4.

[73] *Marvin Roy Feldman v. United Mexican States*, ICSID Case No. ARB(AF)/99/1, Award (December 16, 2002) para. 170.

legitimate regulatory distinctions test and thus not constitute less favorable treatment? Legitimate climate change regulation could be found to be consistent with the nondiscrimination obligations of IIAs on either ground.

Public interference with the right of foreign investors to benefit from support schemes for renewable energy or with their right to GHG emission credits can constitute an unjustified difference in treatment. In the *Nykomb v. Latvia* case, the Tribunal held that Latvia had violated the national treatment standard in the Energy Charter Treaty by refusing to honor a promise of support for low-carbon electricity production on the basis of which Nykomb made its investment. The administrator of the support scheme continued to support low-carbon installations operated by domestic investors, while refusing this payment to Nykomb, which was operating in comparable conditions. The Tribunal held that the host State had failed to justify on the basis of public policy why it refused to pay the promised support to the foreign investor, while continuing to support national investors.[74] This case is consistent with the view that differential treatment should have a basis in a legitimate regulatory distinction.

4.7 Minimum Standard in Customary International Law

NAFTA Article 1105 requires host governments to treat foreign investors and investments in accordance with the minimum standard of treatment in international law. The NAFTA Commission issued an interpretation of this provision that clarifies that "international law" refers to customary international law, which is developed by the common practices of countries, and thus does not include treaty law (including provisions contained in the NAFTA other than Article 1105 and other provisions of Chapter 11).[75] There was some debate in the case law regarding whether this constituted an interpretation or an amendment, but the tribunals decided that this did not matter, since they were bound to follow the decision of the NAFTA Commission regardless. In contrast to nondiscrimination obligations, Article 1105 is framed in absolute terms. The comparative treatment of other investors is not relevant. Article 1105 establishes a minimum standard, under which a Party may not treat foreign investments worse than this standard irrespective of the manner in which the Party treats other investors and their investments.[76]

Following the NAFTA Commission interpretation, several NAFTA Chapter 11 tribunals have sought to define the customary international law standard in 1105. The Tribunal in *Mondev*, for example, emphasized that the application of the customary international law standard does not permit resort to other treaties of

[74] *Nykomb Synergetics Technology Holding AB v. Republic of Latvia*, SCC Case No. 118/2001, Award (December 16, 2003).

[75] *Mexico v. Metalclad* (BCSC) para. 62.

[76] *Mexico v. Metalclad* (BCSC) para. 60.

the NAFTA parties or other provisions within NAFTA.[77] The *ADF* Tribunal noted that recourse to customary international law "must be disciplined by being based on State practice and judicial or arbitral case law or other sources of customary or general international law."[78] The *Loewen* Tribunal observed: "Manifest injustice in the sense of a lack of due process leading to an outcome which offends a sense of judicial propriety is enough, even if one applies the Interpretation according to its terms."[79] The NAFTA Tribunal in *Waste Management* attempted to synthesize the post-interpretation jurisprudence of Article 1105, as:

> [T]he minimum standard of treatment of fair and equitable treatment is infringed by conduct attributable to the State and harmful to the claimant if the conduct is arbitrary, grossly unfair, unjust or idiosyncratic, is discriminatory and exposes the claimant to sectional or racial prejudice, or involves a lack of due process leading to an outcome which offends judicial propriety—as might be the case with a manifest failure of natural justice in judicial proceedings or a complete lack of transparency and candour in any administrative process.[80]

The *Methanex* Tribunal found that Article 1105(1) does not preclude governmental differentiations between nationals and aliens; it does not mention discrimination, while Article 1105(2) does mention it. The NAFTA Commission interpretation confirms this: "A determination that there has been a breach of another provision of the NAFTA, or of a separate international agreement, does not establish that there has been a breach of Article 1105(1)." The *Methanex* Tribunal concluded that, in the absence of a contrary rule of international law binding on the States parties, whether of conventional or customary origin, a State may differentiate in its treatment of nationals and aliens.[81]

As we noted above, in *Mobil Investments Canada Inc. & Murphy Oil Corporation v. Canada*, the 2004 Guidelines required the investors in offshore petroleum projects to pay for research and development in the Province of Newfoundland and Labrador. The investors alleged breaches of the performance requirement prohibition in NAFTA Article 1106 and the minimum standard of treatment in NAFTA Article 1105. The Tribunal held that the 2004 Guidelines violated Article 1106, but not Article 1105.[82] After reviewing the NAFTA jurisprudence on Article 1105, the Tribunal summarized the standard of treatment as follows:

(1) the minimum standard of treatment guaranteed by Article 1105 is that which is reflected in customary international law on the treatment of aliens;
(2) the fair and equitable treatment standard in customary international law will be infringed by conduct attributable to a NAFTA party and harmful to a

[77] *Mondev International Ltd. v. United States*, ICSID Case No. ARB(AF)/99/2, Award (October 11, 2002) paras. 120–1.
[78] *ADF Group Inc.*, Award, para. 184.
[79] *The Loewen Group, Inc. and Raymond L. Loewen v. United States*, ICSID Case No. ARB(AF)/98/3, Decision on Hearing of Respondent's Objection to Competence and Jurisdiction (January 5, 2001) para. 132.
[80] *Methanex v. United States*, Final Award, IV.C.11–12.
[81] *Methanex v. United States*, Final Award, IV.C.14, 17, 25.
[82] *Mobil v. Canada*, Liability and Quantum.

claimant that is arbitrary, grossly unfair, unjust, or idiosyncratic; or is discriminatory and exposes a claimant to sectional or racial prejudice; or involves a lack of due process leading to an outcome which offends judicial propriety.

(3) in determining whether that standard has been violated it will be a relevant factor if the treatment is made against the background of
   (i) clear and explicit representations made by or attributable to the NAFTA host State in order to induce the investment, and
   (ii) were, by reference to an objective standard, reasonably relied on by the investor, and
   (iii) were subsequently repudiated by the NAFTA host State.[83]

The *Mobil* Tribunal explained that customary international law on the treatment of aliens does not require a State to maintain a stable legal and business environment for investments. NAFTA Article 1105 only protects an investor from changes to the rules governing an investment if those changes may be characterized as arbitrary or grossly unfair or discriminatory, or otherwise inconsistent with the customary international law standard. Article 1105 does not prevent a public authority from changing the regulatory environment to take account of new policies and needs, even if some of those changes may have far-reaching consequences and effects, and even if they impose significant additional burdens on an investor. Article 1105 does not provide a guarantee against regulatory change or entitle an investor to expect no material changes to the regulatory framework within which an investment is made. Governments can change, and policies and rules can change. The rules of customary international law only protect against egregious behavior and do not require a legal and business environment to be "set in concrete."[84] Thus, to establish a breach of Article 1105, the Claimants had to establish that (1) clear and explicit representations were made by or attributable to Canada in order to induce the investment, (2) such representations were reasonably relied upon by the Claimants, and (3) these representations were subsequently repudiated by Canada. However, there was no evidence that Canada made representations that there would not be changes to the regulatory regime and no indication of reliance being placed upon such representations. In particular, there was no promise or representation in the underlying regulatory framework to not change an existing benefits plan or to impose a new plan.[85]

Boute notes that support schemes and regulatory frameworks for renewable energy projects and GHG emission reduction projects create incentives that aim to stimulate private investment and that low-carbon investors expect to receive public support in accordance with the schemes existing at the time of investing. Thus, the fair and equitable treatment standard could provide a guarantee of protection against changes to the framework the State has created to attract

[83] *Mobil v. Canada*, Liability and Quantum, para. 152.
[84] *Mobil v. Canada*, Liability and Quantum, para. 153.
[85] *Mobil v. Canada*, Liability and Quantum, paras. 154–9.

low-carbon investments.[86] The fair and equitable treatment standard also requires a tribunal to weigh the investors' legitimate expectations against the legitimate regulatory interests of the host State.[87] However, it is not possible to generalize regarding how a particular category of cases would be addressed, since it depends on the specific domestic regulatory framework, the provisions of the specific IIA[88] and the surrounding circumstances of each case. For example, the two NAFTA investment cases regarding Ontario's FIT program demonstrate how different the facts can be even where they concern the same renewable energy program. Moreover, the *Mobil* Tribunal's statement of the standard in customary international law is consistent with the view that States are entitled to maintain their right to regulate, which includes the right to change the regulatory environment. The real question is whether States design and implement the regulatory framework with customary international law on the treatment of aliens in mind and whether an investor's expectations are justifiable in the circumstances of each case.

In *Pulp Mills on the River Uruguay (Argentina v. Uruguay)*, the Court laid out how customary international environmental law had evolved over time.[89] The Court pointed out that the principle of prevention, as a customary rule, has its origins in the due diligence that is required of a State in its territory. It is "every State's obligation not to allow knowingly its territory to be used for acts contrary to the rights of other States."[90] A State is thus obliged to use all the means at its disposal in order to avoid activities which take place in its territory, or in any area under its jurisdiction, causing significant damage to the environment of another State. This obligation "is now part of the corpus of international law relating to the environment."[91] Moreover, "there are situations in which the parties' intent upon conclusion of the treaty was, or may be presumed to have been, to give the terms used—or some of them—a meaning or content capable of evolving, not one fixed once and for all, so as to make allowance for, among other things, developments in international law."[92] As a result, in *Pulp Mills on the River Uruguay*, the Court found that a provision

> has to be interpreted in accordance with a practice, which in recent years has gained so much acceptance among States that it may now be considered a requirement under general international law to undertake an environmental impact assessment where there is a risk that the proposed industrial activity may have a significant adverse impact in a transboundary context, in particular, on a shared resource.[93]

[86] Boute, "Combatting Climate Change" 637–8.

[87] Boute, "Combatting Climate Change" 649.

[88] For example, some IIAs contain sanctity of contracts clauses, which aim to guarantee by treaty the respect by the host state for the specific contractual obligations it enters into with investors, which could influence the outcome in cases involving contractual claims. Boute, "Combatting Climate Change" 644–7.

[89] *Pulp Mills on the River Uruguay (Argentina v. Uruguay)* paras. 203–19.

[90] *Corfu Channel (United Kingdom v. Albania)* (Merits) ICJ Reports 1949, para. 22.

[91] *Legality of the Threat or Use of Nuclear Weapons* (1996) (Advisory Opinion) ICJ Reports 1996 (I), para. 29.

[92] *Dispute Regarding Navigational and Related Rights (Costa Rica v. Nicaragua)* (2009) (Judgment) para. 64.

[93] *Pulp Mills on the River Uruguay (Argentina v. Uruguay)* (Judgment) para. 204.

Thus, the International Court of Justice has confirmed that the ongoing development of customary international environmental law has to be taken into account in the interpretation of treaty provisions that are subject to evolutionary interpretation.

NAFTA Article 1105 is subject to evolutionary interpretation.[94] Thus, the minimum standard of treatment of foreign investors under customary international law has to be interpreted in accordance with evolving customary international environmental law. The obligation to avoid activities causing significant damage to the environment of another State is likely to encompass regulations to address climate change. Thus, legitimate climate change regulation would not be inconsistent with the minimum standard of treatment in NAFTA Article 1105. To conclude otherwise would create a conflict between customary international investment law and customary international environmental law.

4.8 Compensation for Expropriation

NAFTA Article 1110 requires host governments to provide compensation for expropriation and measures that are tantamount to expropriation. It also requires that expropriations be made for a public purpose, be nondiscriminatory, and be in accordance with due process and Article 1105(1).

In *S.D. Myers v. Canada*, the Tribunal considered that the term "expropriation" in Article 1110 must be interpreted in light of the whole body of State practice, treaties, and judicial interpretations of that term in international law cases. The general body of precedent usually does not treat regulatory action as amounting to expropriation, because expropriations tend to involve the deprivation of ownership rights and regulations a lesser interference. Moreover, an expropriation usually amounts to a lasting removal of the owner's ability to make use of its economic rights. In this case, the trade ban was temporary and there was no expropriation.[95]

International law establishes that the following types of government actions may constitute expropriation: (1) the taking of title to property, in whole or in part; (2) the use of police, administrative, or legal powers to take control of the operation of an investment, or shut the investor out of its rights of control and ownership, without the transfer of title; and (3) creeping expropriation: the use of a series of measures that cumulatively, rather than individually, accomplish the removal of ownership or control of an investment. A more controversial argument is that the diminution of economic value due to a regulation that protects the public interest can be the basis for a finding of expropriation.[96]

Under customary international law, where economic injury results from *bona fide* regulation within the police powers of a State, compensation is not required. Thus, as a general matter, States are not liable to compensate aliens for economic loss incurred as a result of a nondiscriminatory action to protect the public

[94] *Mondev v. United States*, Award. [95] *S.D. Myers v. Canada*, Partial Award, paras. 280–8.
[96] *Methanex Corporation v. United States*, Amicus Curiae Submission para. 82.

interest.[97] However, once an expropriation has taken place, compensation is due even if it is for an environmental purpose.[98] In the context of NAFTA Article 1110, if there is a finding of expropriation, compensation is required, even if the taking is for a public purpose, nondiscriminatory, and in accordance with due process of law and Article 1105(1).[99] Thus, not all government regulatory activity that makes it difficult or impossible for an investor to carry out a particular business is an expropriation under Article 1110.[100] For example, in *Methanex*, the Tribunal concluded that the California methanol ban was made for a public purpose, was nondiscriminatory, and was accomplished with due process. Hence, from the standpoint of international law, the California ban was a lawful regulation and not an expropriation.[101]

Twelve of the treaties surveyed by Gordon and Pohl contain provisions that clarify the understanding of the parties that nondiscriminatory environmental regulation does not constitute expropriation. These clauses state:

> The Parties confirm their shared understanding that: . . . Except in rare circumstances, non-discriminatory regulatory actions by a Party that are designed and applied to protect legitimate public welfare objectives, such as public health, safety, and the environment, do not constitute indirect expropriations.[102]

Arguably, this is just a confirmation of existing customary international law and, contrary to the conclusion that Gordon and Pohl draw, does not mean that States that do not include such provisions may thus be exposed to compensation claims for expropriation that could discourage modifications of environmental regulation or make them onerous.

The cases also have considered what types of property interests are subject to Article 1110. In *Pope & Talbot Inc. v. Canada*, the Tribunal held that "the Investor's access to the U.S. market is a property interest subject to protection under Article 1110."[103] This is consistent with the approach taken in *Cargill v. Mexico*. However, in *Methanex*, the Tribunal observed that items such as goodwill and market share may constitute an element of the value of an enterprise and, in a

[97] *Methanex Corporation v. United States*, Amicus Curiae Submission para. 84; Restatement of the Law Third: The Foreign Relations of the United States:

> A state is not responsible for loss of property or for other economic disadvantage resulting from bona fide general taxation, regulation, forfeiture for crime, or other action of the kind that is commonly accepted as within the police powers of states, if it is non-discriminatory (*Azinian v Mexico* cited in *Marvin Feldman v. Mexico*, Award, para. 105).

[98] *Compañia del Desarrollo de Santa Elena S.A. v. Republic of Costa Rica*, ICSID Case No. ARB/96/1, Award (February 17, 2000); *Metalclad v. Mexico*, Award.

[99] *Marvin Feldman v. Mexico*, Award, para. 98.

[100] *Marvin Feldman v. Mexico*, Award, para. 112.

[101] *Methanex v. United States*, Final Award, IV.D.15.

[102] Gordon and Pohl, "Environmental Concerns in International Investment Agreements" 22; United States Model BIT 2004 Annex B; Canada Model BIT (2004) Annex B.13(1); Belgium/Luxembourg–Colombia BIT (2009); Canada–Czech Republic BIT (1990); Canada–Jordan BIT (2009); Canada–Latvia BIT (2009); Canada–Peru BIT (2006); Canada–Romania BIT (1996); United States–Rwanda BIT (2008); United States–Uruguay BIT (2005).

[103] *Pope & Talbot*, Award in relation to Preliminary Motion, para. 96.

comprehensive taking, these items may figure in valuation. However, they cannot stand alone as the basis for a claim of expropriation.[104]

4.9 Rights of Investors versus Right to Regulate

Some argue that NAFTA Chapter 11 strikes the right balance between non-compensable exercises of regulatory authority and exercises of regulatory authority that amount to expropriation of an investment, achieving sustainable development through the right balance between environmental protection and economic development.[105] It is difficult to determine where to draw the line between these two objectives, other than on a case-by-case basis. In some cases, environmental protection and economic development may be mutually supportive. However, given the economic and environmental consequences of climate change, it seems that *bona fide* climate change regulation should take precedence over investors' rights, though the correct balance likely will have to be decided on a case-by-case basis. It is also important to protect foreign investors from unfair or arbitrary treatment by governments that are motivated by short-term political interests rather than long-term environmental risks. For this reason, we emphasize that we are referring to *bona fide* climate change regulation.

Striking the right balance between the regulatory risks that investors face and the litigation risk that governments face is not the same for all markets. Larger markets can have a greater degree of regulatory risk and still attract foreign investors. In contrast, smaller, less economically attractive markets may need to strike a balance that is more in favor of investors' rights and reduces regulatory risk to a greater degree, in order to attract foreign investment. Larger markets are also a greater source of GHG emissions, so the balance should favor climate change regulations over compensation to foreign investors, in order to limit the risk of regulatory chill and to enhance the right to regulate. Their attractiveness to foreign investors means that large markets should seek to negotiate IIAs that leave adequate regulatory space to combat climate change. However, there may be limits to how precise the provisions should be in this regard. For example, the negotiating history of NAFTA Chapter 11 indicates that there has been no attempt to address directly the problem of how to distinguish legitimate non-compensable regulations having an effect on the economic value of foreign investments, and "regulatory takings" requiring compensation, instead leaving this determination to be made on a case-by-case basis.[106]

[104] *Methanex v United States*, Final Award, IV.D.17.

[105] Sanford E. Gaines, "Protecting Investors, Protecting the Environment: The Unexpected Story of NAFTA Chapter 11" in David L. Markell and John H. Knox (eds.), *Greening NAFTA: The North American Commission for Environmental Cooperation* (Stanford University Press, Stanford 2003) 173, 184, 190–1.

[106] Daniel M. Price, "Chapter 11—Private Party Vs. Government, Investor-State Dispute Settlement: Frankenstein or Safety Valve?" (2000) 26 Canada–U.S. L.J. 107.

The NAFTA Preamble is a relevant context for interpreting the provisions of Chapter 11. According to the Preamble, the NAFTA parties seek to ensure a predictable commercial framework for business planning and investment, in a manner consistent with environmental protection and conservation. At the same time, they seek to preserve their flexibility to safeguard the public welfare, promote sustainable development, and strengthen the development and enforcement of environmental laws and regulations. These aspects of the Preamble support an interpretation of NAFTA Chapter 11 that gives *bona fide* climate change regulation precedence over investors' rights.

The *Methanex* case indicates that science-based regulatory decisions are likely to withstand scrutiny in NAFTA Chapter 11.[107] In that case, the science was preliminary, but there was sufficient scientific evidence of the potentially serious health effects of methanol to support the State regulation. The *Methanex* case is of particular relevance to climate change regulation, since it shows that regulation based on the precautionary principle can survive a challenge by foreign investors. There is sufficient scientific evidence of the potentially serious effects of climate change to justify climate change regulation, even if it also has the effect of diminishing the value of some foreign investments. There are 66 IIAs with general language in preambles that mentions environmental concerns and establishes protection of the environment as a concern of the parties to the treaty.[108]

4.10 Conclusion

International investment law has the potential to have a chilling effect on climate change regulation, by raising issues regarding the risk that climate change regulation will expose host states to claims from foreign investors. Our analysis in this chapter shows that legitimate climate change regulation should not trigger liability to compensate foreign investors. However, this may not eliminate the chilling effect, since it is costly for States to defend against such claims even if they do not succeed. Awards of costs against investors who file such claims may discourage such claims, but this may not be sufficient to overcome the chilling effect in the short term. Moreover, the lack of a system of precedents permits tribunals to reach different conclusions on similar issues, which increases the uncertainty regarding the outcome of litigation. Nevertheless, there is room in international investment law for striking an appropriate balance between the right to regulate climate change and right of foreign investors to seek compensation for arbitrary and discriminatory governmental actions.

[107] This stands in contrast to *Ethyl Corporation v. The Government of Canada*, NAFTA/UNCITRAL, Award (June 24, 1998), a case in which the Canadian government agreed to withdraw a trade ban and publicly concede that there was no scientific basis for the ban. Gaines, "Protecting Investors, Protecting the Environment" 182–3.

[108] Gordon and Pohl, "Environmental Concerns in International Investment Agreements" 9.

5

Intellectual Property Rights for New Plant Varieties

5.1 Introduction

Climate change will increase the importance of new plant varieties that can adapt to changing climactic conditions. This chapter analyzes intellectual property rights for plant varieties. Section 5.2 analyzes Article 27 of the WTO Agreement on Trade Related Aspects of Intellectual Property Rights. Section 5.3 analyzes the UPOV Convention. Section 5.4 analyzes the Convention on Biological Diversity (CBD). TRIPS Article 27.3(b) requires WTO Members to protect plant varieties with patents, a *sui generis* system, or any combination thereof. The UPOV Convention and the CBD represent competing visions of what type of *sui generis* system might be appropriate. However, Sabrina Safrin argues that the combined effect of the two competing systems is to limit genetically based innovations and to create obstacles to international conservation and collaboration.[1]

The appropriate policy choices for WTO Members regarding IPRs for plant varieties will depend on the circumstances of each Member. For developed and middle-income countries with strong biotechnology industries and the financial capacity to pay for climate change adaptation, strong IPRs may be beneficial. They may be less beneficial for developing countries with less financial capacity to pay for adaptation, greater vulnerability to the effects of climate change, and a larger percentage of the population devoted to subsistence farming. However, the circumstances of countries change over time, and the effects of climate change are difficult to predict. As a result, policy choices that are appropriate for a specific country today might not be appropriate tomorrow. Against this backdrop of uncertainty, it is best to maintain flexibility regarding policy choices.

WTO Members do not agree on how to address IPRs for plant varieties. Japan and the United States argue that plant varieties should be protected to allow development of new technological solutions in the field of agriculture.[2] The African Group, Peru, Zimbabwe, and Kenya argue that the protection of plant varieties

[1] Sabrina Safrin, "Hyperownership in a Time of Biotechnological Promise: The International Conflict to Control the Building Blocks of Life" (2004) 98 AJIL 641.

[2] Note by the Secretariat, Review of Article 27.3(b). Paper IP/C/W/369/Rev.1, revised March 9, 2006, para. 45 <http://www.wto.org/english/tratop_e/trips_e/art27_3b_e.htm> (accessed August 20, 2011).

can have an adverse impact on food security, health, and rural development in developing countries, and fails to protect traditional knowledge systems.[3] Developed countries with a strong biotechnology industry generally have favored stronger IPRs for plant varieties. Developing countries generally have favored weaker IPRs, particularly countries with the greatest biodiversity or large numbers of subsistence farmers. Surprisingly, the relevance of climate change to this debate has not played a significant role in WTO negotiations.

5.2 TRIPS Article 27

TRIPS Article 27 refers to patentable subject matter. The first sentence of Article 27.1 requires that patents be available for "any inventions," whether products or processes, and in all fields of technology. The invention must meet each of three criteria: (1) "new"; (2) "involve an inventive step"; and (3) "capable of industrial application." There is no definition of the term "inventions" or the term "new." However, footnote 5 permits Members to deem the terms "inventive step" and "capable of industrial application" to be synonymous with the terms "non-obvious" and "useful" respectively. The first sentence of Article 27.1 also makes the obligation regarding patentable subject matter subject to the provisions of paragraphs 2 and 3.

The second sentence of Article 27.1 requires WTO Members to make patents "available and patent rights enjoyable without discrimination as to the place of invention, the field of technology and whether products are imported or locally produced." This nondiscrimination obligation is subject to Article 27.3. According to the Panel in *Canada—Pharmaceutical Patents*, the term "discriminate" in Article 27.1 extends beyond the concept of differential treatment and is potentially broader than national treatment or most-favored-nation treatment. Discrimination may arise from explicitly different treatment (*de jure* discrimination) or from ostensibly identical treatment which, due to differences in circumstances, produces differentially disadvantageous effects (*de facto* discrimination).[4] TRIPS Article 1.3 requires that treatment be accorded to "nationals" and that both the national treatment and MFN obligations in TRIPS apply to "nationals," not products. Thus, in some cases, Article 27.1 only requires nondiscrimination with respect to inventors, which would require evidence that any inventor of a competing product has been granted a patent in order for a complainant to meet its burden of proof.

Article 27.3(b) permits Members to exclude from patentability "plants and animals other than micro-organisms, and essentially biological processes for the production of plants or animals other than non-biological and microbiological processes." However, Article 27.3(b) requires Members to "provide for the protection of plant varieties either by patents or by an effective *sui generis* system or by any combination thereof." The text in Spanish requires protection of "todas las obtenciones vegetales" (all plant

[3] Note by the Secretariat, Review of Article 27.3(b) para. 46.

[4] Panel Report, *Canada—Patent Protection of Pharmaceutical Products*, WT/DS114/R, adopted April 7, 2000, para. 7.94.

varieties), whereas the English and French versions do not specify the scope of this obligation in this manner. This point has been raised in the Article 27.3(b) negotiations regarding the scope of the subject matter that is covered by the obligation.[5] Thus, Article 27.3(b) initially carves out an exception from the obligations contained in Article 27.1, by creating a right to exclude animals and plants from patentability, but then creates an obligation to provide IPRs for inventors of plant varieties. However, those IPRs do not have to be in the form of patents. Rather, they can be patents, an effective *sui generis* system, or any combination thereof. The obligation regarding IPRs for plant varieties is more flexible than patent obligations for other products and processes. In this regard, it constitutes an exception from the nondiscrimination obligation in Article 27.1. However, Article 27.3(b) also limits the right to exclude plants from some form of intellectual property protection when "plant varieties" are involved.

The general requirements for patentability apply to plant varieties. For example, existing plant varieties cannot be patented because they are not new inventions. However, the general requirements for patentability do not apply to a *sui generis* system of IPRs for plant varieties. The only requirement for a *sui generis* system is that it be "effective." This interpretation is supported by the use of the term "plant varieties" rather than the term "new plant varieties"; the plant varieties need only be "new" to qualify for a patent. The UPOV Convention also requires that a plant variety be new to qualify for breeders' rights. Thus, a *sui generis* system can grant IPRs for traditional knowledge regarding uses of existing plant varieties.[6] This can prevent the granting of IPRs for this use, by way of patents or UPOV breeders' rights, by demonstrating that it is not "new."[7]

The negotiating history of Article 27 shows that the negotiators were aware of the negotiation of the UPOV Convention revisions.[8] The negotiation history also shows that some parties wanted to exclude plants from patentability, while others believed that patents would promote innovation by requiring disclosure of the inventive process.[9] One party argued that plant variety rights were a distinct *sui generis* category of rights regulated by a separate convention and should therefore not be patentable subject matter.[10] The negotiating history indicates that Article

[5] Note by the Secretariat, Review of Article 27.3(b) para. 52.

[6] Thomas Cottier, "The Protection of Genetic Resources and Traditional Knowledge: Towards More Specific Rights and Obligations in World Trade Law" (1998) 1(4) JIEL 555, 583 fn. 68 and accompanying text; Thomas Cottier and Mario Panizzon, "Legal Perspectives on Traditional Knowledge: The Case for Intellectual Property Protection" (2004) 7 JIEL 371.

[7] Adam Masarek, "Treetop View of the Cathedral: Plant Variety Protection in South and Southeast Asian Least-Developed Countries" (2010) 24 Emory International Law Review 433, 462.

[8] Note by the Secretariat, Negotiating Group on Trade-Related Aspects of Intellectual Property Rights, including Trade in Counterfeit Goods, Meeting of the Negotiating Group of July 12–14, 1989, Chairman: Ambassador Lars E. R. Anell (Sweden), MTN.GNG/NG11/14, September 12, 1989.

[9] Note by the Secretariat, Negotiating Group on TRIPS, July 12–14, 1989; Note by the Secretariat, Negotiating Group on Trade-Related Aspects of Intellectual Property Rights, including Trade in Counterfeit Goods, Meeting of the Negotiating Group of October 30–November 2, 1989, Chairman: Ambassador Lars E. R. Anell (Sweden), MTN.GNG/NG11/16, December 4, 1989.

[10] Note by the Secretariat, Negotiating Group on Trade-Related Aspects of Intellectual Property Rights, including Trade in Counterfeit Goods, Meeting of the Negotiating Group of November 1,

27.3(b) represents a compromise that permits flexibility regarding the design of IPRs for plant varieties. However, while the negotiating history indicates that the *sui generis* system some parties had in mind was the UPOV Convention, the final text of Article 27.3(b) does not limit WTO Members to that particular *sui generis* system.[11] In the Article 27.3(b) negotiations the European Communities, Japan, Switzerland, the United States, and Uruguay have argued that the UPOV Convention is an effective *sui generis* system as required by Article 27.3(b) and that its use should be widespread.[12] Brazil, India, Malaysia, Mexico, Singapore, Zambia, Zimbabwe, and the African Group have argued that Article 27.3(b) does not require Members to use the UPOV Convention as a model, although it may be an important point of reference.[13] Brazil, India, Zambia, Zimbabwe, and the African Group argue that Members are free to choose a system based on other models, such as the CBD or the Food and Agriculture Organization (FAO) International Undertaking on Plant Genetic Resources.[14]

The majority of UPOV Members are WTO Members, but the majority of WTO Members are not UPOV Members.[15] Interpreting Article 27.3(b) to require that WTO Members adopt the UPOV Convention to meet the obligation to use an effective *sui generis* system seems unlikely. The language of Article 27.3(b) does not support such an interpretation. Rather, it reflects the lack of agreement on this issue in the Uruguay Round. In addition, the 1978 and 1991 Acts are significantly different in several important respects, so it would not be clear which Act would be the most "effective." While the Vienna Convention permits the interpretation of treaty terms in light of "any relevant rules of international law applicable in the relations between the parties," in these circumstances it would be inappropriate to interpret Article 27.3(b) in light of the UPOV Convention.

Article 27.3(b) was to be reviewed four years after the date of entry into force of the WTO Agreement. This review requirement provides a further indication that

1990, Chairman: Ambassador Lars E. R. Anell (Sweden), MTN.GNG/NG11/27, November 14, 1990.

[11] Cottier, "Genetic Resources" 555 fn. 68.

[12] Note by the Secretariat, Review of Article 27.3(b) para. 61. For an analysis of what amounts to an effective *sui generis* system as required under TRIPS Article 27.3(b), see Srividhya Ragavan and Jamie Mayer, "Has India Addressed Its Farmers' Woes? A Story of Plant Protection Issues" (2007) 20 Georgetown International Environmental Law Review 97.

[13] Note by the Secretariat, Review of Article 27.3(b) para. 62.

[14] Note by the Secretariat, Review of Article 27.3(b) para. 62. For an analysis of the International Treaty on Plant Genetic Resources for Food and Agriculture, see Gregory Rose, "International Law of Sustainable Agriculture in the 21st Century: The International Treaty on Plant Genetic Resources for Food and Agriculture" (2003) 15 Georgetown International Environmental Law Review 583; Harold J. Bordwin, "The Legal and Political Implications of the International Undertaking on Plant Genetic Resources" (1985) 12 Ecology Law Quarterly 1053; Kal Raustiala and David G. Victor, "The Regime Complex for Plant Genetic Resources" (2003) Working Paper No. 14, Program on Energy and Sustainable Development, Stanford University Center for Environmental Science and Policy <http://iis-db.stanford.edu/pubs/20190/pgr_regime_complex.pdf> (accessed September 20, 2011).

[15] As of December 5, 2012, there were 71 UPOV Members, of which 51 were parties to the 1991 Act, 19 to the 1978 Act, and one to the 1961/1972 Act. Members of the International Union for the Protection of New Varieties of Plants <http://www.upov.int/export/sites/upov/members/en/pdf/pub423.pdf> (accessed December 25, 2012).

Article 27.3(b) represented a compromise. However, paragraph 19 of the 2001 Doha Declaration expanded this review to require the TRIPS Council also to examine the relationship between the TRIPS Agreement and the CBD, the protection of traditional knowledge and folklore. The Doha Declaration also requires work on these topics to be guided by the TRIPS Agreement's objectives (Article 7) and principles (Article 8), and to take development issues fully into account.[16] Brazil, India, Kenya, the African Group, Thailand, and the European Union have underlined the flexibility of Article 27.3(b) with respect to the choice of *sui generis* protection and several have argued that the term "effective *sui generis* system" requires further clarification.[17] The United States has argued that there are specific criteria that can be used to judge the effectiveness of a *sui generis* system, whereas some other Members have argued that the issue of whether a *sui generis* system is effective should be left for Members to decide.[18] Given the lack of criteria in the TRIPS Agreement and the ongoing debate among Members, it could be considered inappropriate for a WTO panel or the Appellate Body to resolve this issue by defining the criteria. However, a WTO panel faced with an Article 27.3(b) effectiveness issue might consider the remedies that a WTO Member's legal system provides to institute and enforce judgments in favor of owners.[19] Such an approach could be justified by interpreting the term "effective" in light of the obligations regarding enforcement of IPRs in Part III of the TRIPS Agreement.

The WTO ministers meeting in Hong Kong in 2005 added a separate process of consultations on the relationship between the CBD and TRIPS, chaired by Director-General Pascal Lamy. The adoption of the Nagoya Protocol[20] has added a new wrinkle to WTO discussions regarding the relationship between the CBD and TRIPS and the review of Article 27.3(b).[21] However, WTO Members have not reached any agreement on these issues. As of December 2012, 92 countries had signed the Nagoya Protocol.[22] The experience in the Uruguay Round and the Doha Round both demonstrate the difficulties of WTO Members in reaching agreement on the relationship between TRIPS obligations related to plant varieties, the UPOV Convention, and the CBD.[23]

16 Ministerial declaration, November 20, 2001, WT/MIN(01)/DEC/1, adopted on November 14, 2001, <http://www.wto.org/english/thewto_e/minist_e/min01_e/mindecl_e.htm> (accessed May 20, 2011); Note by the Secretariat, Review of Article 27.3(b).

17 Note by the Secretariat, Review of Article 27.3(b) para. 47.

18 Note by the Secretariat, Review of Article 27.3(b) para. 50.

19 Masarek, "Treetop View of the Cathedral" 460.

20 The Nagoya Protocol was adopted by the Conference of the Parties (COP) to the UN Convention on Biological Diversity (CBD) at its 10th meeting on October 29, 2010 in Nagoya: Nagoya Protocol on Access to Genetic Resources and the Fair and Equitable Sharing of Benefits Arising from their Utilization to the Convention on Biological Diversity (Nagoya Protocol) <http://www.cbd.int/abs/text/> (accessed May 20, 2011).

21 WTO News, "Nagoya gives new context to old views in intellectual property council" March 1, 2011 <http://www.wto.org/english/news_e/news11_e/trip_01mar11_e.htm> (accessed May 20, 2011).

22 Signatories to the Nagoya Protocol <http://www.cbd.int/abs/nagoya-protocol/signatories/> (accessed December 25, 2012).

23 The progress made with respect to the review of Art. 27.3(b) in the TRIPS Council is summarized in Note by the Secretariat, Review of Article 27.3(b). The progress made in the Director-General's consultative process regarding TRIPS and the CBD is summarized in Director-General Pascal Lamy,

Article 27.2 creates a general exception to the patentability obligation of Article 27.1, first sentence. This exception permits Members to

> exclude from patentability inventions, the prevention within their territory of the commercial exploitation of which is necessary to protect *ordre public* or morality, including to protect human, animal or plant life or health or to avoid serious prejudice to the environment, provided that such exclusion is not made merely because the exploitation is prohibited by their law.

This exception does not apply to the nondiscrimination obligation contained in Article 27.1, second sentence. Moreover, this exception would only apply in the case of plant varieties if a Member chooses patents for plant varieties. If a Member instead chooses a *sui generis* system the exception in Article 27.2 would not apply, because no issue of "patentability" would arise.

The effect of Article 27.3(b) is that WTO Members have the right to exclude plant varieties from patentability, provided that they offer a *sui generis* form of intellectual property protection. However, should Members choose not to provide *sui generis* protection, Article 27.3(b) obliges them to provide patent protection for plant varieties. This raises the issue of whether a Member could then invoke Article 27.2 to justify the exclusion of some plant varieties from patentability. The wording of Article 27 suggests that this is possible. Article 27.3 provides additional grounds for non-patentability to those contained in Article 27.2. Article 27.3 begins with the phrase, "Members may also exclude from patentability." The obligation in Article 27.3(b) to provide patents and/or *sui generis* protection appears as an exception to the exception in Article 27.2. There is no indication that Article 27.2 cannot be invoked to justify excluding certain plant varieties from patentability once a Member has chosen a system of patents instead of a system of *sui generis* protection. However, such exclusions would have to meet the criteria of Article 27.2. Nevertheless, if a Member chooses a system of *sui generis* protection, Article 27.2 could not be invoked to exclude certain plant varieties from *sui generis* protection, since Article 27.2 only provides a right to exclude from patentability.

Assuming that a WTO Member has chosen a system of patents instead of a system of *sui generis* protection, which exclusions would meet the criteria of Article 27.2? The party invoking the exception in Article 27.2 has the burden of proof to show that (1) the commercial exploitation of the invention is prevented within its territory and (2) prevention is "necessary to protect *ordre public* or morality, including to protect human, animal or plant life or health or to avoid serious prejudice to the environment."

Report on Issues Related to the Extension of the Protection of Geographical Indications Provided for in Article 23 of the TRIPS Agreement to Products other than Wines and Spirits and those Related to the Relationship between the TRIPS Agreement and the Convention on Biological Diversity, TN/C/W/61 (also circulated as WT/GC/W/633), April 21, 2011 <http://www.wto.org/english/tratop_e/trips_e/art27_3b_e.htm> (accessed May 25, 2011). See also the International Treaty on Plant Genetic Resources for Food and Agriculture (2001) adopted on November 3, 2001 in Rome <http://www.planttreaty.org/content/texts-treaty-official-versions> (accessed April 9, 2013).

The phrase "necessary to protect *ordre public* or morality, including to protect human, animal or plant life or health or to avoid serious prejudice to the environment" echoes the language of GATT Article XX(a) ("necessary to protect public morals") and GATT Article XX(b) ("necessary to protect human, animal or plant life or health"), as well as GATS Article XIV(a) ("necessary to protect public morals or to maintain public order") and GATS Article XIV(b) ("necessary to protect human, animal or plant life or health"). Given the similarity of the language in all of these exceptions, it is reasonable to interpret TRIPS Article 27.2 in light of the relevant GATT and GATS jurisprudence.

The Appellate Body in *US—Gambling* interpreted the term "necessary" in GATS Article XIV(a) to mean the same as the term "necessary" in GATT Article XX.[24] Thus, the term "necessary" has been given the same interpretation in similarly worded exceptions in covered agreements that apply to different sectors (goods and services). While TRIPS addresses a different sector, *US—Gambling* indicates that this is not an obstacle to applying the same interpretation in Article 27.2. WTO panels have diverged in their interpretations of the term "unnecessary" in Article 2.2 of the TBT Agreement and the Appellate Body has not interpreted this term in the same manner as the term "necessary" in GATT Article XX (although the analysis is similar).[25] However, Article 2.2 is an obligation, not an exception, and, unlike TRIPS Article 27.2, its wording is quite distinct from the provisions in GATS Article XIV and GATT Article XX which use the term "necessary." Articles 2.1 and 2.2 of the SPS Agreement also use similar wording to that of GATT Article XX(b) and, like Article 2.2 of the TBT Agreement, set out obligations rather than exceptions. However, the jurisprudence has yet to determine the extent to which the interpretation of GATT Article XX(b) is relevant to the interpretation of the SPS Agreement.[26] As the Appellate Body noted in *US—Stainless Steel (Mexico)*, WTO Members cite WTO jurisprudence in legal

[24] Appellate Body Report, *United States—Measures Affecting the Cross-Border Supply of Gambling and Betting Services (US—Gambling)*, WT/DS285/AB/R, adopted April 20, 2005, para. 291. In *China—Publications*, the panel followed the same interpretation in GATT Art. XX(a): Panel Report, *China—Measures Affecting Trading Rights and Distribution Services for Certain Publications and Audiovisual Entertainment Products (China—Publications)*, WT/DS363/R, adopted January 19, 2010, para. 7.759.

[25] Appellate Body Report, *United States—Measures Concerning the Importation, Marketing and Sale of Tuna and Tuna Products (US—Tuna II (Mexico))*, WT/DS381/AB/R, adopted June 13, 2012, paras. 318–19; Appellate Body Report, *United States—Certain Country of Origin Labelling Requirements (US—COOL)*, WT/DS386/AB/R, adopted July 23, 2012, para. 374. In *US—Clove Cigarettes*, the Panel's approach did not differ significantly from the approach taken in GATT Art. XX(b): Panel Report, *United States—Measures Affecting the Production and Sale of Clove Cigarettes (US—Clove Cigarettes)*, WT/DS406/R, paras. 7.358–7.361. In *US—Tuna II (Mexico)*, the Panel's approach differed from the approach taken in GATT Art. XX(b): Panel Report, *United States—Measures Concerning the Importation, Marketing and Sale of Tuna and Tuna Products (US—Tuna II (Mexico))*, WT/DS381/R, adopted June 13, 2012, para. 7.465. In *US—COOL*, the Panel considered the jurisprudence of GATT Art. XX(b) to be relevant to the interpretation of Art. 2.2 of the TBT Agreement, but did not complete its analysis in this regard: Panel Report, *United States—Certain Country of Origin Labelling Requirements (US—COOL)*, WT/DS384/R, WT/DS386/R, adopted 23 July 2012, paras. 7.670, 7.693, 7.720.

[26] Appellate Body Report, *European Communities—Measures Concerning Meat and Meat Products (Hormones) (EC—Hormones)*, WT/DS26/AB/R, adopted February 13, 1998, paras. 238–9; Panel

arguments in dispute settlement proceedings and take the jurisprudence into account when enacting or amending national legislation.[27] WTO Members also take the jurisprudence into account in trade negotiations. Thus, the interpretation of identical terms should be similar, given the similarities in the language that is used in the GATT, GATS and TRIPS provisions, the fact that all three are exceptions and the similar contexts of these provisions.

The context of TRIPS Article 27.2 is not identical to that of GATS Article XIV and GATT Article XX. The term "*ordre public*" is broader in Article 27.2 because it encompasses measures to "protect human, animal or plant life or health or to avoid serious prejudice to the environment." GATT Article XX(a) and GATS Article XIV(a) do not apply to human, animal, or plant life or health, which are addressed in GATT Article XX(b) and GATS Article XIV(b). GATT Article XX(b) has been interpreted to cover environmental protection measures, as long as the measure in question is designed to protect "human, animal or plant life or health."[28] The reference to the environment in TRIPS Article 27.2 is narrower, since it is restricted to "serious prejudice." However, the phrase "human, animal or plant life or health" should, as in GATT Article XX(b), encompass environmental protection measures designed to protect "human, animal or plant life or health." Environmental measures that do not address these issues would face the stricter threshold of "serious prejudice" in TRIPS Article 27.2. However, taken together, GATT Article XX(a) and (b), GATS Article XIV(a) and (b), and TRIPS Article 27.2 cover substantially the same matters.

There is another difference in the context of TRIPS Article 27.2, compared to that of GATS Article XIV and GATT Article XX. While the language of the chapeau in GATS Article XIV regarding arbitrary or unjustifiable discrimination reflects that of the chapeau in GATT Article XX, TRIPS Article 27.2 contains no comparable language. However, the Article 27.2 exception does not apply to the nondiscrimination obligation of Article 27.1. Thus, Article 27.2 should be interpreted to require compliance with the broad nondiscrimination obligation of Article 27.1,[29] which would have a similar effect to the nondiscrimination requirements of the chapeau in

Report, *United States—Certain Measures Affecting Imports of Poultry from China (US—Poultry (China))*, WT/DS392/R, adopted October 25, 2010, paras. 7.268–7.269, 7.479–7.481.

[27] Appellate Body Report, *United States—Final Anti-dumping Measures on Stainless Steel from Mexico (US—Stainless Steel (Mexico))*, WT/DS344/AB/R, adopted May 20, 2008, para. 160.

[28] Panel Report, *Brazil—Measures Affecting Imports of Retreaded Tyres (Brazil—Retreaded Tyres)*, WT/DS332/AB/R, adopted August 20, 2009, paras. 7.44–7.46. Since GATS Art. XIV(b) uses identical language, the same interpretation should apply to it, given the Appellate Body's interpretation of GATS Art. XIV(a) in *US—Gambling*.

[29] Regarding nondiscrimination in Art. 27, in Panel Report, *Canada—Patent Protection of Pharmaceutical Products (Canada—Pharmaceutical Patents)*, WT/DS114/R, adopted April 7, 2000, para. 7.92 the Panel stated:

> Article 27 does not prohibit bona fide exceptions to deal with problems that may exist only in certain product areas. Moreover, to the extent the prohibition of discrimination does limit the ability to target certain products in dealing with certain of the important national policies referred to in Articles 7 and 8.1, that fact may well constitute a deliberate limitation rather than a frustration of purpose. It is quite plausible... that the TRIPS Agreement would want to require governments to apply exceptions in a non-discriminatory manner, in

GATS Article XIV and GATT Article XX. However, in the case of plants, Article 27.3 operates as an exception to the nondiscrimination obligation in Article 27.1, except in the case of plant *varieties*. Moreover, the complainant would have the burden of proving a violation of the nondiscrimination obligation of Article 27.1, whereas the respondent would have the burden of proof under the chapeau in GATS Article XIV and GATT Article XX.

Another difference between TRIPS Article 27.2, and GATS Article XIV and GATT Article XX, has to do with a linguistic difference. In English and French, TRIPS Article 27.2 uses the same form of the word "necessary" as in GATT Article XX and GATS Article XIV, but in Spanish there is a small variation. The term in Spanish is "necesarias" in GATT and GATS, and "necesariamente" in TRIPS Article 27.2. The text in Spanish suggests that the difference between TRIPS and GATT/GATS is due to grammatical differences in the structure of the provision: the Spanish text uses an adverb applied to a verb, whereas the French and English texts use an adjective applied to a noun. The text in French and English indicates that negotiators intended to convey the same idea in all three agreements.

A final difference in the context of TRIPS Article 27.2, compared that of GATS Article XIV and GATT Article XX, is the final requirement of the former. In TRIPS Article 27.2, the party invoking the exception must also show that the "exclusion is not made merely because the exploitation is prohibited by their law." However, the manner in which Article 27.2 uses this phrase suggests that the inquiry must take place in a broader context than the national law of the Member in question. In addition, the justification for exclusion must be based on one of the permissible grounds for exclusion set out in Article 27. The negotiating history of the TRIPS Agreement appears to confirm this, since an earlier draft listed "law" as one of the permissible grounds for exclusion and was not included in the final text.[30] The patentability of the subject matter in other jurisdictions might support the view that the exclusion of patentability in the respondent's measure is made merely because the exploitation is prohibited by the respondent's law. Thus, this requirement does not appear to alter significantly the subject matter of the Article 27.2 exception, as compared to that of the equivalent provisions in GATS Article XIV and GATT Article XX.

Given the similar wording and context, GATT and GATS jurisprudence suggests the following analysis would be appropriate in TRIPS Article 27.2. First, the party invoking the exception must make a *prima facie* case that the policy goal at issue in its measure falls within the range of policies designed "to protect *ordre public* or morality, including to protect human, animal or plant life or health or to avoid serious prejudice to the environment." Since the latter part of this phrase is illustrative, *ordre public* or morality would include a potentially broad array of policy goals.

order to ensure that governments do not succumb to domestic pressures to limit exceptions to areas where right holders tend to be foreign producers.

30 *Status of Work in the Negotiating Group—Chairman's Report to the GNG*, WTO Doc. MTN. GNG/NG11/W/76.

Once it is established that the policy goal fits the exception, the party would then have to prove that the measure is "necessary" to achieve the policy goal. This analysis takes place in light of the level of risk that a Member has set for itself. To demonstrate that the measure is necessary involves weighing and balancing a series of factors. First, the greater the importance of the interests or values that the challenged measure is intended to protect, the more likely it is that the measure is necessary. GATT Article XX jurisprudence has addressed the importance of human life and health[31] and of environmental protection,[32] and would be relevant at this stage of the analysis. Second, the greater the extent to which the measure contributes to the end pursued, the more likely that the measure is necessary. In *Brazil—Retreaded Tyres*, the Appellate Body observed that a measure adopted in order to attenuate global warming and climate change, the contribution of which is not immediately observable, could meet this requirement under Article XX(b); the same should be true under TRIPS Article 27.2.[33] Third, the less WTO-inconsistent the challenged measure is, the more likely it would be considered necessary. The final issue is whether a WTO-consistent alternative measure which the Member concerned could reasonably be expected to employ is available, or whether a less WTO-inconsistent measure is reasonably available. The weighing and balancing process of the first three factors also informs this analysis.[34]

The party invoking the exception may point out why alternative measures would not achieve the same objectives as the challenged measure, but it is under no obligation to do so in order to establish, in the first instance, that its measure is "necessary." If the other party raises a WTO-consistent alternative measure that, in its view, should have been taken, the party invoking the exception would be required to demonstrate why its challenged measure nevertheless remains "necessary" in the light of that alternative or, in other words, why the proposed alternative is not, in fact, "reasonably available." If the party invoking the exception demonstrates that the alternative is not "reasonably available," in the light of the interests or values being pursued and the party's desired level of protection, it follows that the challenged measure must be "necessary."[35] One alternative that should be considered in the context of TRIPS is to issue a compulsory license under TRIPS Article 31, rather than denying patent protection altogether.

[31] Appellate Body Report, *European Communities—Measures Affecting Asbestos and Asbestos-Containing Products (EC—Asbestos)*, WT/DS135/AB/R, adopted April 5, 2001.

[32] Appellate Body Report, *Brazil—Measures Affecting Imports of Retreaded Tyres (Brazil—Retreaded Tyres)*, WT/DS332/AB/R, adopted August 20, 2009, para. 179.

[33] Appellate Body Report, *Brazil—Retreaded Tyres* para. 151; Panel Report, *China—Measures Related to the Exportation of Various Raw Materials (China—Raw Materials)*, WT/DS398/R, adopted February 22, 2012, paras. 7.536–7.538.

[34] Appellate Body Report, *Korea—Measures Affecting Imports of Fresh, Chilled and Frozen Beef (Korea—Beef)*, WT/DS161/AB/R, adopted January 10, 2001; Appellate Body Report, *EC—Asbestos*; Appellate Body Report, *US—Gambling*; Appellate Body Report, *Dominican Republic—Measures Affecting the Importation and Internal Sale of Cigarettes (Dominican Republic—Cigarettes)*, WT/DS302/AB/R, adopted May 19, 2005; Appellate Body Report, *Brazil—Retreaded Tyres.*

[35] Appellate Body Report, *US—Gambling* paras. 310–11.

In the Doha Round negotiations, WTO Members have debated the kinds of exceptions that would be appropriate in a *sui generis* system of protection. This debate also provides a preview of the types of measures that might be justified under Article 27.2 or under other exceptions set out in TRIPS Articles 30 and 31. Suggested limitations and exceptions can be categorized in two groups: (1) exceptions regarding experimental use and the right to use a protected variety for further breeding aim to benefit other breeders and to promote further innovation (breeders' exemption); and (2) exceptions that aim to ensure food security and preserve the integrity of rural communities include exceptions to the benefit of farmers, non-commercial use of plant varieties, and the system of seed saving and exchange as well as the selling among farmers (farmers' privilege). Compulsory licenses might serve both objectives. While some Members agree that the breeders' exemption allows breeders to freely use plant varieties protected by plant breeders' rights in their breeding activities, others note that the scope of the exemption is unclear, particularly regarding compensation due to right holders. Some Members have argued that farmers' privilege allows farmers to replant on their own holdings propagating material of protected plant varieties that they have harvested on their own holdings, with the issue of remuneration left to the national legislator. Other Members have argued that farmers' privilege should not be limited to saving and replanting the material only on a farmer's own holdings. Rather, the privilege should provide greater latitude in the case of subsistence and small farmers, as long as the commercial interests of plant breeders are protected. However, there is disagreement regarding whether certain categories of commercial farmers in developing countries should benefit from farmers' exceptions.[36]

5.3 UPOV Convention

This Section will compare IPRs under TRIPS to IPRs under the UPOV Convention. The UPOV Convention established the International Union for the Protection of New Varieties of Plants (UPOV).[37] The 1961 UPOV Convention has been revised in 1972, 1978, and 1991. In the review of TRIPS Article 27.3(b), WTO Members have debated the merits of the various UPOV conventions and their relationship to the TRIPS Agreement. Some advocate the 1991 Act of the UPOV Convention (UPOV 1991) as the most appropriate system and level of protection. Others argue that, since UPOV 1991 does not permit limits on the eligibility for protection of varieties by species of plant, newly developed varieties of species of plant that would not have been eligible for protection under the 1978 Act are now eligible for protection under the 1991 Act.[38]

[36] Note by the Secretariat, Review of Article 27.3(b) paras. 55–6, 61.

[37] International Convention for the Protection of New Varieties of Plants of December 2, 1961, as Revised at Geneva on November 10, 1972, on October 23, 1978, and on March 19, 1991 <http://www.upov.int/en/publications/conventions/1991/act1991.htm> (accessed May 15, 2011).

[38] Note by the Secretariat, Review of Article 27.3(b) para. 64.

The UPOV Council has adopted a series of Explanatory Notes on the UPOV Convention.[39] The Preambles in these Explanatory Notes contain this proviso: "The only binding obligations on members of the Union are those contained in the text of the UPOV Convention itself, and these Explanatory Notes must not be interpreted in a way that is inconsistent with the relevant Act for the member of the Union concerned." However, this proviso is silent regarding the effect of these Explanatory Notes on the interpretation of the provisions of the UPOV Convention. The interpretation of the Convention is not the same as the interpretation of the Explanatory Notes. Thus, while the Explanatory Notes do not create binding obligations, they are relevant to the interpretation of the binding obligations, as well as the rights, set out in the Convention itself, by virtue of Articles 31 and 32 of the Vienna Convention on the Law of Treaties.

UPOV 1991 obliges contracting parties to "grant and protect breeders' rights" (Article 2) and to apply UPOV 1991 "to all plant genera and species" (Article 3). The minimum period of protection is 20 years from the date of the grant of the breeder's right and, for trees and vines, 25 years (Article 19). Thus, the minimum period of protection for trees and vines is 5 years longer than for patents under Article 33 of the TRIPS Agreement. Contracting parties must provide national treatment to nationals of other contracting parties insofar as the grant and protection of breeders' rights (Article 4).

In contrast, the length of protection varies among *sui generis* systems. Thailand grants rights for a maximum of 12, 17, or 27 years, depending on the type of plant involved. India grants protection for 15 or 18 years, also depending on the type of plant variety involved. Some argue that the length of protection that IPRs confer should vary with the particular technology.[40] For example, 20 years of patent protection for pharmaceuticals confer a shorter term in practice, due to the length of time required to obtain marketing approval for the product.

Article 1(vi) of UPOV 1991 defines the term "variety" as:

> a plant grouping within a single botanical taxon of the lowest known rank, which grouping . . . can be defined by the expression of the characteristics resulting from a given genotype or combination of genotypes, distinguished from any other plant grouping by the expression of at least one of the said characteristics and considered as a unit with regard to its suitability for being propagated unchanged.

The TRIPS Agreement does not define the term "plant varieties." WTO Members have debated whether and how to define several terms in Article 27.3(b).[41] Some suggest that the 1991 UPOV definition inform the definition of the term "plant varieties" in Article 27.3(b).[42] Some critics argue that biotechnology advances,

[39] Explanatory Notes on Conditions and Limitations Concerning the Breeder's Authorization in Respect of Propagating Material under the UPOV Convention, UPOV/EXN/CAL/1, October 21, 2010 <http://www.upov.int/explanatory_notes/en/> (accessed April 9, 2013).

[40] Masarek, "Treetop View of the Cathedral" 463–4.

[41] Note by the Secretariat, Review of Article 27.3(b). The terms debated are plants and animals, micro-organisms, and non-biological and micro-biological processes.

[42] Note by the Secretariat, Review of Article 27.3(b) para. 68.

particularly molecular marker technologies, are making the very concept of plant variety obsolete.[43]

UPOV 1991 Article 5 provides that the breeder's right shall be granted where the variety is new, distinct, uniform, and stable. This provision is more detailed than the patentability requirements of TRIPS Article 27.1 (new, inventive step and usefulness) and adapted to the specific situation of plant varieties. However, advances in biotechnology also make these UPOV criteria more difficult to apply in practice, since the analysis must take place at the molecular level, and molecular information can be used to avoid detection of IPR infringements more easily.[44] As noted above, TRIPS Article 27.3(b) does not have a novelty requirement for *sui generis* systems.

Article 14, subject to Articles 15 and 16, requires that the following acts in respect of the propagating material of the protected variety obtain the authorization of the breeder: (i) production or reproduction (multiplication), (ii) conditioning for the purpose of propagation, (iii) offering for sale, (iv) selling or other marketing, (v) exporting, (vi) importing, and (vii) stocking for any of the foregoing purposes. The same acts with respect to harvested material obtained through the unauthorized use of propagating material of the protected variety also require the authorization of the breeder. Contracting parties may add other acts to this mandatory list. These provisions also apply to varieties which are essentially derived from the protected variety, where the protected variety is not itself an essentially derived variety (Article 14(5)(a)(i)), varieties which are not clearly distinguishable in accordance with Article 7 from the protected variety (Article 14(5)(a)(ii)) and varieties whose production requires the repeated use of the protected variety (Article 14(5)(a)(iii)). Again, advances in biotechnology make these criteria more difficult to apply in practice.[45]

Article 15(1) provides compulsory exceptions to the breeder's right: (i) acts done privately and for non-commercial purposes; (ii) acts done for experimental purposes; and (iii) acts done for the purpose of breeding other varieties (the breeder's exemption), and, except where the provisions of Article 14(5) apply (varieties which are essentially derived from the protected variety), acts referred to in Article 14(1) to Article 14(4) in respect of such other varieties.

The breeder's exemption is also affected by technological advances. New "reverse breeding" techniques shorten the time needed to create new varieties, thereby shortening the *de facto* period of exclusivity enjoyed by the original breeder. This has led to proposals to phase in the breeder's exemption according to the time it takes to reverse breed new varieties.[46] Such proposals are comparable to proposals to increase the patent term for pharmaceuticals to account for the time required to obtain marketing approvals and appear designed to extend monopolies that would limit access.

[43] Mark D. Janis and Stephen Smith, "The Protection of Rights in Plant Varieties: Technological Change and the Design of Plant Variety Protection Regimes" (2007) 82 Chicago-Kent Law Review 1557, 1577–8.

[44] Janis and Smith, "The Protection of Rights in Plant Varieties" 1583–92.

[45] Janis and Smith, "The Protection of Rights in Plant Varieties" 1596–7.

[46] Janis and Smith, "The Protection of Rights in Plant Varieties" 1604–5.

The Explanatory Notes on Article 15(1)(i) set out two examples of acts "possibly" not falling within the scope of the exception for acts done privately and for non-commercial purposes: non-private acts for non-commercial purposes and private acts which are undertaken for commercial purposes.[47] The Explanatory Notes on Article 15(1)(i) also set out examples of acts "possibly" falling within the scope of the exception for acts done privately and for non-commercial purposes: the propagation of a variety by an amateur gardener for exclusive use in his own garden and the propagation of a variety by a farmer exclusively for the production of a food crop to be consumed entirely by that farmer and the dependents of the farmer living on that holding.[48] According to this explanation of Article 15(1)(i), if I am an amateur gardener, I propagate a protected variety of roses in my garden, my neighbor invites me over for dinner and I bring my neighbor a bouquet of those same roses, I am in violation of Article 15(1)(i). However, it seems a bit of a stretch to consider such an act a "commercial purpose."

Regarding Article 15(1)(ii), the Explanatory Notes simply state that "[t]he breeder's right does not extend to the use of the protected variety for experimental purposes."[49] No definition of the term "experimental purposes" is provided in the Explanatory Notes or the Convention itself. However, this term should not cover "acts done for the purpose of breeding other varieties," since those acts are subject to the breeder's exemption in Article 15(1)(iii). To interpret Article 15(1)(ii) otherwise would make Article 15(1)(iii) redundant, contrary to the principle of effective treaty interpretation. Regarding the breeder's exemption in Article 15(1)(iii), the Explanatory Notes state, "there are no restrictions on the use of protected varieties for the purpose of breeding new plant varieties."[50]

Article 15(2) provides an optional exception to the breeder's right:

> Each Contracting Party may, within reasonable limits and subject to the safeguarding of the legitimate interests of the breeder, restrict the breeder's right in relation to any variety in order to permit farmers to use for propagating purposes, on their own holdings, the product of the harvest which they have obtained by planting, on their own holdings, the protected variety or a variety covered by Article 14(5)(a)(i) or Article 14(5)(a)(ii).

The Explanatory Notes suggest that the farmer's exemption in Article 15(2) was aimed at those crops where, for the Member of the Union concerned, there was a common practice of farmers saving harvested material for further propagation, based on a recommendation of the Diplomatic Conference of 1991.[51] However, there are two aspects to note about this explanation. First, the Diplomatic Conference note uses the phrase "should not be read," which indicates that this interpretation of Article 15(2) is not mandatory. Second, the Diplomatic Conference note would constitute negotiating history under Article 32 of the Vienna

[47] Explanatory Notes on Exceptions to the Breeder's Right under the 1991 Act of the UPOV Convention, UPOV/EXN/EXC/1, adopted by the Council at its forty-third ordinary session on October 22, 2009, paras. 5 and 6 <http://www.upov.int/explanatory_notes/en/> (accessed April 9, 2013).

[48] Explanatory Notes para. 7.

[49] Explanatory Notes para. 8.

[50] Explanatory Notes para. 9.

[51] Explanatory Notes paras. 14–15.

Convention on the Law of Treaties and only be relevant as a supplementary means of interpretation.[52]

The Explanatory Notes also suggest that the farmer's exemption in Article 15(2) relates to selected crops where the product of the harvest is used for propagating purposes, for example small-grained cereals where the harvested grain can equally be used as seed. They suggest further that it may be inappropriate to introduce the exception for agricultural or horticultural sectors, where it has not been a common practice for the harvested material to be used as propagating material, based on the same recommendation of the Diplomatic Conference of 1991.[53]

Regarding the phrase "reasonable limits and safeguarding of the legitimate interests of the breeder," the Explanatory Notes commentary expands the breeder's right and limits the farmer's exemption beyond what the text of Article 15(2) suggests. They suggest considering several factors to determine whether the implementation of the farmer's exemption introduces limits to the breeder's right that are reasonable and whether the national legislation safeguards the legitimate interests of the breeder:

(1) type of variety (whether there has been a common practice of farmers saving harvested material for further propagation);
(2) size of farmer's holding, crop area, or crop value (to determine the extent to which a farmer is permitted to use farm-saved seed and with a different level of remuneration to breeders);
(3) proportion or amount of harvested crop (to determine the maximum percentage of the harvested crop or maximum acreage which the farmer may use for further propagation and/or the level of remuneration);
(4) changes in the level of harvested material used for further propagation, evolution of farming practices, and breeding and propagation methodologies, and economic developments to justify limiting the level of farm-saved seed to those levels which had been common practice before the introduction of plant variety protection; and
(5) remuneration to safeguard the legitimate interests of the breeders.[54]

The Explanatory Notes go on to emphasize that the farmer's exemption does not extend to propagating material which was produced on the holding of another farmer.[55] They then assert that the use of the words "within reasonable limits and subject to the safeguarding of the legitimate interests of the breeder" means that the

[52] However, there is some disagreement among experts regarding the role of Art. 32: John H. Jackson, *The Jurisprudence of GATT and the WTO: Insights on Treaty Law and Economic Relations* (Cambridge University Press, Cambridge 2000) 145 fn. 37 (arguing that preparatory history is to be used only when the means specified in Article 31 do not resolve an interpretative problem); Stephen M. Schwebel, "May Preparatory Work be Used to Correct Rather than Confirm the 'Clear' Meaning of a Treaty Provision?" in Jerzy Makarczyk (ed.), *Theory of International Law at the Threshold of the 21st Century* (Springer, Berlin 1996) 541 (arguing that preparatory work must be given a greater role in treaty interpretation than the words of Art. 32 might suggest).

[53] Explanatory Notes para. 15.

[54] Explanatory Notes paras. 19–23.

[55] Explanatory Notes para. 24.

farmer's exemption should be implemented "in a way which does not undermine the incentives provided by the UPOV Convention for breeders to develop new varieties."[56]

The Explanatory Notes then suggest that contracting parties consider several factors in deciding if and how to implement Article 15(2): the impact on breeding, the costs and mechanisms required for implementation, and the overall economic impact on agriculture. They also suggest consultation with the interested parties[57] and introducing the farmer's exemption in a manner that facilitates modifying the legislation over time to adapt to the evolution of farming practices and breeding and propagation methodologies, as well as economic developments.[58]

WTO Members have debated UPOV provisions regarding the farmers' exemption. Some argue that UPOV 1978 allows farmers to save, exchange, and, to a limited degree, sell seeds of protected varieties, whereas UPOV 1991 gives the government discretion as to whether to permit farmers to save seeds for use on their own holdings and makes this exception subject to "reasonable restrictions" and the protection of the "legitimate interests" of the breeder. Moreover, the exception only applies to material that has been harvested on the same holdings and not to propagated material (just "the product of the harvest"). Thus, this exception does not benefit farmers in the case of GMO varieties that do not produce viable seeds (terminator technology).[59] Many Members, particularly developing countries, therefore argue that UPOV 1991 would have a negative impact on food security in developing countries and create dependence on foreign commercial breeders for seeds.[60]

Article 16(1) limits exhaustion of the breeder's right when the acts involve: (1) further propagation of the variety or (2) an export of material of the variety, which enables the propagation of the variety, into a country which does not protect varieties of the plant genus or species, except where the exported material is for final consumption purposes. Thus, Article 16(1) gives contracting parties less discretion than TRIPS Article 6 gives to WTO Members. TRIPS Article 6 leaves WTO Members free to determine their own regime with respect to exhaustion of patent rights, subject to national treatment and most-favored-nation treatment.[61] Some WTO Members argue that UPOV 1991 limits exhaustion of the right to sell or otherwise market plant varieties made within the national territory of the contracting party concerned, in contrast to Article 6 of the TRIPS Agreement, which leaves the issue of exhaustion of intellectual property rights to the discretion of each Member. For the foregoing reasons, several WTO Members argue that legislation based upon UPOV 1978 provides effective *sui generis* protection for plant variety rights for the purposes of Article 27.3(b). However, other Members

[56] Explanatory Notes para. 25. [57] Explanatory Notes para. 26.

[58] Explanatory Notes para. 27.

[59] Jason A. Barron, "Genetic Use Restriction Technologies: Do the Potential Environmental Harms Outweigh the Economic Benefits?" (2008) 20 Georgetown International Environmental Law Review 271.

[60] Note by the Secretariat, Review of Article 27.3(b) para. 65.

[61] Declaration on the TRIPS Agreement and Public Health, para. 5(d).

argue that the farmers' exemption can be justified under Article 27.3(b) as an exception to plant variety protection or under Article 30 of the TRIPS Agreement as an exception to patent protection, for farming activities are limited to small, subsistence farms which customarily reuse seed because they lack access to financial resources for new seeds every growing season, or where commercial activities of farmers are limited geographically.[62]

Article 17 of UPOV 1991 limits restrictions on the exercise of the breeder's right to reasons of public interest. When the restriction "has the effect of authorizing a third party to perform any act for which the breeder's authorization is required," the contracting party must take all measures necessary to ensure that the breeder receives equitable remuneration. Article 9 of the 1978 UPOV Convention also limits restrictions on the exercise of the breeder's right to reasons of public interest. When the restriction "is made in order to ensure the widespread distribution of the variety," the Member State of the Union must take all measures necessary to ensure that the breeder receives equitable remuneration.

What would qualify as "public interest" justifications for restricting a breeder's right? At the very least, if a WTO Member issues a compulsory license in accordance with TRIPS Article 31, such compulsory licenses should qualify. Since TRIPS Article 27.3(b) permits WTO Members to choose between patents and *sui generis* protection for plant varieties, there should be no inherent conflict between UPOV and TRIPS. The negotiating history of TRIPS indicates that the negotiators were aware of UPOV 1991, as well as the fact that many countries might prefer UPOV 1978 to UPOV 1991, when they drafted Article 27.3(b). This interpretation would be consistent with the presumption against conflicts between treaties.

Neither UPOV 1978 nor UPOV 1991 defines "equitable remuneration," as in the case of TRIPS Article 31(h), which requires "adequate remuneration" according to the circumstances of each case taking into account the economic value of a compulsory license. The TRIPS agreement sheds some light on this issue, since Article 1.1 permits WTO Members to determine freely the adequate method of applying TRIPS provisions according to their own legal system and practice. Another issue is what measures are considered "necessary to ensure that the breeder receives equitable remuneration."

The balance of rights between producers and users of biotechnology seems more balanced in the TRIPS Agreement than in UPOV 1991. First, TRIPS provides the flexibility of choosing patents or a *sui generis* system of IPRs. Second, even if patents are chosen, rather than a *sui generis* system, UPOV is less balanced. UPOV 1991, particularly when read together with the Explanatory Notes, favors the interests of plant breeders over the interests of farmers. This is particularly true with respect to the interpretation of the farmer's exemption in the Explanatory Notes, which, if accepted, would be more restrictive than the exception for compulsory licensing of patents in TRIPS Article 31.

[62] Note by the Secretariat, Review of Article 27.3(b) para. 66.

5.4 Convention on Biological Diversity

The CBD contains provisions regarding use of traditional knowledge and access to genetic resources, in particular with respect to prior informed consent and the equitable sharing of benefits derived from traditional knowledge and access to genetic resources. WTO Members have raised two general issues regarding the relationship between the TRIPS Agreement and the CBD: (1) whether there is conflict between the TRIPS Agreement and the CBD and (2) whether something needs to be done to ensure that the two instruments are applied in a non-conflicting and mutually supportive way.[63] As Cottier has noted in his seminal article on the protection of genetic resources and traditional knowledge, the reference in TRIPS Article 27.3(b) to effective *sui generis* protection is sufficiently broad to encompass systems to protect innovations based on traditional knowledge as well as existing knowledge through a system of "traditional intellectual property rights."[64]

Some Members argue that there is a conflict between the TRIPS Agreement and the CBD, for two reasons. First, the TRIPS Agreement provides for the appropriation of such genetic resources by private parties in a way that is inconsistent with the sovereign rights of countries over their genetic resources as provided for in the CBD, by requiring that certain genetic material be patentable or protected by *sui generis* plant variety rights and by not preventing the patenting of other genetic material. Second, the TRIPS Agreement provides for the patenting or other intellectual property protection of genetic material without ensuring that the provisions of the CBD, including those relating to prior informed consent and benefit sharing, are respected. Similar points have been made about the relationship between the TRIPS Agreement and the provisions of the CBD relating to the traditional knowledge of indigenous peoples and local communities.[65] Some Members argue that the TRIPS Agreement needs to be amended to avoid patents being granted on existing traditional knowledge or genetic resources subject matter, without the prior informed consent of the source countries, while others argue that no amendment is required to achieve these objectives.[66] These concerns may seem exaggerated, since there is no obligation to implement the CBD in a manner that

[63] Note by the Secretariat, The Relationship between the TRIPS Agreement and the Convention on Biological Diversity: Summary of Issues Raised and Points Made, IP/C/W/368/Rev.1, February 8, 2006, para. 6.

[64] Cottier, "Genetic Resources." Professor Cottier's article provides an eloquent argument in favor of a better balance of rights and obligations regarding biotechnology, IPRs genetic resources and traditional knowledge, as well as concrete suggestions on how that might be achieved. Therefore, we will limit this discussion to some of the legal issues that arise between TRIPS and the CBD.

[65] Note by the Secretariat, The Relationship between the TRIPS Agreement and the Convention on Biological Diversity para. 15.

[66] Note by the Secretariat, The Relationship between the TRIPS Agreement and the Convention on Biological Diversity paras. 8–15; Director-General Pascal Lamy, Report on Issues Related to the Extension of the Protection of Geographical Indications Provided for in Article 23 of the TRIPS Agreement to Products other than Wines and Spirits and those Related to the Relationship between the TRIPS Agreement and the Convention on Biological Diversity, TN/C/W/61 (also circulated as WT/GC/W/633), April 21, 2011, para. 27.

would conflict with WTO obligations. However, the CBD serves as an alternative model for national legislation relating to plant varieties and it may prove difficult to implement effectively in a manner that is consistent with WTO obligations (regarding export restrictions for example).

CBD Article 8(j) requires contracting parties to encourage the equitable sharing of the benefits arising from knowledge, innovations, and practices of indigenous and local communities embodying traditional lifestyles relevant for the conservation and sustainable use of biological diversity. Since the obligation merely is to "encourage" and is "subject to national legislation," this provision contains no real obligation to equitably share benefits from traditional knowledge.[67]

CBD Article 15(1) recognizes that the authority to determine access to genetic resources rests with the national governments. WTO Members are limited by their obligations under the Covered Agreements. For example, export restrictions would have to comply with GATT Article XI or be justified under GATT Article XX(b) or (g). Export restrictions that provide preferential access to natural resources to domestic firms would be difficult to justify under Article XI:2(a) and Article XX.[68] Providing preferential access to nationals could be inconsistent with non-discrimination obligations in GATT Articles I or III,[69] or nondiscrimination obligations in TRIPS Articles 3, 4, or 27. However, the CBD does not oblige national governments to take any particular course of action in their legislation. It should be possible to implement its provisions in a manner that is consistent with WTO obligations. For example, evenhanded measures that restrict exports in order to preserve biodiversity could be justified under GATT Article XX(g).[70] However, it is not clear that a *sui generis* system that discriminates between nationals and foreigners could meet the effectiveness requirement of TRIPS Article 27.3(b).

CBD Article 15(4) provides that access, where granted, shall be on mutually agreed terms and subject to the provisions of Article 15. CBD Article 15(5) provides that access to genetic resources shall be subject to prior informed consent of the contracting party providing such resources, unless otherwise determined by that party. As Professor Cottier has noted, this contractual approach has its limitations, but a number of countries have designed national legislation to implement the CBD nevertheless.[71]

CBD Article 15(7) requires each contracting party to take legislative, administrative, or policy measures, as appropriate, and, in accordance with Articles 16 and 19, with the aim of sharing in a fair and equitable way the results of research and development and the benefits arising from the commercial and other utilization of

[67] For a summary of the WTO debate on this issue, see Note by the Secretariat, The Protection of Traditional Knowledge and Folklore: Summary of Issues and Points Made, IP/C/W/370/Rev.1, 9 March 2006.

[68] Panel Report, *China—Raw Materials* and Appellate Body Report, *China—Raw Materials.*

[69] Mitsuo Matsushita, Thomas J. Schoenbaum, and Petros C. Mavroidis, *The World Trade Organization: Law Practice and Policy* (2nd edn., Oxford University Press, Oxford 2006) 714–15.

[70] Bradly J. Condon, "Climate Change and Unresolved Issues in WTO Law" (2009) 12 JIEL 895, 911–13.

[71] Cottier, "Genetic Resources."

genetic resources with the contracting party providing such resources, upon mutually agreed terms. The wording of this provision does not set out any meaningful obligation regarding equitable sharing of the benefits from genetic resources, unless the source party uses its authority to determine access to the genetic resource to negotiate access to the benefits that may arise. It is not clear what "as appropriate" means. It is not clear what "aim" means. The obligation to share benefits is subject to reaching an agreement on "mutually agreed terms." This amounts to a mere obligation to negotiate, rather than an obligation to reach a particular set of terms. Thus, these CBD provisions regarding the equitable sharing of the benefits do not really have any teeth unless the source party uses its authority to determine access to the genetic resource to enhance its bargaining position. However, developing countries that are not a source of the genetic material would lack such bargaining power.

It is not clear how the CBD provisions regarding informed consent can be made effective. Some WTO Members have expressed concern that the TRIPS Agreement allows the granting of patents for inventions that use genetic material without requiring that the provisions of the CBD in relation to prior informed consent and benefit sharing are respected. Some have proposed national solutions to this issue, including legislation on access and benefit sharing and contracts, outside the intellectual property system, to directly regulate the conduct in question.[72] Others have suggested disclosure requirements for patent applicants as a supplementary measure to national legislation and contracts, including in international forums other than the WTO. A proposal has been made to amend the TRIPS Agreement to oblige WTO Members to require that an applicant for a patent relating to biological materials or to traditional knowledge provide the following information, as a condition of acquiring patent rights: (i) the source and country of origin of the biological resource and of the traditional knowledge used in the invention; (ii) evidence of prior informed consent from the authorities under the relevant national regime; and (iii) evidence of fair and equitable benefit sharing under the relevant national regime.[73]

Paragraphs 1–5 of CBD Article 16 must be read together, since these paragraphs contain cross references that explicitly require that they be interpreted and applied consistently with each other. Article 16(1) recognizes that technology includes biotechnology, and that both access to and transfer of technology among contracting parties are essential elements for the attainment of the objectives of the Convention. It provides that each contracting party undertakes (subject to the provisions of this Article) to provide and/or facilitate access for and transfer to other contracting parties of technologies that make use of genetic resources.

Article 16(2) provides that access to and transfer of technology to developing countries shall be provided and/or facilitated under fair and most favorable terms, including on concessional and preferential terms where mutually agreed, and, where necessary, in accordance with the financial mechanism established by Articles 20

[72] Note by the Secretariat, The Protection of Traditional Knowledge and Folklore, paras. 28–9.
[73] Note by the Secretariat, The Protection of Traditional Knowledge and Folklore, para. 71.

and 21. Article 16(2) also provides that access to and transfer to developing countries of technology subject to patents and other intellectual property rights shall be provided on terms which recognize and are consistent with the adequate and effective protection of intellectual property rights. Thus, while paragraph 1 contains a mere undertaking regarding access to technology, paragraph 2 contains an obligation to provide such access in a manner that protects intellectual property rights. The access of developing countries to biotechnology takes a back seat to intellectual property rights and depends on access to financing. Concessional and preferential terms of financing are subject to reaching a mutual agreement. In effect, developing countries are required to pay for biotechnology that is subject to intellectual property rights and negotiate further for favorable financing of those purchases. Even this uncertain access to favorable financing is subject to a necessity test, which could depend on the financial and economic situation of the developing country that seeks the financing. Finally, the application of paragraph 2 must be consistent with paragraphs 3, 4, and 5 of Article 16. When read in light of Article 15, these provisions of Article 16 mean that the source party must use its authority to determine access to the genetic resource to negotiate favorable terms before intellectual property rights are granted. However, developing countries that are not a source of the genetic material would lack such bargaining power.

Article 16(3) requires each contracting party to take legislative, administrative, or policy measures, as appropriate, with the aim that contracting parties, in particular those that are developing countries, which provide genetic resources, are provided access to and transfer of technology which makes use of those resources, on mutually agreed terms, including technology protected by patents and other intellectual property rights, where necessary, through the provisions of Articles 20 and 21 and in accordance with international law and consistent with paragraphs 4 and 5. Thus, even where developing countries are the source of the genetic resources that are used to create biotechnologies, their access depends on negotiating financing and is subject to a needs test. The requirement that access be in accordance with international law is vague. However, it likely includes IPR treaties to which the developing country in question is a party, such as the TRIPS Agreement or the UPOV Convention. Once again, the source party must use its authority to determine access to the genetic resource to negotiate favorable terms before intellectual property rights are granted. However, developing countries that are not a source of the genetic material would lack such leverage.

Article 16(4) requires each contracting party to take legislative, administrative, or policy measures, as appropriate, with the aim that the private sector facilitate access to, joint development, and transfer of technology referred to in paragraph 1 for the benefit of both governmental institutions and the private sector of developing countries. Since this paragraph requires each contracting party in this regard to abide by the obligations included in paragraphs 1, 2, and 3, it implies an obligation to protect intellectual property rights. It might also be read to encourage the use of the compulsory licensing provisions of TRIPS Article 31, for example in the case of biotechnology that is subject to patents. However, Article 31 requires that compulsory licensing of patents predominantly serves the domestic market of the WTO

Member that issues the license. This means that the measures taken under CBD Article 16(4) would not include obliging the private sector to transfer biotechnology to developing countries. However, it might include other measures, such as tax incentives. The effectiveness of such incentives would be doubtful if the private sector would stand to lose more by facilitating access to, joint development, and transfer of technology to developing countries than it would gain through the tax incentives. Thus, it is doubtful that Article 16(4) will result in the adoption of any meaningful measures, absent political will.

Under Article 16(5), the contracting parties recognize that patents and other intellectual property rights may have an influence on the implementation of the CBD and requires them to cooperate in this regard, subject to national legislation and international law, in order to ensure that such rights are supportive of and do not run counter to its objectives. An obligation to cooperate is not an obligation to reach a particular outcome. WTO Members have been trying to cooperate in this regard in the context of the WTO negotiations on TRIPS Article 27.3(b) and the Director General's consultations on the relationship between the TRIPS Agreement and the CBD, but have failed to reach any agreement.

CBD Article 19 governs handling of biotechnology and distribution of its benefits. Article 19(1) requires each contracting party to take legislative, administrative, or policy measures, *as appropriate*, to provide for the effective participation in biotechnological research activities by those contracting parties, especially developing countries, which provide the genetic resources for such research, and *where feasible* in such contracting parties. Article 19(2) requires each contracting party to take all *practicable* measures to promote and advance priority access on a fair and equitable basis by contracting parties, especially developing countries, to the results and benefits arising from biotechnologies based upon genetic resources provided by those contracting parties. Such access shall be on mutually agreed terms. Again, to make these provisions effective, the source country would have to use its bargaining power when it decides whether or not to grant access to genetic resources.

It is difficult to see how legal conflicts could arise between the CBD, on the one hand, and the TRIPS Agreement or UPOV Convention, on the other hand, with respect to biotechnology. The CBD contains no meaningfully binding legal obligations regarding protection of traditional knowledge or sharing of the benefits of biotechnology that is subject to intellectual property rights, with the notable exception of cases in which the biotechnology in question was developed using the genetic resources of a particular source country. Even those obligations are subject to compliance with the TRIPS Agreement or UPOV Convention, as the case may be, and subject to the source country using its bargaining power regarding access to the genetic resources. Unless the source country uses this bargaining power to negotiate later access to any technology that may be developed as a result of access to the genetic resource, its access will be subject to the intellectual property rights. There is no obligation in the CBD to make the granting of intellectual property rights subject to proving prior informed consent of countries that are the source of genetic material. Even if there were such an obligation, it would only benefit the source countries. Thus, while the CBD preamble speaks of awareness "that conservation and sustainable use of biological diversity is of critical importance for meeting the food, health and other needs of the growing world population, for which purpose access to and sharing of both genetic resources and technologies are essential," it does not provide legally binding obligations to achieve these goals.

The provisions of the CBD could be relevant to determining compliance with TRIPS Article 27.2. TRIPS Article 27.2 provides a right to exclude from patentability, not an obligation to do so. Lack of informed consent or other inconsistencies with the CBD, particularly with respect to use of traditional knowledge, might justify exclusion from patentability on public policy grounds, particularly in developing countries. However, such an approach could be undermined by the use of terminator technologies or the refusal by multinational holders of intellectual property rights to introduce technologies into countries that exclude biotechnology from patentability.

5.5 Conclusion

TRIPS Article 27.3(b) requires Members to provide IPRs for plant varieties either by patents or by an effective *sui generis* system. The UPOV Convention and the CBD are alternative models for *sui generis* IPRs, though not the only alternatives. The CBD emphasizes the rights of countries to control access to genetic resources and to protect traditional knowledge, but is flexible regarding how that is done. One way to protect traditional knowledge is via a registry, which can then serve as evidence that the products derived from traditional knowledge are not new inventions and therefore not patentable nor susceptible to IPR protection under the UPOV Convention.

In essence, the CBD and the UPOV Convention set out competing visions of what is the most appropriate form of protection for plant IPRs. TRIPS Article 27.3(b) reflects the lack of consensus on this issue, just as the CBD and the UPOV Convention do. Some WTO Members have already limited their options, by choosing the 1978 or 1991 version of the UPOV Convention, or by entering into Free Trade Agreements in which they agree to adopt one of these versions of the UPOV Convention. Others have implemented a CBD-based model for plant IPRs.[74] The CBD, with less strict obligations and more flexibility than the UPOV Convention, leaves countries with more options should they wish to change their national system in future.

While the UPOV Convention sets out clearer, more binding obligations than CBD, either can serve as a model for national policy regarding plant IPRs. The real issue is which model Members will choose to follow and why, since they can implement a mandatory scheme at the national level, regardless of whether CBD is mandatory. It is beyond the scope of this chapter to determine whether UPOV or

[74] Cottier and Panizzon, "Legal Perspectives on Traditional Knowledge."

CBD will better serve the goal of food security in the varying national contexts. It may be that UPOV suits countries with large investments in biotechnology, through their multinational firms or public vehicles. It may be that CBD better suits mega-diverse countries with important stores of traditional knowledge. It may be that some other *sui generis* system better suits developing countries with large numbers of subsistence farmers and greater susceptibility to climate change. It may be that one size does not fit all. However, the focus of this chapter is on understanding the nature of the obligations in TRIPS Article 27.3 and the relationship between WTO law, UPOV, and CBD so that each country can make a more informed decision in light of its own circumstances.

Climate change, technological change, and economic change are converging to make current intellectual property laws, and current debates regarding IPRs and access to technology, obsolete. In this environment, making the right policy choices is increasingly difficult. The right of WTO Members to choose between patents, the UPOV Convention, or some other form of *sui generis* intellectual property protection for new plant varieties provides the flexibility that is necessary to adapt policies as climate change alters the environment. Moreover, the flexibility inherent in Article 27.3(b) facilitates avoiding conflicts between TRIPS obligations and the obligations contained in other instruments, such as the UPOV Convention and the CBD. Thus, the impasse in the negotiations under Article 27.3(b) may prove to be a good thing. Indeed, the impasse itself may be a sign that reducing existing flexibilities through the adoption of a new multilateral regime is not the best course of action.

6

Economics of Climate Change

6.1 Introduction

Economics is helpful in different aspects of the climate change debate. First, it helps us to formalize the idea and the potential solutions by formulating climate change as a market failure. Second, economic ideas can be used to address climate protection by examining costs and benefits. Third, economic ideas help us to identify a set of mitigation and adaptation strategies. Fourth, economics is critical in designing climate change mitigation and adaptation policy instruments such as taxes and tradable permits. Fifth, it helps us rationalize the international architecture of climate policies. Finally, it helps us understand the political economy at work behind domestic and international climate policy negotiations.[1]

There are two separate paradigms for studying environmental issues: ecological and economic. The ecological paradigm, based on the science of ecology, stresses the health and survival of ecosystems. The economic paradigm emphasizes environmental economics—the application of economic theory to environmental issues—and emphasizes maximizing the welfare of humans (even if this means harming the environment). Contrary to popular belief, they are *not* incompatible. Bonds et al. show that ecological problems give rise to a reduction in biodiversity which in turn impacts per capita income.[2] Thus, ecological and economic concerns are intertwined.

This chapter begins with an examination of environmental economics. We then apply the concepts of environmental economics to the problem of climate change. In earlier chapters, we have made the case for using unilateral measures to address climate change and to create incentives for multilateral action. In this chapter, we highlight the fact that unilateral measures represent only a partial solution because they will not achieve the desired level of emissions reductions. However, they can still be used to create incentives for multilateral action, which is what is needed to effectively address the risks associated with climate change. In Chapter 2, we criticized the division of countries into developed and developing as an ineffective approach to addressing climate change. In this chapter, we critique this approach in

[1] These ideas were set forth by Jason Shogren in "Climate protection: What insight can economics offer?" in Anthony D. Owen and Nick Hanley (eds.), *Economics of Climate Change* (Routledge, London 2004).

[2] Matthew H. Bonds, Andrew P. Dobson, and Donald C. Keenan, "Disease Ecology, Biodiversity, and the Latitudinal Gradient in Income" (2012) 10 PLOS Biology e1001456.

economic terms, and further develop the idea of an index that would categorize countries more appropriately.

6.2 Environmental Economics and Climate Change

First we introduce the concept of externality. An economic transaction takes place through markets. Buyers (consumers) and sellers (producers) get together in the market. When a transaction takes place in a market, the buyer pays, and the seller receives, the price owed for the product. When producers undertake production, their production activities can have "side effects" on third parties who are not part of the market transaction between the consumer and the producer. In economic parlance, such side effects are called "externalities."

Consider the following example. A coal power plant generates electricity for a community. The community is the consumer and the plant is the producer. In the market for electricity, the community pays for the power it consumes. Coal power stations have rotating machines to convert the heat produced by burning coal into mechanical energy. The prime mover is a steam or gas turbine. There are byproducts released directly into the atmosphere or into river or lake water, or indirectly into the atmosphere using a cooling tower with river or lake water used as the cooling medium. The flue gas from the combustion of coal is discharged into the air. This gas contains carbon dioxide (CO2), water vapor, and more corrosive substances such as nitrogen oxides (NOx), sulfur oxides (SOx), mercury, and fly ash. While the price of electricity is paid for by the community, these other byproducts generate an externality that not only affects the neighboring community but also other communities with access to the river or to the air. These other communities bear the cost of the production of electricity indirectly, outside the market, without receiving any benefits. Therefore, from a societal point of view, the price paid for the electricity by the community typically does not take into account these emissions that cause harm through these toxic emissions.

This general principle is illustrated in Figure 6.1a. When consumers and producers are operating in the market alone, the producer considers only private marginal cost. In the context of the example of electricity production, this would mean that the producer does not take into account all the effluence such as NOx, SOx, mercury, and fly ash he produces in the atmosphere while producing electricity. We have assumed that the marginal cost of producing electricity goes up with higher volume of electricity production. It is well known that there are economies of scale in electricity production at low volumes of production. However, cost per unit eventually goes up.[3] This is represented by the upward slope of the marginal private cost line. The consumers in the aggregate are represented by the marginal (social and private) cost line—the demand curve. It is downward sloping as a higher price per unit would induce the consumers to use less electricity. The market equilibrium is reached at the

[3] Laurits R. Christensen and William H. Green, "Economies of Scale in U.S. Electric Power Generation" (1976) 84 Journal of Political Economy 655.

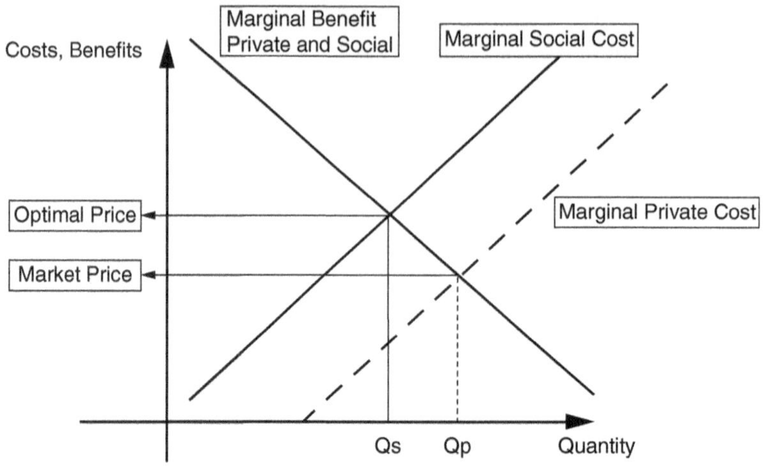

Figure 6.1a Cost–benefit analysis when social and private costs differ

"market price" when the equilibrium quantity of electricity produced with no government intervention in the market is Qp (quantity in the private market).

The additional cost of the effluence comes about through the impact of NOx, SOx, and other pollutants on the environment and human health. If, through government intervention, the social cost of the impact of the effluence is taken into account, the production would become more costly for the electric utility company. The "marginal social cost" line represents such an outcome: the cost of producing electricity at every level of output will go up. From a social point of view, taking the externality into account, the optimal output is Qs—socially optimal output. The price the consumers would pay in that case would be what is labeled as the "optimal price."

Accounting for all externalities, valuing them, and calculating the marginal social cost line is not a trivial matter. But if the uncertainty regarding the social cost through different channels is small, this calculation can be performed. Note that Qs represents a level of output at which there will be less effluence than the private equilibrium at Qp—but not zero effluence. Therefore, Qs represents *a compromise* between society's desire for electricity and society's desire for a clean environment (in this case, clean water and clean air). From an economic point of view, a tradeoff exists between goods output and an unpolluted environment. It is rare to produce electricity with zero pollution. Even extremely clean production of electricity through photovoltaic cells generates pollution during the production of those cells themselves. Therefore we must decide how much pollution to tolerate if we wish to have the benefits of electricity production. In doing so, we must measure the social costs of pollution and balance those costs against the social benefits of electricity production.

Such externalities lead to what economists call a market failure—the outcome generated by the market is not socially optimal. This example illustrates the case of a negative externality. There are examples of positive externalities as well. Economic theory offers solutions to such problems through internalization of externalities.

How would such internalization work? Suppose the government imposes a tax based on the output. The marginal social cost now becomes the private cost. The object of this tax is not just to penalize the electric utility company. It is to send an economic message: companies that create less pollution will pay less tax. Indirectly, it will create an incentive for the electric utility companies to generate electricity through cleaner technologies that produce less effluence.

Figure 6.1b illustrates how such an effluence tax would work. The socially optimal output is now reached through the market itself. Qs is the new market equilibrium with the tax. It is usually possible to find ways to produce electricity that produces less effluence. Electricity that can be produced with less effluence will be taxed less. In that case, the social optimum will be achieved not through any reduction of production, but by shifting to a different production technology. The tax would still achieve the purpose of creating an economic incentive to achieve a socially beneficial economic shift. Arthur Pigou proposed this solution to "internalize" the total costs of an activity into the market. In honor of Pigou, it is often called a Pigovian tax.

It is challenging to design a tax that will achieve a socially optimal output that the society desires. We cannot be sure that the tax will hit the optimal quantity target we are trying to achieve.

The goal is to arrive at the socially optimal level of effluence production (Qs, in our example). Determining the correct level of Pigovian tax to reduce the effluence to the socially optimal level can be very difficult. For example, if a rising level of affluence causes ecological problems in addition to economic problems (such as the disappearance of an endangered species), it could be difficult to put a price on the ecological impact. But, suppose, after all such economic and ecological considerations, we come up with an optimal level of production such that all the considerations are taken into account with our optimal level. So, in our example, we have

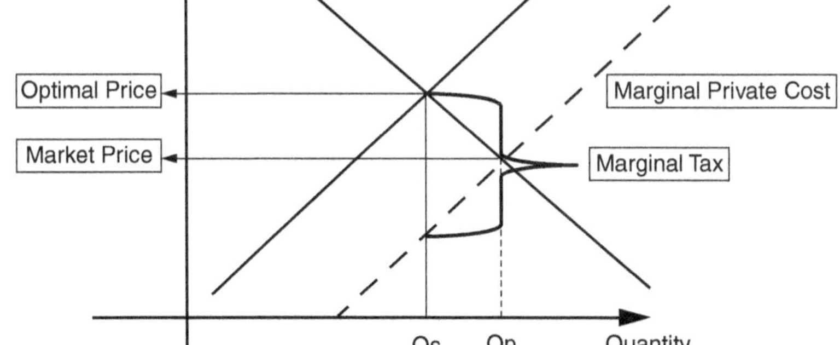

Figure 6.1b How a tax works to internalize an externality

determined that Qs is the level that would not wipe out the endangered species that we are concerned about. It is possible to design a market that will solve the optimal output problem as long as it is possible to monitor effluence production.

Suppose the government has determined that to achieve a socially optimal output of electricity Qs, the necessary effluence reduction is Es. The government then issues permits that allow all the firms together to produce a quantity of effluence of Es. The permits may be distributed to existing firms. They can be sold by auction to firms that are producing electricity for the community. This system will reach the same level of effluence as a command and control regulatory system. However, in a command and control regulatory system, every company has to comply with environmental regulation regardless of cost. Under the auctioning system, a company could choose to increase its level of effluence if it is able to buy credits from another firm. If a producer has a superior technology which produces very little effluence, it does not need as many permits to produce effluence. It can therefore sell the additional permits to companies that produce more effluence. The buyers of these permits can postpone costly controls until they can switch to a new technology that produces less effluence. In this system, the price of an effluence permit is determined by the effluence market through supply and demand. The government needs to monitor the activities of the companies in order to ensure compliance. It would be possible to tighten effluence standards over time by reducing the number of available permits according to revised goals each year. In this system, the role of the government is reduced to the necessary monitoring and sanctioning, but the role of the government is not eliminated. With command and control regulation, the government has to resort to micro management, which requires more government resources than a system based on permit trading.

The price of the permits will depend on two critical elements. All other things being equal, if fewer permits are available, the price of the permits will rise. On the other hand, if technological progress makes cutting effluence cheaper, then the price of the permits will fall.

In the case of a negative externality, the main problem that arises is often called the "tragedy of the commons." The main problem was illustrated by Garrett Hardin:

> The tragedy of the commons develops in this way. Picture a pasture open to all. It is to be expected that each herdsman will try to keep as many cattle as possible on the commons. Such an arrangement may work reasonably satisfactorily for centuries because tribal wars, poaching, and disease keep the numbers of both man and beast well below the carrying capacity of the land. Finally, however, comes the day of reckoning, that is, the day when the long-desired goal of social stability becomes a reality. At this point, the inherent logic of the commons remorselessly generates tragedy. As a rational being, each herdsman seeks to maximize his gain. Explicitly or implicitly, more or less consciously, he asks, "What is the utility to me of adding one more animal to my herd?" This utility has one negative and one positive component. 1) The positive component is a function of the increment of one animal. Since the herdsman receives all the proceeds from the sale of the additional animal, the positive utility is nearly +1. 2) The negative component is a function of the additional overgrazing created by one more animal. Since, however, the effects of overgrazing are

shared by all the herdsmen, the negative utility for any particular decision-making herdsman is only a fraction of –1. Adding together the component partial utilities, the rational herdsman concludes that the only sensible course for him to pursue is to add another animal to his herd. And another; and another. . . . But this is the conclusion reached by each and every rational herdsman sharing a commons. Therein is the tragedy. Each man is locked into a system that compels him to increase his herd without limit—in a world that is limited. Ruin is the destination toward which all men rush, each pursuing his own best interest in a society that believes in the freedom of the commons.[4]

In summary, it is easy to see the relevance of negative externality and the tragedy of the commons to the context of climate change. The emissions of greenhouse gases (principally CO2, N2O, SO2, and CH4) caused by human activities contribute substantially to the rising temperature in the long run around the globe. This process, in turn, contributes to climate change. Climate change does not necessarily mean simply a rise in the temperature. It also means a rise in the variability of the temperature. These effects produce more frequent extreme weather-related events, such as drought, floods, hurricanes, and tornados, and contribute to rising sea levels. All of these weather-related events have an impact on the existing ecological systems across the globe. These effects, in turn, will give rise to extreme poverty, migration, and other disruptive forces. If there were no short-term gains from these effects for any country, then it would be easy to get all countries around the table to find a solution. Unfortunately, the tragedy of the commons works with a vengeance in the short run. If a country can get away with adding to the greenhouse gases, which would give them business or economic growth advantage in the short run, they have every incentive to pursue that path. Like the herdsmen in the story above, countries would be willing to add to greenhouse gases for short-term gains, in spite of their long-term effects.

In economic terms, climate change is an externality with the following characteristics.

First, it is global. This feature gives rise to the tragedy of the commons problem. One country or one region cannot reduce the greenhouse gas emissions of other regions or countries. There is no simple way to deal with that problem because it is not possible to ban greenhouse gases from traveling from one part of the globe to another. It may be possible to create incentives for other countries, for example with unilateral trade restrictions or foreign aid that is conditional upon emissions reductions. However, as we saw in Chapter 3, trade restrictions may violate WTO law. Moreover, trade barriers do not just impose costs on the exporting country. They also impose costs on the importing country, where importers are affected by the increased cost of inputs. In addition, few domestic markets are large enough for trade barriers to have an economic impact that would be sufficient to create an adequate incentive to reduce emissions.[5] In the case of foreign aid, as we will see in Chapter 8, foreign aid that is conditioned upon the use of inputs from the donor

[4] Garrett Hardin, "The Tragedy of the Commons" (1968) 162 Science 1243.

[5] Bradly J. Condon, *Environmental Sovereignty and the WTO: Trade Sanctions and International Law* (Transnational Publishers, Ardsley NY 2006) 224–9.

country also may violate WTO law. Moreover, foreign aid costs money for donor countries, too. Even if we resolve the problems of cost and WTO consistency of trade barriers and foreign aid, they remain partial solutions only because they will not achieve the desired level of emissions reductions.

Second, there is substantial uncertainty about the impact of greenhouse gas emissions. We do know that human activities are contributing to the global stock of greenhouse gases. We can estimate that. However, the link between rising greenhouse gases and rising temperature under natural conditions is uncertain. Knutti and Hegerl reviewed the nature of this uncertainty in great detail.[6] Their results are summarized in Figure 6.2.

Figure 6.2 shows the following. Consider the top part of the graph. Suppose the carbon dioxide equivalent (CO2e) level rises to 400 parts per million (ppm) in the atmosphere. The consequence will be a rise in the temperature (relative to the pre-industrial level). However, the exact rise cannot be predicted with precision. What we can say is that the temperature will rise between 0.5°C and 5.0°C. However, the chance of an increase in that range of 0.5°C and 5.0°C is not uniform. The consequence of a 400 ppm rise in CO2e will have a 90 percent chance of a temperature rise between 1°C and 3°C. Similarly, suppose the CO2e level rises to 650 ppm in the atmosphere. Then the temperature will rise between 1.5°C and 6°C. The probability that it will rise between 2°C and 6°C will now be 90 percent. How likely are these increases in CO2e? Upton gives us some idea.[7] Suppose the

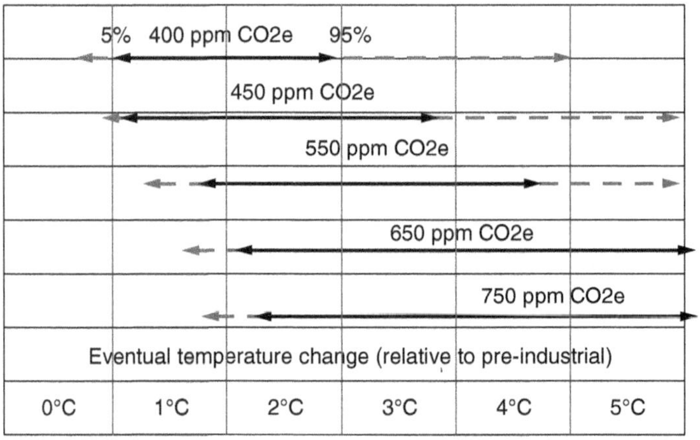

Figure 6.2 Distribution of rising temperature as a consequence of rising CO2 levels

[6] Reto Knutti and Gabriele C. Hegerl, "The Equilibrium Sensitivity of the Earth's Temperature to Radiation Changes" (2008) 1 Nature Geoscience 735 <http://www.nature.com/ngeo/journal/v1/n11/abs/ngeo337.html> (accessed March 15, 2013).

[7] Simon Upton, "Environmental Outlook to 2050" Online slide presentation (June 27, 2012) slide 8 <http://www.slideshare.net/EPAIreland/speaker-6-simon-upton> (accessed March 15, 2013).

so-called "business as usual" policy continues, where no action is taken to reduce the emissions. Then, by 2050, the atmosphere will have between 500 and 600 ppm of CO2e, and by 2060 the atmosphere will have between 600 and 700 ppm of CO2e. Thus, we can estimate the future concentrations of greenhouse gases and we can estimate the probability of a range of temperature increases, but we cannot say precisely what the temperature increase will be, how that temperature increase will vary from one part of the planet to another, and what the ecological and economic effects will be.

Third, the problems of climate change are long-term problems. John Maynard Keynes once said: "In the long run we are all dead."[8] In particular, the effects of more CO2e in the atmosphere do not show up immediately. They take decades to appear. There may be feedback effects. Specifically, small changes in temperature can lead to bigger changes in temperature in the future through other mechanisms such as rise in the sea level. The dire consequences of rising greenhouse gases in the early parts of the twenty-first century may well happen in the early parts of the twenty-second century. Most humans alive now will be dead by then. It is therefore difficult to motivate today's citizens and voters to take action that will be costly for them, when the benefits of averting *possible* disasters are far into the future.

Fourth, the damages caused by climate change are potentially large and irreversible. For example, small island countries such as Kiribati, Maldives, and Tuvalu are most at risk to be completely under water within a few decades. Other countries such as Barbados or Bangladesh will not be completely under water but they will have to resettle a large number of people within their own borders away from the densely populated coastal areas. Thus, while we cannot predict with absolute certainty how and when these damages will occur, the risks are great.

The 2013 *Draft Climate Assessment Report* encapsulates the nature of this uncertainty as follows:

> A significant issue in studying and preparing for global climate change is the fact that changes in human, social, and physical systems do not always occur gradually. Same changes may occur in a relatively predictable way, while others involve unexpected break-points or thresholds beyond which there are irreversible changes or changes of higher magnitudes than expected based on previous experience. These "tipping points" are very hard to predict, as there are many uncertainties associated with understanding future conditions. These uncertainties come from a number of sources, including insufficient data associated with low probability/high consequence events, models that are not yet able to represent the interactions of multiple stresses, incomplete understanding of physical climate mechanisms related to tipping points, and a multitude of issues associated with human behavior, risk management, and decision-making.[9]

[8] John M. Keynes, *A Tract on Monetary Reform* (MacMillan Publishers, London 1924) ch. 3, 92 (accessed December 30, 2012). The context of the quote is somewhat different. Keynes was discussing current affairs back in the 1920s: "The long run is a misleading guide to current affairs. In the long run we are all dead. Economists set themselves too easy, too useless a task if in tempestuous seasons they can only tell us that when the storm is past the ocean is flat again."

[9] National Climate Assessment Development Advisory Committee, *Draft Climate Assessment Report* (January 11, 2013) 13 <http://ncadac.globalchange.gov/> (accessed January 12, 2013).

Many developing countries are especially vulnerable to climate change. They will face rising water in coastal regions and thus need to relocate people. For example, Bangladesh will need to move at least 20 million people if the sea level rises by just three feet. A large number of rural poor will suffer due to falling agricultural yields that will lead to falling incomes of the marginal farmers and rural agricultural workers. Rising disruptive rainfall will lead to more malnutrition and disease. Migration and conflict will rise as a consequence of climate change.[10]

Developed countries are not immune either. Regions that experience high levels of water shortages today will experience worse shortages in the decades to come. In particular, such problems will become more pronounced in Southern Europe and in the South Western United States (such as Arizona, California, Nevada, and Texas). Sea level rises could force the relocation of people from populous cities such as New York, Osaka, and Tokyo. Insurance against hurricanes, floods, and drought will become more expensive and, in some cases, nonexistent. We analyze insurance issues in Chapter 7.

The foregoing analysis points to the importance of internalizing the externalities associated with greenhouse gas emissions. However, these negative externalities are not being internalized in the price of goods and services. Indeed, as we will see in Chapter 8, the opposite is occurring. Countries around the world, both developed and developing, spend far more money subsidizing the production and consumption of fossil fuels than the amount of money developed countries have promised to mobilize by 2020 to combat climate change.

6.3 Calculating Effects of Climate Change with Macroeconomic Models

There are two types of methods for calculating the economic effects of climate change. The first can be called an "enumerative approach."[11] In this approach, estimates of the "physical effects" of climate change are obtained from natural sciences based on some climate models, impact models, and laboratory experiments. Each physical impact is then assigned a price and added up. For agricultural products, results from models are used to predict the effect of climate on crop yield. Then, market prices are used to value the change in output as a consequence of climate change. The effect of the rise in sea level is calculated as the additional cost of coastal protection and land lost. These estimates are taken from the engineering studies. For goods and services that are not sold in the markets, such as the impact on health, different methods are needed. In this method, epidemiological results are

[10] Rafael Reuveny, "Climate Change-Induced Migration and Violent Conflict" (2007) 26 Political Geography 656.

[11] This name was first suggested by W.R. Cline, "The costs and benefits of greenhouse gas abatement. A guide to policy analysis" in OECD (ed.), *The Economics of Climate Change* (OECD, Paris 1994).

used to calculate the aggregate impact by simply assigning some cost to additional lives lost or damage done to human bodies.

The second method can be called a "statistical approach."[12] The economic impact of climate change is calculated based on estimates using observed variations. The observations are typically made within a given country or region over a span of time. This impact is then extrapolated to examine what would happen in other countries or regions if and when similar events take place. Thus, physical modeling of the underlying process of the economic phenomenon is completely eliminated.

Each calculation method has its own advantages and disadvantages. The enumeration method is based on actual physical phenomena. Thus, they are realistic. However, when they are extrapolated to the future on a much larger canvas, the errors get magnified. Perhaps more importantly, they do not take into account changes in human behavior. Changes in climate conditions will induce policy responses. The model does not take into account such changes. They are mechanical models.

The statistical models avoid such traps. These models are based on actual differences observed in the real world. The problem arises with these models because of biases arising from omitted variables. Specifically, if some changes in the relevant variables take place, such as income, almost always there are other variables that we have not measured that would have an impact on the income level, not just the climate variable. In addition, these models often rely on cross section data that do not exhibit large variation. On the other hand, the questions for which we want answers are spread out over decades or centuries. Hence, the changes in climate variables are much larger. Once again, if our statistical models have errors, they simply get amplified.

6.3.1 The Nordhaus model

William Nordhaus has been at the forefront of modeling the economic impact of CO2 emissions. In 1979, he devoted an entire chapter to this topic in his Cowles Commission Monograph.[13] In a chapter entitled, "Strategies for the Control of Carbon Dioxide," he expounded a method that would model the economic impact of different levels of CO2 production. He also calculated the (shadow) price of carbon for controlling CO2 emissions at different levels. However, his model was not yet a fully fledged economic growth model, which he later developed.

Nordhaus developed the Dynamic Integrated Climate Economy (DICE) model, which takes into account CO2 in a full general equilibrium economic model (see Appendix to this chapter for a full description of the model). In the usual economic model, the economic agent solves an intertemporal problem of maximizing a utility function u that depends on her consumption every year of her life

[12] This nomenclature is owed to Richard S.J. Tol, "The Economic Effects of Climate Change" (2009) 23 Journal of Economic Perspectives 29.

[13] William Nordhaus, *The Efficient Use of Energy Resources* (Cowles Foundation for Research in Economics, Yale University Press, New Haven 1979).

subject to certain lifetime constraints such as income during her working life. Formally, the utility function depends on consumption c at every period: $u(c_t)$ where t is a time period index. However, future utility is discounted at some rate (ρ). Nordhaus introduced an additional environmental factor T (i.e. temperature) so that the utility function at every period depends on two components: c and T. Thus, the utility function at any time t is written as $u(c_t, T)$. While higher consumption gives higher utility, a higher value of T *reduces* the utility.[14] In addition, there is a production function for the economy. This function, in the usual economics literature, depends on labor and capital only.[15] Formally, the output Y in the economy depends on two factors: capital K and labor L ($Y = F(K, L)$). The output increases when either capital or labor increases. The capital comes from savings that are not consumed in any period. In its simplest form: $K = Y - C$. This process links the utility to the production function in the model. In the Nordhaus model, the output $Y = F(K, L, T)$. The output now depends on an additional environmental factor T. Moreover, any increase in T leads to a decrease in output. The DICE model has been used in many studies since Nordhaus introduced it. Nordhaus himself introduced more detailed and disaggregated versions of it.[16]

In the latest version of the model, called the Regional Integrated model of Climate and the Economy (RICE) model,[17] Nordhaus considers four different scenarios:

(1) Baseline: No climate-change policies.

(2) Optimal: Climate-change policies maximize economic welfare with full participation starting in 2010 and no climatic constraints.

(3) Limit temperature to 2°C: The optimal policies are taken subject to a further constraint that global temperature would not increase more than 2°C above the 1900 average.

(4) Copenhagen Accord: High-income countries implement deep emissions reductions (similar to those included in the current US proposals), with developing countries following in the next two to five decades.

(5) Copenhagen Accord with only rich countries: This policy assumes that high income countries implement deep reductions as in case (4), but developing countries do not participate (as in the current Kyoto Protocol).

Nordhaus concludes as follows:

> Using a discount rate of 5 percent, the optimal program raises the present value of world income by USD 8.05 trillion, or 0.35 percent of discounted income. This is the equivalent

[14] William Nordhaus, "An Optimal Transition Path for Controlling Greenhouse Gases" (1992) 258 Science 1315.

[15] See e.g. Robert Solow, *Growth Theory: An Exposition* (Oxford University Press, Oxford 1988).

[16] William Nordhaus, *A Question of Balance: Weighing the Options on Global Warming Policies* (Yale University Press, New Haven 2008).

[17] William Nordhaus, "Economic Aspects of Global Warming in a Post-Copenhagen Environment" (2010) 107(26) Proceedings of the National Academy of Sciences 11721 <http://www.pnas.org/content/107/26/11721.full> (accessed March 15, 2013).

to an annuity of USD 402 billion per year at a 5 percent discount rate. Imposing the 2°C temperature constraint is quite costly, reducing the net benefit by almost half, because of the difficulty of attaining the 2°C target. The Copenhagen Accord with phased in participation of developing countries has substantial net benefits, but lack of participation reduces these substantially.

He goes on to suggest that charging a carbon price to USD 100 per ton upfront, as suggested by some,[18] is counterproductive. It should be raised gradually over the decades:

The estimated 2°C-limiting carbon price for 2010 is estimated to be USD 64 per ton carbon (2005 prices), whereas the effective globally average carbon price today is around USD 5 per ton.

Nordhaus shows net costs and benefits to different countries and regions for complying with the Copenhagen Accord (2009). He demonstrates that large developing countries like India and China will benefit from it but there will be a net cost to the developed world.[19] That does not mean all developed countries will incur a net cost. Canada, for example, could expand landmass under agricultural production due to warmer months. The effect could be a net gain from climate change.[20]

6.3.2 The Stern model

The British economist Nicholas Stern was asked by the British Treasurer to do a "review" of the impact of climate change on the economy and what can be done to mitigate the problem in economic terms. Stern produced a report that amounted to "a call to action."[21] It argued that, by incurring modest costs today (in the order of one to two percent of the gross world product), it is possible to reduce future high cost if no action is taken at all (in the order of ten to twenty percent of the world gross product). After reviewing the evidence, the Stern Report concluded that, given the 2005 rate of emissions, it is impossible to stabilize carbon emissions at a level any lower than 450 ppm. On the other hand, any level of carbon emissions above 550 ppm will be catastrophic.

The basis for calculating the impact of climate change in the Stern Report is the PAGE2002 model.[22] The model itself is not all that different from the DICE/RICE models used by Nordhaus. However, the difference in the outcomes and the consequent recommendations arise from the way the future is "discounted" in the

18 For example, in the documentary produced by Al Gore, *An Inconvenient Truth* (2006).

19 Nordhaus, "Economic Aspects of Global Warming."

20 Afshin Amiraslany, *The Impact of Climate Change on Canadian Agriculture: A Ricardian Approach*, PhD thesis, University of Saskatchewan, 2010.

21 Kenneth Arrow, "Global Climate Change: A Challenge to Policy" Economists' Voice, June 2007.

22 PAGE2002 is a Microsoft Excel-based model to be used with the risk management program called @RISK as an add-in. The model calculates the impact at regional and global level of climate change. It also has a component to calculate social costs of different greenhouse gases. It calculates the costs of abatement and adaptation. It belongs to the class of Integrated Assessment Models.

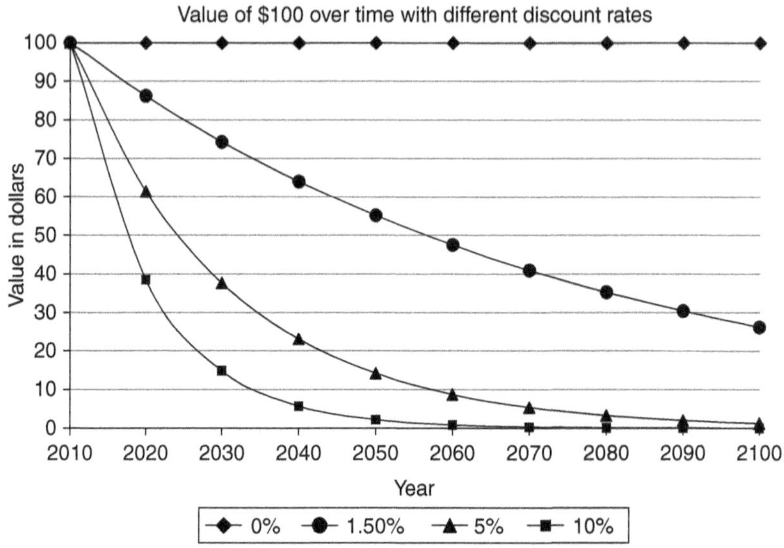

Figure 6.3 The impact of discounting

Stern Report. Let us consider the simple question of how discounting the future matters with a graphical illustration.

Suppose we start with USD 100 in different decades starting with 2010. How much would USD 100 in 2030 be worth in 2010? The answer depends critically on the discount rate used to bring that USD 100 in 2030 to 2010. Figure 6.3 illustrates this point. If we do not discount the future at all (that is, if our discount rate is 0 percent), then (and only then) USD 100 in 2100 has the same value in 2010. However, with a discount rate of 1.5 percent per year, the *present value* in 2010 of USD 100 at future dates keeps falling. For example, USD 100 in 2060 is worth less than USD 48 in 2010 at that discount rate of 1.5 percent. At a discount rate of 10 percent per year, the value of USD 100 in 2020 is less than USD 39 in 2010. With discount rates of 10 and 5 percent, the value of USD 100 in 2090 is virtually zero in 2010. Even with a low discount rate of 1.5 percent per year, in 80 years, almost 70 percent of the value is lost. This exponential decay is a characteristic of compounding.

The discount rate for consumption, ρ, can be expressed by the simple formula, $\rho = \delta + g\eta$, where δ is the social rate of time preference, g is the projected growth rate of average consumption, and η is the elasticity of the "social weight" attributed to a change in consumption. The parameter η is there to account for the following: as consumption grows, the marginal unit of consumption may be considered as having less *social* value. It is similar to diminishing marginal private utility of private consumption. The appropriate value to assign to η is not without controversy, but a value in the range of 2 or 3 seems reasonable. The Stern Report puts the value of η at 1,[23] but this level does not seem compatible with other evidence.

[23] Nicholas Stern, *The Economics of Climate Change* (Cambridge University Press, Cambridge 2007).

There is a big disagreement surrounding the appropriate value for δ, the social rate of time preference. This parameter allows for discounting the future simply because it is the future, even if future generations were no better off than we are. Therefore, the Stern Report adopts a value of *zero* for δ—the social rate of time preference. The reason is that Stern did not want to treat the welfare of future generations with any less weight than the current generation. This approach has been underpinned by theoretical argument in terms of generational equity questions.[24] It is also consistent with the principle of sustainable development. Thus, in the Stern Report, the value of ρ ends up being small, making discounting of future values low relative to the Nordhaus model. Therefore, it is not surprising that the Stern Report advocates strong action now.

6.3.3 The Weitzman model

While uncertainty and risk are discussed in both the Nordhaus and Stern models, they take the form of certainty equivalence. Essentially, the way to model risk with a certainty equivalence approach is to add more "cost" to account for more risk. For example, in a model of a project evaluation, if a project has no risk compared with another risk, the investor in the project with risk will demand a higher rate of return on investment. However, the essence of the model does not change. The relationship between the temperature change and CO2 rise is not without uncertainty. This uncertainty is illustrated in Figure 6.4.

Figure 6.4 shows the following. It shows that a 3°C temperature rise and a 550 ppm CO2 level happen with a probability of 0.99 (99 percent), a rise of 5°C and 650 ppm CO2 level happen with a probability of 0.58 (58 percent), and so on. Thus, there is a great deal of uncertainty associated with any change in temperature that corresponds to a given level of CO2 concentration.

Martin Weitzman proposed the following theoretical argument when there is uncertainty about different future states of the world: if catastrophic events have higher probabilities than standard exponentially decaying probability distributions, then the standard economic models (of calculating costs and benefits) cannot be applied.[25] This result has become known as the "Weitzman's Dismal Theorem." This is the consequence of the so-called fat right tail of a probability distribution and catastrophic losses. Weitzman's own summary of the result and its relevance for climate change was:

> The burden of proof in climate-change cost-benefit analysis (CBA) is presumptively upon whoever calculates expected discounted utilities without considering that structural uncertainty might matter more than discounting or pure risk. Such a middle-of-the-distribution modeler should be prepared to explain why the bad fat tail of the posterior-predictive

[24] John Quiggin, "Equity between Overlapping Generations" (2012) 14 Journal of Public Economic Theory 273.

[25] Martin Weitzman, "On Modeling and Interpreting the Economics of Catastrophic Climate Change" (2009) 91 Review of Economics and Statistics 1.

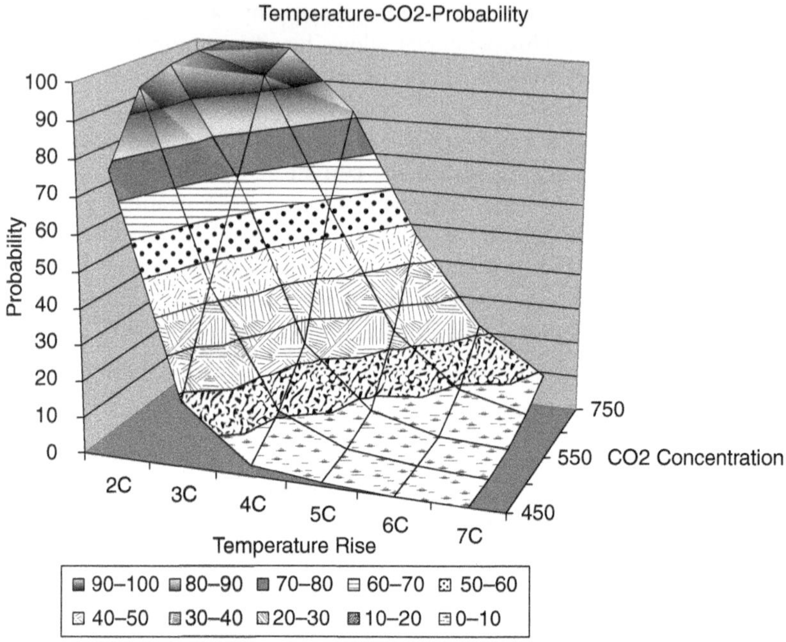

Figure 6.4 Uncertainty in the relationship between rise of CO2 and rise of temperature

probability distribution function is not empirically relevant and does not play a very significant—perhaps even decisive—role in climate-change CBA.[26]

To understand the significance of events of large magnitudes, we can consider the example of the distributions of large earthquakes. It is well known that the magnitude of large earthquakes has a fat tail (a power law distribution). Thus, there were four earthquakes in recorded history that were of the magnitude equal to or greater than the 2011 Japanese earthquake that resulted in a devastating tsunami. With a probability distribution with a fat tail, such events are not that rare. If we were to apply the same event and pretend that the distribution is normal, then it would be equivalent to running into a person 30 feet tall.[27] Weitzman's approach has been criticized by some economists—notably by Nordhaus.[28] More recently, a new approach has been suggested to reconcile Nordhaus's model with Weitzman's.[29]

[26] Weitzman, "On Modeling and Interpreting the Economics."

[27] William Nordhaus, "Economic Policy in the Face of Severe Tail Events" (2012) 14 Journal of Public Economic Theory 197.

[28] Nordhaus, "Economic Policy in the Face of Severe Tail Events."

[29] Masako Ikefuji et al., "Weitzman meets Nordhaus: Expected Utility and Catastrophic Risk in a Stochastic Economy-Climate Model" Discussion Paper No. 825, The Institute of Social and Economic Research Osaka University, Osaka, Japan (December 2011) <http://econpapers.repec.org/paper/dprwpaper/0825.htm> (accessed March 15, 2013).

6.3.4 The Llavador–Roemer–Silvestre North–South model

The Llavador–Roemer–Silvestre study is the only one that directly tackles the issue of climate change as a North–South problem.[30] In their model, they simulate North by the current situation in the US, and the South by the current situation in China. They assume that the goal of this stylized world is to stabilize emissions at 450 ppm of CO2. They assume that in this world there will be a convergence of North and South in terms of sustainable growth (to keep the CO2 from rising above 450 ppm of CO2) at the rate of 1 percent gross domestic product (GDP) growth per year. In order to achieve that, they find that the transition paths require a drastic reduction of the share of emissions in the North, large investments in knowledge in both the North and the South, as well as very large investments in education in the South. To sustain the economic growth rate in the North, some resources must be transferred from South to North during the transition. Thus, there is a tradeoff during the transition: the North cuts emissions but requires resources *from* the South. Thus, the standard argument that developed countries have to reduce CO2 emission *and* give resources to the developing countries is not possible in terms of a resource allocating growth maximizing strategy of the world as a whole.

6.3.5 Critiques of macroeconomic models of climate change

The macroeconomic models that we have elaborated on (such as the Nordhaus or Stern or Weitzman models) rely heavily on a mechanical model where individuals do not modify their behavior based on experience or learning. However, we know that people do not behave the same way under all conditions. The impact of climate change is very different depending on the level of development of a country. Dell, Olken, and Jones (2010)[31] show the following:

(1) Higher temperatures substantially reduce economic growth in poor countries (countries with less than the median per capita income in the world) but have no effect in OECD countries. There is negligible impact from changing precipitation.
(2) Higher temperatures reduce growth rates in poor countries, rather than just the level of output.
(3) Higher temperatures affect the poor countries in a number of different dimensions by reducing agricultural output, industrial output, and aggregate investment. In addition, higher temperatures also increase political instability.

[30] Humberto Llavador, John E. Roemer, and Joaquim Silvestre, "North–South Convergence and the Allocation of CO2 Emissions" Paper Presented at the European Economic Association & Econometric Society, 2011 Parallel Meetings, Oslo, Norway, August 26, 2011 <http://www.eea-esem.com/EEA-ESEM/2011/m/viewpaper.asp?pid=432> (accessed March 15, 2013).

[31] Melissa Dell, Benjamin F. Jones, Benjamin A. Olken, "Temperature and Income: Reconciling New Cross-Sectional and Panel Estimates" (2009) 99(2) American Economic Review 198.

In the context of environmental economics, the inequality within a country looms larger.[32] Standard neoclassical models of economic growth (maximizing intertemporal utility function subject to intertemporal budget constraints) may not be suitable for studying inequality within a country. Thus, it requires a different type of modeling.[33]

There is a huge difference between most developed and most developing countries in terms of the importance of the agricultural sector. For most *developing* countries, agriculture still constitutes a large sector. Thus, any impact on agriculture will affect a large section of the population. For example, in India, half the population is engaged in the agricultural sector. Therefore, climate change impacts on agriculture will affect half of India directly. On the other hand, in a country like the US, where less than 2 percent of the population directly relies on agriculture, any negative impact of climate change on agriculture will be far less disruptive. The energy sector will be affected in all countries. So will the housing market. All of them will add up to macroeconomic effects of significant magnitude.

The needs of small island nations such as Vanuatu or Maldives are much more immediate. These island nations could be wiped off the map with a relatively small rise in the sea level. On the other hand, other nations such as Bangladesh will be hit by the rise of frequency and severity of flooding (either through the rise of the sea level or by the change in the pattern of rainfall across the country).

Jason Sampson[34] has designed a climate disaster vulnerability index (CDVI). He created a cylindrical map of the world (rather than Eckert type IV or type VI). It shows that those who are likely to be the most vulnerable to climate change are populations living in low-latitude, hot regions of the world, like central South America, the Arabian Peninsula, and much of Africa (all developing countries), and relatively less so in India and China—the two most populous nations on the planet.

6.4 Climate Sensitivity Index of Countries

As we have argued in Chapter 2, dividing the world into developed and developing countries, and using that categorization as a basis for addressing climate change, is too simplistic. The effects of climate change will not be distributed on this basis, the financial and technological endowments of countries are not distributed on this basis, and neither is static. It makes more sense to categorize countries on a scale, based on objective criteria that are in accordance with the economic impacts of climate change and the capacity to address the cost of mitigation and adaptation.

These criteria could be used to create an index of countries that evolves as conditions change, what we shall call the "Climate Sensitivity Index" (CSI). Our

[32] World Bank, *World Development Report 2006: Equity and Development* (World Bank, Washington 2006).

[33] Daron Acemoglu and Melissa Dell, "Productivity Differences Between and Within Countries" (2010) 2(1) American Economic Journal: Macroeconomics 169.

[34] Jason Samson et al., "Geographic Disparities and Moral Hazards in the Predicted Impacts of Climate Change on Human Populations" (2011) 20 Global Ecology and Biogeography 532.

proposal is to include (1) current per capita annual GHG emissions; (2) proportion of population who are subsistence farmers; (3) population of the country who are poor in absolute terms. Other criteria such as the per capita GDP, some measure of technological capacity, and historical GHG emissions can also be added.

The WTO recognizes as least-developed countries those countries that have been designated as such by the United Nations. There are currently 50 least-developed countries on the UN list. A country like Maldives does not qualify as a least-developed country as it graduated from the list in 2011. Nevertheless, 0.49 meters rise in the sea level will leave most of the islands of that country under water. In addition, a rise in sea level of 1.46 meters will submerge the country entirely.[35]

Least-developed countries are officially designated as such by the United Nations General Assembly on the basis of a number of criteria, including: low national income (per capita GDP under USD 900 for countries now joining the list); low levels of human development (a combined health, nutrition, and education index); and economic vulnerability (a composite index based on indicators of instability, inadequate diversification, and the handicap of small size). The population of countries that meet all the other criteria for admission into the category must not exceed 75 million inhabitants.

This definition provides an inadequate measure for countries that need help in the ambit of climate change. Consider the case of the Maldives. Using the foregoing criteria, the Maldives would not qualify as it "graduated" from the list of least-developed countries in 2011. Therefore, we propose that a different index be used (somewhat similar to the Human Development Index) to categorize countries, and to determine their need for help, on a country-by-country basis. Like the list of least-developed countries, the construction of the index would have to be revised on an annual basis so that a country does not stay in one category or another indefinitely.

It is important to emphasize that we are not proposing that this index be used regarding other matters such as eligibility for general foreign aid. Nor are we proposing that this index be used to justify the erosion of core nondiscrimination obligations in WTO law and international investment agreements. Rather, we propose that this index be used for the narrow purpose of assistance with climate change mitigation and adaptation. The current methods of defining developing and least-developed countries in international law are inadequate. The legal rights and obligations that flow from the concepts of common but differentiated responsibilities in international environmental law and special and differential treatment for developing countries under WTO law are far from clear. However, we are by no means suggesting this index as a general solution to these highly controversial issues.

Here, we discuss in some detail the following criteria: (1) current per capita annual GHG emissions; (2) proportion of population who are subsistence farmers; (3) per capita GDP in the bottom 20 percent of the population.

[35] James G. Titus, "Rising Sea Levels: The Impact They Pose" (1986) 12 EPA Journal 17, 18. See further Alasdair Edwards, "The Implications of the Sea-Level Rise for the Republic of Maldives" Report to the Commonwealth Expert Group on Climate Change and Sea Level Rise, June 1989 <http://www.islandvulnerability.org/slr1989/edwards1989.pdf> (accessed March 15, 2013).

6.4.1 Current per capita annual GHG emissions

The emissions have to be above some threshold rate that will be determined by a world body. Once a country is above that threshold rate, it will qualify for climate change technology transfer or other technical assistance. This criterion is aimed at mitigation.

Because this has to be done impartially on a global scale, we need to decide which world body should decide on the threshold. It could be done through the UNFCCC or the World Bank. As we will see in Chapter 8, the World Bank has begun to be more proactive in aligning its policies with climate change mitigation and adaptation, and it is well-placed to formulate policy advice on these issues.

Suppose the per capita GHG emissions in a country is denoted by r. Suppose also one has agreed upon a world threshold to be t(r). For example, the threshold rate could be 350 ppm CO2. The criterion used would be as follows. A country would qualify if its actual rate of emissions were larger than the agreed-upon threshold value (to use a formula, a country would qualify if $r > t(r)$ regardless of the economic development or any other criterion). A high rate of emissions would override every other criterion. The rationale for using this method is simple. It does not exclude countries that may not qualify because of otherwise mitigating factors such as a relatively high level of per capita income. How should the threshold be determined? This should depend on the importance of the threshold in question. The threshold should be determined by a level such that with a higher rate, the problem of climate change would rapidly escalate. For example, the forthcoming Fifth Assessment Report of the IPCC considers that a rise of CO2e beyond 650 ppm would be catastrophic.[36]

6.4.2 Proportion of a country's population who are subsistence farmers

One needs to include a country if there are a significant proportion of subsistence farmers. This criterion is aimed at adaptation. We will see in Chapter 8 that climate change is going to make farmers with less than 3 hectares of land particularly vulnerable because they will be hard-pressed to buy genetically modified seeds that would be able to withstand more variable weather as a consequence of climate change. This criterion requires a threshold value below which a country would qualify. Once again, the threshold would be determined by a world body. In this case, the most suitable organization would be the Food and Agricultural Organization, although the World Bank may play a role with respect to finance issues. If a country has a proportion of subsistence farmers above an agreed threshold, it would automatically qualify. Let us denote the proportion of subsistence farmers by pci and the corresponding threshold per capita income by t(pci). Expressed in symbols,

[36] See the Intergovernmental Panel on Climactic Change <http://www.ipcc.ch/> (accessed December 29, 2012).

the criterion is written as follows: If the proportion of subsistence farmers is above some threshold, pci > t(pci), then the country automatically qualifies.

6.4.3 Proportion of a country's population who are poor

We need to have a mechanism to take into account the poverty level in a country. One possible candidate could be the average level of income in the country. Unfortunately, the average income does not do the job because it papers over the inequality in income among the population.

There are different measures of inequality that are potential candidates. One commonly used measure is the Gini coefficient. The most common geometric definition of the Gini coefficient is based on the Lorenz (or concentration) curve. It represents cumulative income share as a function of cumulative population share. If a population share is always exactly equal to a share in overall income then there is a situation of perfect equality. However, the Gini coefficient is not affected by a multiplicative factor. If everyone's income increases ten-fold, the Gini coefficient is not affected. Technically, the Gini coefficient is a relative measure of inequality. To take a concrete example, Uganda and the United States have approximately the same Gini coefficient of income distribution. Both are around 38 percent. The lower the number, the more equal the income. Conversely, the higher the number is, the higher the inequality is going to be. Thus, a zero (percent) Gini coefficient means everybody in the economy has the same income. On the other hand, a 100 percent value Gini coefficient implies that one person has all the income in the country and everyone else has zero income. Of course, in real life, neither extreme is observable. In real life, it ranges from around 25 percent (for countries such as Belgium, Finland, and the Czech Republic) to over 60 percent (such as Brazil or Sierra Leone). Thus, the Gini coefficient would be meaningless as a measure of inequality that can be compared across countries at a given point in time.

A more appropriate measure of inequality is to include people who are poor in the country in absolute terms. One possibility would be to consider a threshold of some proportion of people who are below some absolute measure of poverty level. The rationale is simple. If there are many people below some absolute poverty level, they cannot afford to pay for adaptation or mitigation activities. A simple measure (available for most countries around the world) is the proportion of people in the country who live on USD 1 or less a day. So, the criterion would be the following: if the proportion of population (p) with USD 1 income exceeds some threshold t(p), then the country would automatically qualify. In symbols, if p > t(p), then a country automatically qualifies.

Thus, there are three possible criteria that could be used for determining the countries that qualify. The three are combined to arrive at a single criterion. If a country qualifies using any of the above threshold criteria, it should qualify. The following criterion, which includes all three measures using a compact notation, can be used: If maximum {r – t(r), pci – t(pci), p – t(p)} > 0, the country qualifies. This criterion ensures that: (1) if the emissions rate (r) is above the predetermined threshold (t(r)), then the country qualifies; (2) if the subsistence farmers in the population (pci) is above the predetermined threshold (t(pci)), then the country qualifies; and (3) if the proportion of population who are very poor (p) is above certain threshold (t(p)), then the country qualifies.

Although the measure above is useful, it is not entirely satisfactory. Suppose a country has all the above problems but it fails each threshold criterion by some amount and therefore fails to qualify. Clearly, we will need a method of "adding" each "score" to come up with an aggregate value that reflects the issue in all three dimensions. There are two ways of achieving this, which are discussed below.

Let max(r) be the country with the maximum emissions rate. Let max(pci) be the country with maximum proportion of subsistence farmers. Let max(p) be the country with the maximum proportion of people below USD 1 per day per capita income. We construct the following absolute index (Absolute Climate Sensitivity Index or Absolute CSI):

$$\text{Absolute CSI for a country} = [r/\max(r) + pci/\max(pci) + p/\max(p)]/3$$

The rationale for the formula is as follows. If a country hits the maximum emissions rate, maximum subsistence farmers level and maximum number of poor people in the pool of all countries, the CSI will hit a maximum of 1. We can set a predetermined value of the Absolute CSI such that any country with the value of the index above that level would qualify.

Since this measure will never hit zero, some people might consider this measure unsatisfactory. We can adjust that by considering a modified version that measures different dimensions in relative terms. Thus, we construct the Relative Climate Sensitivity Index:

$$\text{Relative CSI for a country} = [I(r) + I(pci) + I(p)]/3$$

where, $I(r) = [r - \min(r)]/[\max(r) - \min(r)]$, $I(pci) = [pci - \min(pci)]/[\max(pci) - \min(pci)]$ and $I(p) = [p - \min(p)]/[\max(p) - \min(p)]$.

To see why we take such ratios, consider the first one: I(r). If, for a given country, the emissions rate r is the highest among all countries, then the index I(r) = 1. On the other hand, if the emissions rate r is the lowest among all countries, I(r) = 0. Similarly, if the proportion of subsistence farmers (pci) is the highest among all countries, then the index I(pci) = 1. On the other hand, if the proportion of subsistence farmers pci is the lowest among all countries, then the index I(pci) = 0. If the USD 1 a day population p is the highest among all the countries in question, then I(p) = 1, whereas if it is the lowest, then I(p) = 0. Thus, the relative CSI is a measure bounded by 0 and 1 as two extremes. By construction, the relative CSI could touch the limits for the best (in that case, it will touch 0) and the worst (in that case, it will touch 1) case scenarios. It should be noted that the worst outcome country in terms of Absolute CS Index may not be the worst outcome country in terms of the Relative CS Index. Thus, it is quite probable that in the list of all countries we will never observe the extreme value 1 for the Relative CS Index.

Similarly the best outcome country in the RCSI measure may not be the best outcome country in the ACSI measure. Thus, we might not observe the extreme value 0 in a sample of countries.

With the Relative CSI (RCSI), the criterion should specify a threshold (t): if the RCSI > t, the country should qualify under the composite measure for the most favorable level of treatment available. To incorporate this measure into our overall criterion, we propose the following:

If maximum {r – t(r), pci – t(pci), p – t(p), RCSI - t} > 0, then a country should qualify for the most favorable level of treatment.

The use of our index facilitates policies for sustainable economic growth in these types of countries. That process should eventually push them over the threshold value of the index so that they no longer qualify. Without such measures, these countries will be caught in a vicious circle and, therefore, will never generate the level of purchasing power needed to create a market where sustainable development becomes possible.

The index allows us to redress this balance of incentives through a mechanism that uses objective standards recognized by international bodies that represent all interested parties. It can be used to determine which countries should receive multilateral climate financing for adaptation and mitigation in a manner that takes into account the changing circumstances of countries over time.

6.5 Conclusion

The term "global" has been overused. Unlike in many other contexts, the environment is truly a "global public good." No single nation can capture for itself a substantial part of the benefits from its own emissions reductions. Thus, it raises unprecedented problems of international coordination and institutional design that require a global solution. In this chapter, we have explored the concept of externality in the context of climate change. Then, we described various macroeconomic models that have been used to model such externality. The way forward is to implement market-based solutions to the problem of mitigation and adaptation to climate change. These solutions have come from different market experiments, such as cap and trade in the case of sulfur, which can be used to control the emission of CO2. On the other hand, insurance market solutions such as the use of catastrophic bonds can be used to help adapt to changing climate. We analyze these issues in the next chapter.

Appendix to Chapter 6

A social planner maximizes a utility function of the economy subject to production constraints.[37]

[37] The model in this appendix is taken from Samuel Fankhauser and Richard S.J. Tol, "On Climate Change and Economic Growth" (2005) 27 Resource and Energy Economics 1.

Maximize

$$\int_{0}^{\infty} u(c, T)e^{(n-\rho)t}\,dt$$

Subject to

$$\dot{K} = F(K, L, T) - cL - \delta T(K)$$

And

$$\dot{L} = nT(L) \text{ with } L_0 = 1$$

Here u denotes the utility function of the representative agent.

The utility function depends on two things: c is the per capita consumption of the agent and T—a climate variable. We assume that higher consumption gives higher utility and higher T gives lower utility. Formally, we assume:

$$\partial u(c, T)/\partial c \geq 0$$
$$\partial u(c, T)/dT \leq 0$$

The function F is the production function which depends on three factors, capital K, labor L, and the climate variable T.

K, the capital—depreciates at rate δ

The agent discounts future utility at the rate of ϱ.

L is labour supply growing at rate n, starting from an initial, normalized level of 1 (hence $L_0 = 1$); growth in labor supply is exogenous in this formulation and basically reflects changes in population.

This is the standard Ramsey Model of Economic Growth. The only twist to include climate change comes from the addition of a factor T (for temperature) that enters negatively in the utility function and the production function of the economy.

7

Climate Change and the Financial Industry

7.1 Introduction

There are two aspects of climate change: mitigation and adaptation. Financial and insurance markets can play a critical role in both. The principal contribution the financial industry can have to mitigation is through cap-and-trade markets. These markets can reduce the growth of new emissions, they can encourage better mitigation strategies for countries, and they can bring innovations to the targeted industries that are not related to the emissions themselves. The insurance industry can contribute to both mitigation and to adaptation. Insurance and reinsurance companies take a long-term view of the types of risks associated with climate change, can measure those risks, and can create incentives for their clients to mitigate and adapt to those risks.

In this chapter, we first analyze how the lessons learned from SO2 markets in the US can be applied to CO2 markets. Then, we will discuss two CO2 markets—one compulsory and one voluntary. Finally, we will discuss the role of insurance and reinsurance markets in the context of adaptation and mitigation strategies for climate change.

7.2 Mitigation Lessons from SO2 Market

The traditional approach to reduce pollution had been "command and control" regulation until the SO2 market came along. In the command and control method, a government determines a pollution target and decides on how much each polluting agent has to reduce pollution. This is usually done by setting a uniform emissions rate for a class of emitters (such as a fixed rate for electric utility company per ton of coal used) or by mandating a specific type of pollution-control equipment (such as a scrubber, regardless of the technology being used). This "one size fits all" approach ignores the heterogeneity of types of technologies that exist in the industry and, as a result, the cost of compliance would vary considerably across plants of different vintages.

Why did the Environmental Protection Agency (EPA) target the electric utilities for their SO2 emissions? Electric utilities accounted for about 70 percent of SO2 emissions in 1990. Coal-fired electric generation units accounted for 96 percent of this total, and oil-fired units accounted for the remainder. The other 30 percent of emissions were accounted for by a wide variety of industrial/commercial/residential boilers, smelters, paper mills, and other process oriented sources.[1]

The SO2 market operating in the US since 1995 became the first cap-and-trade market to operate successfully in the world. There are lessons to be learned from that for the CO2 market. Therefore, we take a closer look at the SO2 market.

In 1990, the Clean Air Act Amendments established the Title IV Acid Rain Program (ARP). The amendment (in the Title IV) of the Clean Air Act mandated requirements for the control of acid deposition—also known as acid rain.[2] The Clean Air Act Amendments of 1990 set a goal of reducing annual SO2 emissions by 10 million tons below 1980 levels. To achieve these reductions, the law required a two-phase tightening of the restrictions placed on fossil fuel-fired power plants.

How does one go about creating such markets?

First, to set the parameters, the legislators have to come together to pass certain laws. The political economy of such rule-making is messy. Any new legislation produces winners and losers. In the case of SO2 it was no different. There were states that produced coal and were net exporters to other states. There were states that produced and used coal for power generation but did not export either to other states. In addition, there were states that did neither but were affected by acid rain for being located in the neighborhood. West Virginia, a large coal producer, was a clear loser.[3]

Second, regulators have to set up regulatory parameters based on the laws enacted. They have to anticipate whether such regulations would produce perverse reactions, or produce unintended negative consequences somewhere else. Essentially, they have to produce regulation to properly internalize the externality we discussed in Chapter 6. Third, once the regulation is set forth, institutions have to be built for permit exchange. There has to be constant vigilance to prevent fraudulent activities. Additional legal institutions might be necessary for such new activities to take place.

Finally, a new legion of traders has to be trained and the buyers and the sellers of the permits have to be educated. At every stage, a number of lawyers will have to be involved to make sure all the transactions are admissible.

Phase I of the process began operating in 1995. It was targeted to 263 units at 110 coal-burning electric utility plants located in 21 Eastern and Midwestern states of the United States. An additional 182 units were added to Phase I of the program as substitution or compensating units, bringing the total of affected units to 445 in Phase I.

[1] Environmental Protection Agency, "National Air Pollution Emission Trends 1900–1998" (2000) <http://www.epa.gov/ttn/chief/trends/trends98/trends98.pdf> (accessed January 3, 2013).

[2] The following document sets out quantitative requirements of sulphur dioxide, nitrous and nitric oxides, and quantitative requirements, along with permit requirements a decade into the future: Clean Air Act, Title IV (Acid Deposition Control) <http://www.epa.gov/airmarkets/progsregs/arp/docs/title4.pdf> (accessed April 5, 2013).

[3] For an in-depth discussion of the political economy of SO2, see Paul L. Joskow and Richard Schmalensee, "The Political Economy of Market-Based Environmental Policy: The U.S. Acid Rain Program" (1998) 41 Journal of Law and Economics 37.

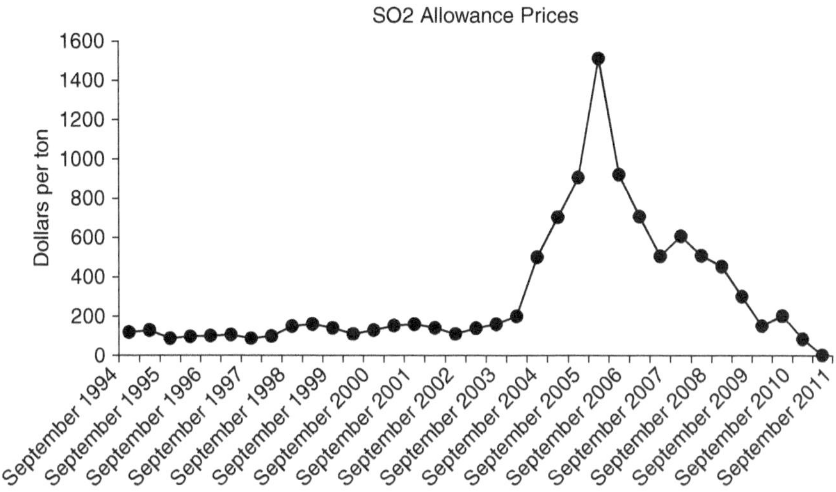

Figure 7.1 Prices of the traded permits in SO2 market 1994–2012
Source: Environmental Protection Agency

The idea of the market was simple. The EPA would auction permits (each representing one ton of pollution). Then, private companies would trade these permits in a specially created product at the Chicago Board of Trade. Once the permits were auctioned, they were resold in the private market. The price was determined by demand and supply not just in the spot market but also by the expected future value of each permit. Thus, expectation plays a vital role in the market for permits (see Figure 7.1 for the market prices).

In the first 12–18 month period, there were a handful of traders. The price was in the \$250–\$300 per ton range. By the end of 1994, the price had dropped below \$150 per ton and the volume of private trades exceeded the volume offered in the EPA auction. The prices had fallen to about \$100 per ton by 1998, and private trading of allowances for more than 5 million tons per year had eclipsed the EPA auction by a factor of 15 to 1. The costs of trading allowances were 2 percent of the prevailing price.

The prices of traded permits were very close to one another. For example, the spread between average bids and lowest winning bids at EPA auctions was 1–3 percent (of the price) and trading in the private market appeared to be similarly concentrated around a single price.[4] The program also spurred innovation in the technology of power plants. All the new plants installed scrubbers to reduce SO2 emissions.

[4] Paul L. Joskow, Richard Schmalensee, and Elizabeth M. Bailey, "The Market for Sulfur Dioxide Emissions" (1998) 88(4) American Economic Review 669.

The technology also had a spillover effect. It reduced other harmful emissions such as NOx, mercury, lead, and microscopic particles. Another innovation was in the mining techniques for extracting lower-sulfur coal seams. These were known methods but not in wide use. Another low-cost modification came from blending low-sulfur and high-sulfur coal. The industry standard did not allow large-scale blending for it was believed that the boilers would not function correctly. That belief turned out to be wrong. This allowed large-scale blending without costly modifications to the equipment. This new practice was propelled by the SO2 cap-and-trade program.

The ARP also generated innovations in the management practices of the utilities. It encouraged the utilities to seek to streamline the fluctuation in the input cost through activities in the futures markets for coal and oil. As a result, the utility companies were able to reduce their overall cost of production.

Phase II began in 2000 with more stringent annual emissions limits imposed on large, higher emitting plants. It also set restrictions on smaller, cleaner plants fired by coal, oil, and gas, encompassing over 2,000 units in total. The program incorporated existing utility units serving generators with an output capacity of greater than 25 megawatts and all new utility units.

In Table 7.1, we reproduce the winners of the permits auctioned by the EPA in 1995. There are names that we expect for such bids, such as Duke Power and Virginia Power. They are companies that were buying permits for their own use. However, there are names of several law schools and law societies, and the Pollution Retirement Center. They were not buying these permits for profit or speculation. They were participating in the market for the explicit purpose of reducing the number of permits that other market players could have so that in the aggregate there would be less pollution. In the end, they did not make much of a difference in the market as they were very minor players. They did not affect the market price or quantity in any significant way. However, this practice highlights the possibility that in other circumstance, other entities can enter the market and change the market dynamics.

In 2003, following the success of SO2 trading, the NOx Budget Trading Program (NBP) started in nine states. The NBP was a cap-and-trade program that required emissions reductions from power plants and industrial plants in the Eastern US during the summer months. The program was to last until 2008. Meanwhile, the Bush Administration tried but failed to tighten the SO2 emissions through the Clear Skies Act. It died at the committee level in the Congress.

The Bush Administration then came up with the Clean Air Interstate Rule (CAIR) in 2005. The purpose was to lower the SO2 emissions to a level 70 percent below the 2003 level. The CAIR tried to achieve this by reducing the cap by two-thirds in some of the states that were not part of the original ARP—it intended to include 28 Eastern States plus the District of Columbia by replacing the entire ARP with CAIR. The target was interstate transport of pollution from upwind states to downwind states.[5] This action had a clear impact on the SO2 price in the market. It

[5] Testimony of Brian McLean, Director, Office of Atmospheric Programs Office of Air And Radiation US Environmental Protection Agency before the Committee on Environment and Public Works Subcommittee on Clean Air and Nuclear Safety United States Senate (July 29, 2008) <http://www.epa.gov/airmarkets/progsregs/cair/docs/testimony.pdf> (accessed January 11, 2013).

Table 7.1 Winning bidders for the 1995 SO2 permits

Bidder's Name	Quantity	Percent of Total	Cost
Duke Power Company	17,750	35.1%	USD 2,332,500
PECO Energy Company	8,000	15.8%	USD 1,061,000
Cantor Fitzgerald Brokerage, L.P.	8,000	15.8%	USD 1,058,000
Virginia Power	6,000	11.9%	USD 800,000
Canterbury Coal Company	4,000	7.9%	USD 520,000
Detroit Edison Company	2,952	5.8%	USD 386,712
Allowance Holdings Corporation	2,160	4.3%	USD 280,800
Hoosier Energy REC, Inc. Ratts Unit 2SG1	500	1.0%	USD 68,000
Marine Coal Sales Company	500	1.0%	USD 67,500
National Healthy Air License Exchange	135	0.3%	USD 18,350
Sam Peltzman Revocable Trust	50	0.1%	USD 7,050
INHALE/Glens Falls, NY Middle School	21	<0.1%	USD 3,171
CATEX Vitol Electric Inc.	12	<0.1%	USD 1,584
University of Michigan Environmental Law Society	5	<0.1%	USD 1,000
Environment Law Students Association	2	<0.1%	USD 410
Hamline University School of Law	2	<0.1%	USD 350
New England School of Law	1	<0.1%	USD 350
Electric Software Products David Gloski	1	0.01%	USD 300
Electric Software Products Alexander Long	1	0.01%	USD 300
Thomas M. Cooley Environment Law Society	1	<0.1%	USD 200
Duke University School of the Environment	1	<0.1%	USD 176
Michael S. Hamilton	1	<0.1%	USD 170
Pollution Retirement Center	1	<0.1%	USD 160
L.J. O'Callaghan, Sr.	1	<0.1%	USD 153
University of Maryland School of Law	1	<0.1%	USD 150
TOTAL	50,600	100.0%	USD 6,676,386

Source: <http://www.epa.gov/airmarkt/trading/1995/95splob.html> (accessed January 11, 2013).

shot up nearly threefold. However, the entire structure of the CAIR was based on the so called "good neighbor policy" interpretation of the Clean Air Act Clause §110(a)(2)(D)(i)(I).[6]

[6] A discussion is available in EPA, Status of CAA 110(a)(2)(D)(i)(I) SIPs Final Rule Technical Support Document (July 2011) <http://www.epa.gov/airtransport/pdfs/StatusSIPs.pdf> (accessed January 11, 2013).

Not surprisingly, a number of states (principally Michigan, Minnesota, and North Carolina) were opposed to the CAIR. In July 2008, the United States Court of Appeals for the District of Columbia Circuit declared that the CAIR cap-and-trade method was fundamentally flawed, concluding that the EPA focused on region-wide emission reductions and did not adequately factor in each state's significant contribution to air pollution issues. It declared that the methods for determining SO2 and NOx pollution were not objective. The Court held that the EPA lacked authority to remove the Acid Rain Program allowances through CAIR.

To salvage the situation, the EPA finalized the Cross-State Air Pollution Rule (CSAPR) to replace CAIR. This rule responded to the court's concerns and fulfilled the "good neighbor" provision of the Clean Air Act by addressing the problem of air pollution that is transported across state borders. The court decision of 2008 and the subsequent maneuver by the EPA to resurrect the SO2 market through other mechanisms failed. Predictably, the price of SO2 permits went down to virtually zero. The introduction of the CSAPR did not fare much better. In another lawsuit, the CSAPR was also struck down in August 2012.[7] The Courts decided that the EPA failed to show that the "downwind states" are different from "upwind states." In addition, the EPA did not give the affected states enough time to come up with their own solution.

In spite of the recent setback suffered by the EPA with respect to SO2 emissions, there is no doubt about the overall success of the program. We will discuss three dimensions of success.

First, the program managed to reduce the SO2 emissions much faster than what was originally stipulated by the EPA. This dimension is illustrated by Figure 7.2. Overall SO2 emissions were reduced from about 23,000 tons per year in 1990 to less than 7,000 tons per year in 2011. One common criticism of this reduction examines the reduction in emissions between 1970 and 1990. During that period too there was a reduction in SO2 emissions. Therefore, it is argued that the reductions during the two decades after 1990 were going to happen with or without the Acid Rain Program.

To understand the impact of the ARP, we need to examine exactly from where such a reduction in SO2 emissions came. Recall that ARP was directed squarely at the power utilities. In Figure 7.2 we examine two separate sources—the utilities and others. In 1990, the utilities accounted for two-thirds of all the SO2 emissions. The electric utilities were the principal source of reductions during the following two decades. The rest did not have much of a reduction. This evidence gives us a clear indication that the SO2 reduction was not simply a "natural" reduction with better technologies from all sources. The ARP was indeed the catalyst for such a change.

[7] *EME Homer City Generation, LP v. EPA and others*, D.C. Cir. No. 11-1302 (December 30, 2011) <http://www.cadc.uscourts.gov/internet/opinions.nsf/19346B280C78405C85257A61004DC0E5/$file/11-1302-1390314.pdf> (accessed January 13, 2013).

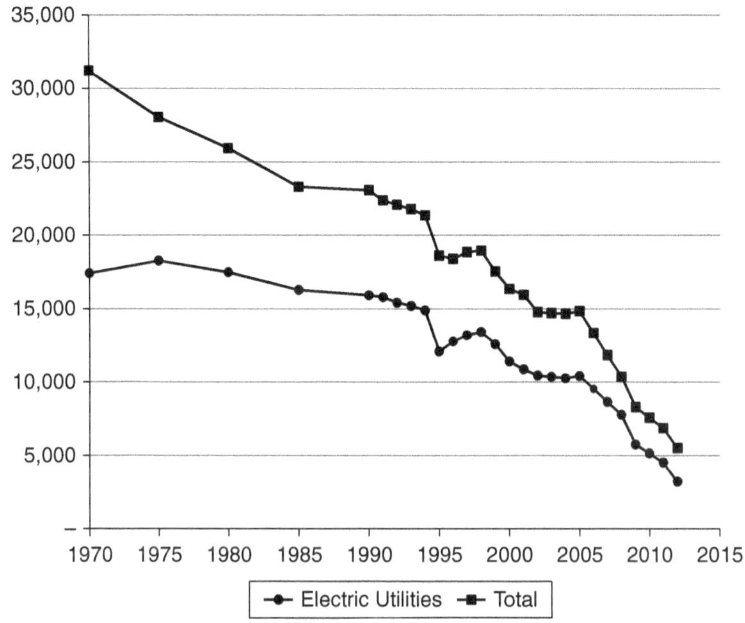

Figure 7.2 Thousands of tons of SO2 emitted 1970–2012

Second, the EPA has documented the geographical distribution of sulfate deposition in the US between 1989 and 2011.[8] From Illinois, Indiana, Ohio, Pennsylvania, all the way up to Maine, the sulfate deposit has been reduced at least by half in most states. In some states, the reduction has been over 70 percent. There is no doubt that this reduction was caused by the Acid Rain Program.

Third, as a consequence of the reduction in pollution in the Midwestern and Northeastern states of the United States, where nearly 40 percent of the US population lives, large benefits have been accruing due to: (1) health impact; (2) improved land value for agriculture; and (3) ecological impact. The value of such benefits has been calculated by the EPA. We demonstrate the impact in Figure 7.3.

Figure 7.3 shows monetized value of cumulative benefits coming from the SO2 reduction program. The value is measured in terms of reduction of lost lives, reduction of respiratory diseases, improved recreational facilities, and ecological improvement, among other things. Estimated median value of such gains for 2000 was 700 billion dollars, rising to 1,300 billion dollars in 2010, and to 2,000 billion dollars in 2020. They also estimated the minimum and maximum gains for each of those years as we illustrate in Figure 7.3. These are extremely large gains. To get an

[8] See Three-Year Wet Sulfate Deposition 1989–1991 and 2009–2011 <http://www.epa.gov/airmarkets/progress/ARPCAIR11_02/wet_sulfate.jpg> (accessed April 1, 2013).

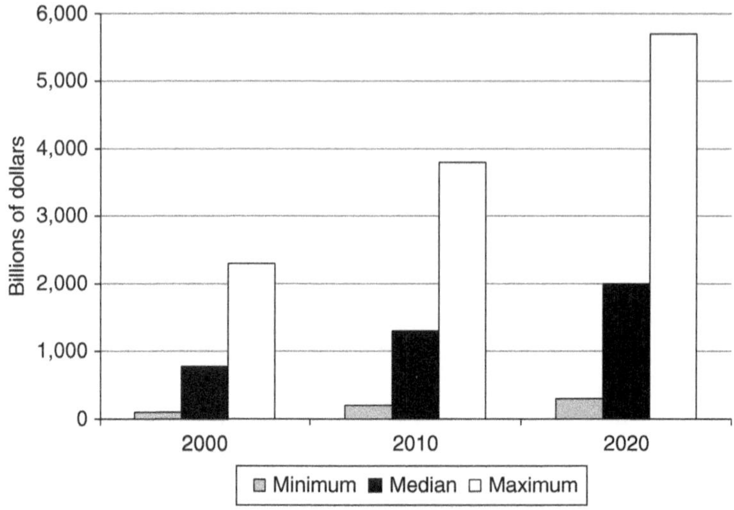

Figure 7.3 Monetized benefits 2000–2020

idea of the magnitude, the US GDP for the year 2011 was estimated at 15,000 billion dollars.

The SO2 experiment shows that cap-and-trade can actually work in real life. While the gains in terms of morbidity and mortality were not well understood at the time when the plan was implemented, it became clear later. These additional benefits eclipsed the original benefits posited when the implementation was being debated in the US Congress.

7.3 CO2 Markets

This brings us to the question of CO2 emissions and cap-and-trade markets for CO2. The first serious stab at reducing CO2 emissions came with the Kyoto Protocol. The Kyoto Protocol set binding targets for 37 industrialized countries and the European Community for reducing greenhouse gas (GHG) emissions. The goal was to reduce the GHG emissions of these countries in 2008–2012 to a level of 5 percent below the level of emissions of 1990. The detailed rules for the implementation of the Protocol were adopted at COP 7 in Marrakesh in 2001 and came into effect on February 16, 2005. Under the UNFCCC, the countries needed to meet the emissions standards with national measures. However, the Protocol created flexibility. As we noted in Chapter 2, it allowed each country to meet the targets with additional measures: (1) cap-and-trade, (2) Clean Development Mechanism, and (3) Joint Implementation (JI). Below we discuss the CDM and the JI in brief. The most important element of the Kyoto Protocol, the cap-and-trade markets, is discussed in greater detail at the end.

Under the Clean Development Mechanism, emissions-reduction projects in developing countries can earn certified emission reduction credits. Industries in the developed countries can use these saleable credits to meet a limited part of their emissions reduction targets under the Kyoto Protocol.

One key element of the CDM is the mechanism for verifying compliance. Some are easy to verify. For example, one of the projects funded for the CDM was the Delhi Metro. It was fairly easy to calculate the benefits of CO2 accrued due to the project. Similarly, another project funded was the 2012 extension of the Mexico City Metro. For these kinds of projects, compliance is not a problem.

There are, however, potential problems with the CDM. It has been argued by some advocates of small-scale developments that the CDM would allow large companies in developed countries to impose projects that are not in the best interests of host countries. The CDM now requires that host countries verify that CDM projects contribute to their own sustainable development. Such certification processes are also not without problems. Since developing country government officials are more susceptible to bribes, it is possible that such processes can be corrupted easily.

Joint Implementation means one country in Annex I[9] can implement a project in another country in Annex I to reduce CO2 emissions to earn credits for reducing emissions. The aim of the JI is to encourage clean energy technology use in the so-called "transition economies"—the countries of the former "Eastern Bloc" (countries such as the Russian Federation and Ukraine that lack modern technology for power generation and other areas).

The logic behind the Kyoto Protocol can be illustrated by the following (see Figure 7.4). In 2010, total emissions in the world were about 43 gigatons of CO2. By 2050, they are slated to grow to over 70 gigatons per year if nothing is done. The emissions would not change much in the European Union over the next forty years. They would go up marginally in the US and the rest of the OECD. However, the main growth of the emissions would come from the BRICS (Brazil, Russia, India, China, and South Africa) and from the rest of the world if nothing is done. Therefore, to make any headway into a possible future reduction, two things have to happen. First, there has to be an absolute reduction in emissions in the developed world, which accounts for nearly 40 percent of the emissions today but contains less than 20 percent of the world's population. Second, the developing countries need to have the technology of the developed world today in order to reduce their emissions.

In this section, we will discuss two separate markets: one compulsory (European Union Emissions Trading System (EU ETS)) and one voluntary (Chicago Carbon Exchange (CCX)). As a direct result of signing of the Kyoto Protocol, the European

[9] Annex I countries as of the end of 2012 are: Australia, Austria, Belarus, Belgium, Bulgaria, Canada, Croatia, Czech Republic, Denmark, Estonia, Finland, France, Germany, Greece, Hungary, Iceland, Ireland, Italy, Japan, Latvia, Liechtenstein, Lithuania, Luxembourg, Malta, Monaco, Netherlands, New Zealand, Norway, Poland, Portugal, Romania, Russian Federation, Slovakia, Slovenia, Spain, Sweden, Switzerland, Turkey, Ukraine, United Kingdom, and the United States of America.

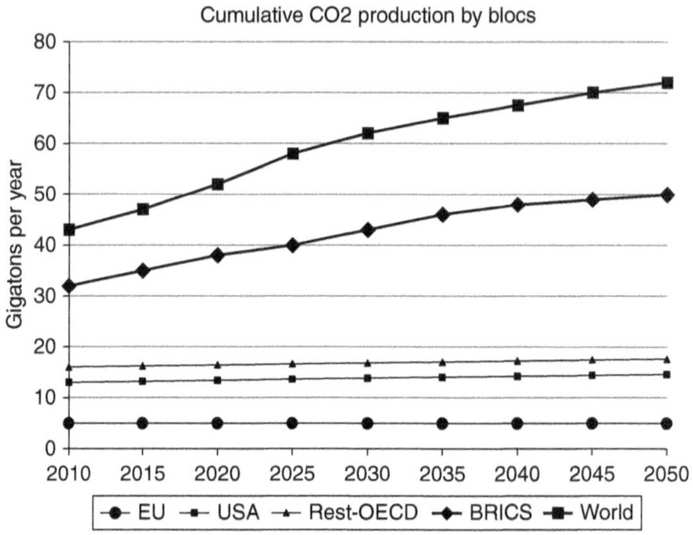

Figure 7.4 CO2 emissions up to 2050 by blocks of countries

Union set up an emissions trading system for 11,300 energy-intensive installations across the 27 Member States. In the Kyoto Protocol, the EU has set its emissions reduction target to 8 percent by 2012 compared with the level observed in 1990. This goal has been modified into emissions reduction objectives at the Member State level following the so-called "Burden Sharing Agreements." In 2003, Directive 2003/87/EC established a scheme for greenhouse gas emission allowance trading within the Community and amending Council Directive 96/61/EC creating the European Union Emissions Trading Scheme. It explicitly recognized the following six gases as greenhouse gases: Carbon dioxide (CO2), Methane (CH4), Nitrous Oxide (N2O), Hydrofluorocarbons (HFCs), Perfluorocarbons (PFCs), and Sulfur Hexafluoride (SF6). The scheme covered about half of all GHGs at the EU level.[10]

There were three phases of the market envisioned in the plan. Phase I went from 2005 to 2007—the so called warm-up period prior to the introduction of the Kyoto Protocol. Phase II went from 2008 to 2012. It is concomitant to the Kyoto Protocol. Phase III is supposed to go from 2013 to 2020. Phase III corresponds to the objectives of the EU "Energy-Climate Package" introduced in January 2008 to reduce emissions by 20 percent by 2020 along with increasing energy efficiency by 20 percent and increasing the share of renewable resources in the energy mix to 20 percent. This plan became known as the "20-20-20 target."

[10] Directive 2003/87/EC establishing a scheme for greenhouse gas emission allowance trading within the Community and amending Council Directive 96/61/EC (October 13, 2003) <http://eur-lex.europa.eu/LexUriServ/LexUriServ.do?uri=OJ:L:2003:275:0032:0032:EN:PDF> (accessed January 13, 2013).

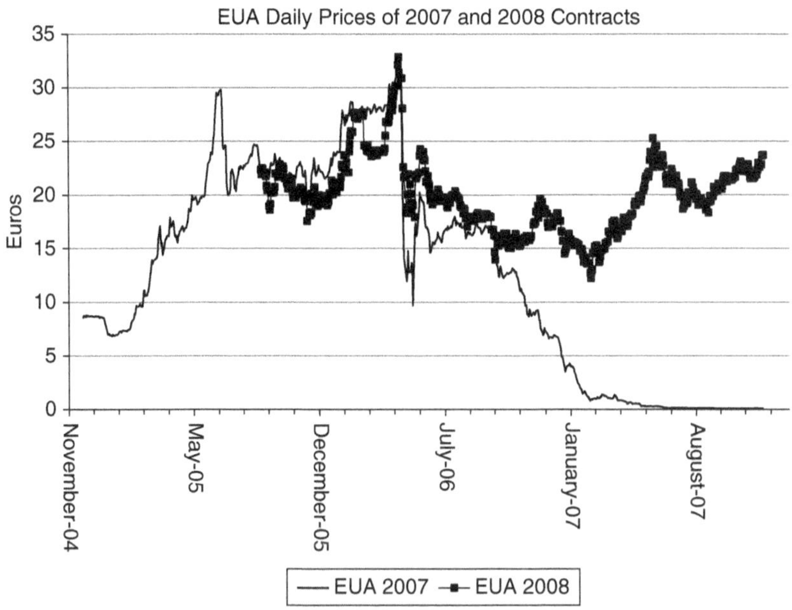

Figure 7.5 EUA 2007 and EUA 2008 prices in the EU ETA market

The allowance for one contract exchanged on the EU ETS corresponds to one ton of CO2 released into the atmosphere. It is called a European Union Allowance (EUA). For the Phase I, 2.2 billion allowances per year were distributed (2005–2007). During the Phase II (2008–2012), 2.08 billion allowances per year were distributed.

Figure 7.5 displays the movement of daily EUA prices between 2005 and 2008 for the contracts expiring in 2007 and 2008. The first notable feature of the prices is that they were volatile. The price covered an enormous range over a relatively short period of time. The second feature is that the EUA 2007 and EUA 2008 prices moved in tandem and close to each other over a year before the dramatic fall of the EUA 2007 price. Since EUA 2007 contracts did not have much value after 2007, it was expected that the prices would diverge at the end. But when they were both valuable, both the prices were close. In economic terms, these products were close substitutes for a while and therefore had very similar prices. This phenomenon is called the "law of one price" in economics.[11] It is often used as an acid test for understanding the pricing of a product. As we will see below, the price in other

[11] Law of one price has been tested in different markets. In the commodities markets, there is some evidence that the law of one price holds: John Pippenger and Llad Phillips, "Strictly Speaking, Law of One Price Works in the Commodities Markets" (2007) Working Paper, University of California, Santa Barbara. However, there are well documented anomalies in the financial markets: Owen A. Lamont and Richard H. Thaler, "Law of one price in the financial markets" (2003) 17 Journal of Economic Perspectives 191.

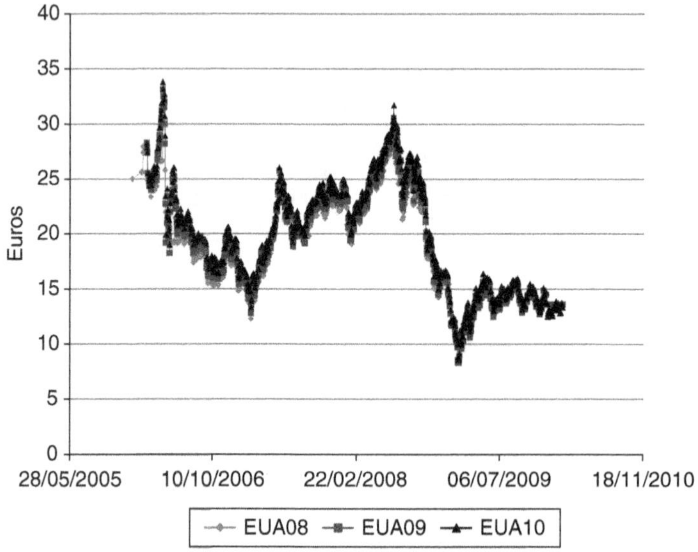

Figure 7.6 Co-movement prices of different contracts in the EU ETA market

parts of the world for one ton of carbon has not been the same. This illustrates that the price is critically dependent not just on the current laws governing the market but also future expectations about the legal status of the market. For example, we saw how the successful challenge of CAIR in the courts caused the collapse of the SO2 market in the US.

Each year, the EU ETS issued more permits with expiry dates and then they were traded in the markets. In Figure 7.6, we examine the co-movement of three such contracts over time: EUA2008, EUA2009, and EUA2010. Once again, the prices show closeness that shows us that they were being treated very similarly in the marketplace.

In addition, we performed correlation analysis between the EUA2009 and EUA2010 prices. It shows that the correlation between the prices is 0.9988. It can also be seen when we plot these two prices together in Figure 7.7.

The EU ETS market was a *compulsory* market—the companies in the EU were forced to participate. In addition, in the US, another experiment was conducted by a *voluntary* market called the Chicago Carbon Exchange. This was the brainchild of Robert Sandor.[12] He took his cue from an observation of the economist Ronald Coase.[13] Coase observed that while economists have argued that economic agents would not pay for public goods with positive externalities (such as the maintenance of a lighthouse), in fact, the Trinity House, a private consortium, has been building lighthouses in England for five centuries and making profits.

[12] Robert Sandor, *Good Derivatives* (John Wiley & Sons, Hoboken NJ 2012).
[13] Ronald H. Coase, "Lighthouse in Economics" (1974) 17 Journal of Law and Economics 357.

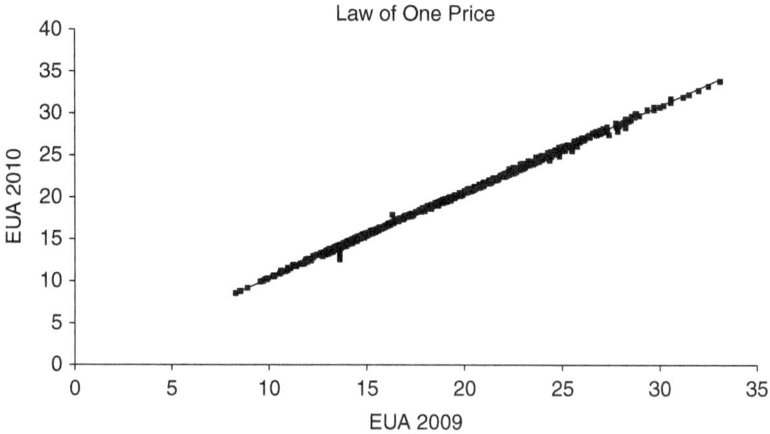

Figure 7.7 Law of one price in EU ETA

While this was a voluntary market, over time it managed to capture 17 percent of the value of the companies listed in the Dow Jones Index. It represented 25 percent of the power industry, and during the peak year, it managed to reduce more CO2 than France did (which operated under the compulsory market of the EU ETS).[14] The CCX not only managed to do business in the US market, but it also helped to generate CO2 reductions in far-flung places like Kerala, India.

In a number of villages in Andhra Pradesh and Kerala, the CCX helped develop collection of waste from cows that were used for capturing methane to be used for cooking. For doing it, families received a financial incentive of about USD 2 per month. As a result, the amount of GHG reduced more than offset the "cost" of carbon capture by other means, as wood burning became unnecessary. It also generated a bonus. In those villages, small school-age girls were used to fetch wood and sticks to burn for cooking. Their time was freed up, allowing them to attend school. Moreover, the use of methane burning stoves reduced the air pollution inside the huts, improving the health of those villagers.[15]

Sandor sets out the requisites for a successful launch of financial products. First, the financial products have to be traded in a well-defined exchange. Second, the trades have to be completely transparent to all the market participants. Third, they have to be regulated. Here, he noted that it does not have to be regulated by the government. The exchange can be self-regulated (such as the lighthouses in England). Finally, there should be a central clearing system in the market to minimize counterparty risk. He notes that these conditions are sufficient for a market to work, but all conditions are not necessary. Products such as interest-rate and foreign-exchange swaps are not centrally cleared, but they work well.[16] He also

[14] Sandor, *Good Derivatives.*
[15] CCX Project 1021, Biogas Digester Project, Angamaly, Kerala, India, 2003–2008.
[16] Sandor, *Good Derivatives.*

insists that there has to be a standard product for the market to work well. Without a well-defined product, the market would collapse. There is also a need for evidence of ownership of the product. If property rights are not well established, the market will not work.

Sandor explained why a volunteer market like the CCX would work. There is no single reason why all the parties came to trade. Some companies traded because they had business to conduct in the EU. That was the case with the Ford Motor Company. Baxter Laboratories traded in the market because they wanted to create goodwill. They wanted to be part of the green market. Other companies came in because they anticipated that sooner or later the market will be implemented across the entire nation. They wanted to have a practice run. In Figure 7.8, we display the entire experiment between 2003 and 2011 with volumes that were traded in the CCX, along with the prices at which they were traded.

The EU ETS and CCX markets are not the only two in the world. There are a few others. For example, within the US, California started operating a market on November 14, 2012. Its stated goal is ambitious: California wants to reduce the 1990 level of CO2 emissions by 80 percent by 2050. California's plan faces numerous challenges. California imports about 30 percent of its electricity from neighboring states. Consequently, the California Air Resources Board (CARB) crafted the cap-and-trade program to account for the emissions of out-of-state generators in an attempt to ensure it does not lead to greater emissions from power plants elsewhere (the so-called carbon leakage problem). That has resulted in a complex set of rules applying to "first deliverers of electricity"—companies in

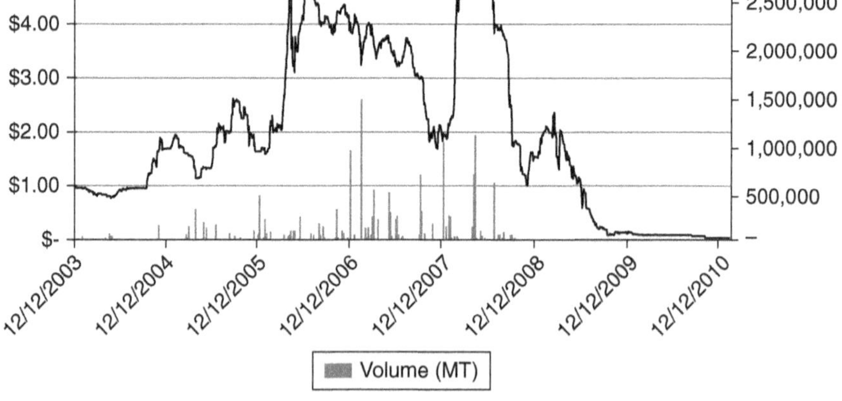

Figure 7.8 Chicago Climate Exchange market price and volume

California that receive power from sources outside the state. Under those rules, first deliverers must account for the greenhouse gas emissions of power sources located outside California and comply with the cap-and-trade scheme accordingly.

They are also prohibited from "resource shuffling"—or making changes that reduce the emissions reportable to CARB, but do not actually lower the total level of emissions they produce. For example, if a firm in Arizona now supplies electricity from a coal-fired plant to California and from a wind farm to Nevada, it might switch these around, rerouting electricity from the coal-fired power plant to Nevada and from the wind farm to California. The rules are designed to prevent such occurrences. However, there are a number of challenges pending in the courts against these rules. It is also being challenged in the federal courts for violating the US Constitution because it is seen as an impediment to interstate commerce.

The Regional Greenhouse Gas Initiative (RGGI) is another mandatory program to reduce greenhouse gas emissions in the US. The RGGI is a cooperative effort among nine eastern states of the US: Connecticut, Delaware, Maine, Maryland, Massachusetts, New Hampshire, New York, Rhode Island, and Vermont. Together, these states have capped and will reduce CO2 emissions from the power sector by 10 percent by 2018. Originally, New Jersey was also part of the RGGI. In 2012, New Jersey pulled out of the program after the Governor argued that it was not an effective way of capping emissions.

The Australian government started pricing carbon by imposing a carbon tax on July 1, 2012. The ultimate policy goal is to create a market-based emissions trading scheme in three to five years. The tax began at a fixed price of AUD 23 per metric ton of CO2e. Due to political pressure, the floor price later was dropped to AUD 15 for 2015. The long-term goal is to reduce emissions to 5 percent below 2000 levels by 2020. Critics have argued that the policy to cut emissions is regressive: low-income households carrying a relatively higher burden. To reduce the burden for the low-income households, the government has introduced a rebate and refund system.

Imposing a carbon tax (as they did in Australia) raises a question about supply: how will the existing supply be affected by a carbon tax? This question has been studied in Brito and Curl. They show that, in the US, if the price of CO2 is USD 10 or below per metric ton, it affects very few power units. However, with a CO2 price between USD 30 and USD 40, over 70 percent of the production units will shut down. Thus, indefinitely raising the price of carbon in the short run can generate a huge shortage. Therefore, any carbon tax has to be announced well in advance, as it takes several years to build electrical power generating plants.[17]

Unlike the quantitative target of a cap-and-trade system, Australia has introduced a tax on the price of carbon. As we discussed in Chapter 6, in theory both types of measures achieve the same goal of reducing consumption. However, with a price mechanism it is much more difficult to reduce the quantity consumed by some

[17] Dagobert L. Brito and Robert F. Curl, "Economics of Pricing the Cost of Carbon Dioxide Restrictions in the Production of Electricity," James Baker Institute, Rice University Working Paper July 2010, 13 Fig. 3.

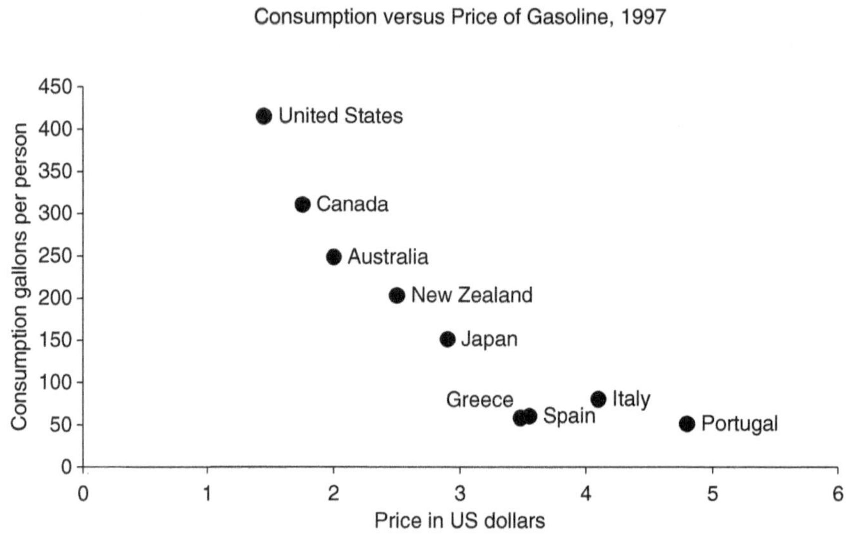

Figure 7.9 Price and consumption of gasoline per capita in OECD

given amount in the long run, since the short-run and long-run demand curve for energy consumption vary substantially. The short-run price elasticity is much lower than the long-run price elasticity because substitution of one form of energy by another is more difficult to achieve in the short run. We illustrate the relationship between price and quantity of energy related product in Figure 7.9. It shows that a higher (equilibrium) price in the market is associated with a lower (equilibrium) level of consumption.

Australia has also announced that by 2015 it will introduce a cap-and-trade system and that it will link the market to the EU ETS market.[18] In fact, in November 2012, Australia's Securities and Investments Commission granted the first 11 licenses to trade emission permits in Australia. However, the opposition party has vowed that it will dismantle the whole carbon tax and related policies if it gets elected in the next Federal election (due by 30 November 2013).

In our discussion above, we observe that the price of carbon differs in different countries. The law of one price does not seem to hold. There is a clear reason for it. The future viability of carbon markets is not clear in many places. There are potentially fatal lawsuits that could invalidate the operations in some markets (such as California). There are legislative threats coming from the opposition in Australia. But, in Australia, we will not have a market-determined price of carbon

[18] On April 16, 2013, in a surprise move, the European Parliament voted down a proposal to continue the support of the ETS. It might cause the collapse of the entire ETS cap and trade mechanism. This may not doom the cap and trade globally. As we have noted in Chapter 1, there is a US–China joint initiative in the offing. The collapse of the ETS may encourage the European Union to back the new US–China initiative.

until 2015. Moreover, if the opposition gets into power that prospect may be doubtful. Progress in multilateral treaty negotiations also is doubtful, with a very limited number of countries agreeing to extend their participation in the Kyoto Protocol (see Chapter 2). All of this makes the same ton of CO2 emissions different in different jurisdictions. This explains why prices are not (yet) equalizing across countries and regions.

In two other OECD countries, carbon taxes and cap-and-trade markets have been implemented: Ireland and Norway. Ireland has imposed a straight carbon tax (like Australia), but the main purpose has been to use the money to bridge the budget deficit. It has generated results on both counts. It has produced taxes to the tune of 25 percent of the budget deficit. It has also reduced carbon emissions by 18 percent compared with 2007. Norway, on the other hand, has closely followed the cap-and-trade model of the EU ETS.

China has proposed pilot programs of cap-and-trade in five provinces and two cities for 2013: Beijing, Shanghai, Tianjin, Chongqing, Shenzhen, Guangdong, and Hubei. Some of them are cities with large industries and others are states with large industrial hubs. In addition, in April 2013, China announced a joint initiative with the US. The indications are that China is examining both the Australian carbon tax scheme and the SO2 market in the US.

Chan et al. perhaps provide the best summary of what we can learn from the SO2 market for the CO2 case:

> Ironically, the cap-and-trade model seems especially well suited to addressing the problem of climate change, in that emitted GHGs are evenly distributed throughout the world's atmosphere. Emissions reductions anywhere make identical contributions to helping alleviate the problem, and there are no pollutant concentration hot spots. The sheer number and variety of GHG-emissions sources heightens the practical difficulty of developing a comprehensive and effective command-and-control approach and magnifies the cost savings that could be achieved by enlisting the market to find the least costly abatement options.
>
> It is difficult to achieve an international agreement to limit GHG emissions, however, for precisely the same reasons—many countries, hosting many emissions sources, must agree to take action. Given the complexities of either developing a domestic U.S. cap-and-trade system for GHGs or obtaining congressional approval for an international system in which the United States participates, it is likely that at least as much bipartisan collaboration would be required as was evident in the . . . process [to enact the Clean Air Act Amendments that established the Acid Rain Program]. Instead, we have much less.
>
> The stakes for a broad-based GHG policy—economic, political, and environmental—are much higher than they were for SO2 policy in 1990. While the debate over federal policy to address climate change is currently in hiatus, the lessons of the SO2 allowance-trading program will prove useful and relevant to future deliberations about climate change policy when the time arrives for serious reflection.[19]

If the fragmented CO2 markets of China, Australia, the EU, RGGI, and California band together, they can produce a vast market for CO2. In that market, they would

[19] Gabriel Chan et al., "The SO2 Allowance-Trading System and the Clean Air Act Amendments" (2012) 65 National Tax Journal 419.

compete for clean energy technology and buy up permits to drive up the demand for clean air technology. The experience with the SO2 market in the US suggests that such an approach would be an effective means to achieve emissions reductions and clean energy technology dissemination and adoption. It would be even more effective with more players. Indeed, if the countries that account for the majority of GHG emissions could negotiate such an agreement, it would represent a big step forward.

7.4 Adaptation, Mitigation, and Insurance

In our discussions so far, we have had very little to say about adaptation strategies. For the governments of many countries, so far, adaptation strategies are nonexistent. This state of affairs is not just for the developing countries. For example, in the "Status of State Climate Adaptation Plans" in the US, most of the states have no plans at all.[20] This is the status in a developed country. Not surprisingly, in the developing countries, adaptation is not even on the radar of the policymakers.

At the international level, the Adaptation Fund was established to finance concrete adaptation projects and programs in developing countries which are signatories to the Kyoto Protocol and which are particularly vulnerable to the adverse effects of climate change. Essentially, the Adaptation Fund finances projects and programs to help developing countries adapt to the negative effects of climate change. It has funded projects in two dozen countries with a total commitment of 166 million dollars, with about a quarter of it spent by the end of 2012.[21] The Adaptation Fund is financed mainly from the share of proceeds on the CDM project activities. The share of proceeds amounts to 2 percent of certified emission reductions (CERs) issued for a CDM project activity.

In summary, there is very little that is being done either at the level of international institutions or at the state and local government level. This is one area where private insurance and reinsurance companies can make a substantial contribution by teaming up with governments. Most individuals and companies avoid risks. Insurance companies, on the other hand, create business by accepting risks.

The four trillion dollar insurance industry is at the vanguard of climate change. The reason is obvious. Many risks that insurance and reinsurance companies cover are related to weather. Flood insurance, hurricane insurance, windstorm damage insurance, crop insurance (along with more recent synthetic insurance policies with features of put options) are staples of the insurance and the reinsurance business. At least half of the world's population lives in regions highly exposed to natural disasters, but only a small fraction of them are insured. Insured losses alone from weather-related disasters have jumped from USD 5 billion per year in the period

[20] See e.g. National Climate Assessment and Development Advisory Committee Report, 2013, 990 Fig. 28.1. It shows that only 15 states out of 50 in the US have drawn up any plan at all.

[21] See Adaptation Fund, Funded Projects <http://www.adaptation-fund.org/funded_projects> (accessed January 14, 2013).

between 1970 and 1989 to USD 27 billion annually over the last two decades. Events such as Hurricane Katrina pushed the annual cost of catastrophes to over USD 100 billion in 2005.[22] While the actual losses from Katrina were over USD 113 billion, less than half of it was insured. More recently, in October 2012, Hurricane Sandy caused losses over USD 71 billion.[23]

In theory, any risk whose probability distribution and loss magnitude can be quantified can be priced and insured. So, it is possible to insure against natural catastrophic losses. The problem is that such risks are changing over time and the volatility of such risks is rising. That makes forecasting future losses a challenge. Many such risks are insured on a year by year basis. But if the underlying variability of such risks changes, that is, events that were once in a hundred years become once in thirty, it may take the insurers a long time to find that out. When they do find out, it may make them insolvent.

Consider crop and flood insurance products. They are related to the amount and the variability of rainfall. Over time, rainfall in a given area is changing. It is not a simple matter of more or less rain—the entire distribution of rainfall is moving in unexpected ways. In Figure 7.10, we examine the rainfall in the Sahel region of Africa. It shows (the kernel density) of the monthly rainfall distribution over a period of 113 years. The figure shows that rainfall in that region has become sparser

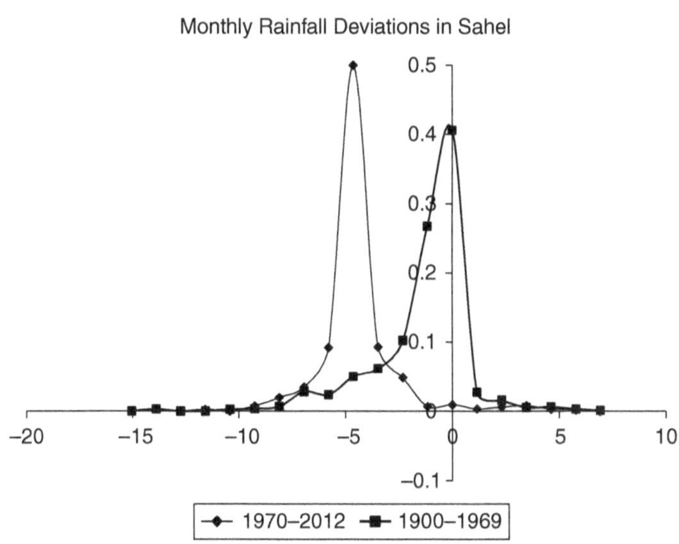

Figure 7.10 Distribution of rainfall 1900–2012

[22] Swiss Re, "Weathering Climate Change: Insurance Solutions for More Resilient Communities" (2010) <http://media.swissre.com/documents/pub_climate_adaption_en.pdf> (accessed January 14, 2013).

[23] See Hilary Russ, "New York, New Jersey put $71B price tag on Sandy" MSN News, November 27, 2012 <http://news.msn.com/us/new-york-new-jersey-put-dollar71b-price-tag-on-sandy> (accessed January 14, 2013).

(the median of the distribution has moved to the left). Moreover, the shape of the distribution itself has changed dramatically making extreme events more likely.

Most of these extreme events are not insured by ordinary insurance companies but by reinsurance companies. Naturally, large global reinsurance companies like Munich Re and Swiss Re have taken great interest in this field.[24] They take the IPCC Reports very seriously. It provides a market test: they have a financial stake in climate change in the long run. Thus, they take climate change more seriously than politicians, many of whom, by the nature of their business, have very short-term interests.

Specific actions have been suggested by many governments. For example, the October 2010 report on climate change from the White House suggests that the government needs to "facilitate the incorporation of climate change into insurance mechanisms."[25] Particularly, insurance markets may have insufficient capital to cover increasing catastrophic losses, especially if rates cannot track climate risks. It also encourages public-private partnerships to produce an open-source risk assessment model.

For the developing countries, the stakes are higher. For some countries, like the small island nations around the globe (the Maldives, for example), the rise of sea water level is critical to survival. Low lying countries will have higher likelihood of inundations. They are the most vulnerable ones. Other countries will face challenges as well. For example, many large cities in Latin America have grown without any plan. There is urban poverty and precarious living conditions. Climate change will cause rising sea levels, more violent storms, flooding, and extreme temperatures. There will be problems of water supply, food will be scarcer and there will be threats to health and sanitation.[26]

The lack of maturity of capital markets brings additional challenges for the developing countries. Routine synthetic financial products (such as rainfall insurance) are simply not available in most developing countries. This is where international multilateral organizations such as the World Bank, International Monetary Fund, Inter-American Development Bank, and Asian Development Bank can play an important role in providing private–public partnerships in the form of microinsurance and other innovative products that are normally not available in such markets.

[24] See Munich Re, "Munich Re*Newables*: Our Contribution to a Low-Carbon Energy Supply" (2009) <www.munichre.com/publications/302-05955_en.pdf> (accessed January 12, 2013) and Swiss Re, "Managing Climate and Natural Disaster Risk" (2012) <http://www.swissre.com/rethinking/climate_and_natural_disaster_risk/> (accessed January 13, 2013).

[25] See The White House Council on Environmental Quality, Progress Report of the Interagency Climate Change Adaptation Task Force: Recommended Actions in Support of a National Climate Change Adaptation Strategy, October 5, 2010 <http://www.whitehouse.gov/sites/default/files/microsites/ceq/Interagency-Climate-Change-Adaptation-Progress-Report.pdf> (accessed January 12, 2013).

[26] This section draws on Raquel Szalachman, "Potential Impacts of Climate Change in Latin American Cities" Working Paper Series of the Geneva Association, Études et dossiers No. 356 (January 2010) <http://www.genevaassociation.org/PDF/Working_paper_series/(1)_GA_E&D_356.08_SZALACHMAN_Climate_change,Economic_development,Vulnerability.pdf> (accessed January 13, 2013).

There are many ways that insurance companies can help to address climate change.[27]

First, insurance companies can help with the understanding of the climate change problem. Insurers are beginning to share their expertise in data collection, catastrophe modeling, and risk analysis to track trends in weather related data. These activities can be used to address problems posed by climate change. Insurers are building forward-looking risk models that take climate change into account. For example, in Figure 7.10, the movement of the rainfall index over the past century can be used to create a rainfall index for the future. For planning and for insurance, such modeling will not produce unexpected losses. Such modeling is essential for the survival of the insurance industry itself. If a hundred-year flood becomes a ten-year flood, insurance companies will have insufficient capital to confront such risks in the long run.

Second, insurance can promote loss prevention through risk mitigation. Managing risks and controlling losses is central to any insurance business. The insurance industry has been setting up fire departments and advocating building codes for natural and man-made hazards for a long time. For example, after the Great Fire of London in 1666, Nicholas Barbon started an insurance company to offer protection against fire. He did two other innovative things:

(1) He started differential rates for different building material used: 2 percent of the rent if the houses were made of bricks and 5 percent of the rent if they were made of wood.
(2) He also hired the watermen from the river Thames to work as part-time fire fighters.

The first of these acts encouraged the building of brick houses in London. The second act created an eventual fire department for the city. More recently, insurance companies have been giving rebates for households with fire insurance if they do not use halogen lamps. For the insurance companies, it reduces the fire hazards and it reduces the use of electricity substantially for households, thereby reducing the GHG emissions.[28]

Third, insurance companies can encourage risk reducing behavior. Insurance contracts can design policy exclusions to instill behaviors that reduce greenhouse gas emissions and appropriate efforts to prepare for the impacts. For example, so called pay-as-you-drive insurance products are offered by insurers that recognize the link between accident risk and distance driven. In this case, the insurance company can do this cost-effectively today because it can use the GPS in the cars to monitor activities with very little cost. It encourages people to drive less. This, in turn, co-benefits the environment.

[27] This section draws heavily on Eric Mills, "Weighing the Risks of Climate Change Mitigation Strategies" (2012) 68 Bulletin of the Atomic Scientists 67.

[28] Erik Page et al., "New Energy Efficient Torchieres Ready For Hot Torchiere Market," Proceedings of the Conference on Energy Efficiency in Household Appliances, November 5–8, 1997, Florence, Italy.

Fourth, insurance companies are promoting innovative products. Insurance companies are offering special rates for "green buildings" and products that cover risks associated with energy efficiency or renewable energy projects. This encourages the construction of more efficient buildings in the future as well as retrofitting old buildings to become more energy efficient.

Fifth, insurance companies are offering climate protection improvements. Insurance companies that also have banking operations are engaged in financing projects that (1) improve resilience to the impacts of climate change and (2) contribute to reducing emissions. Several companies are providing preferential mortgage rates for energy efficient appliances and home upgrades. Some companies are offering "Clean Car Credit" financing for low-emission vehicles.

Sixth, the so-called carbon risk disclosure requirement is being encouraged by insurance companies that insure business entities. The carbon risk disclosure requires a company to disclose information related to risk factors and calls for management discussion and analysis. This issue got a boost with the 2009 position paper of the Securities and Exchange Commission, which stated that businesses "should consider the impact of existing climate change legislation and regulation, international accords or treaties on climate change, indirect consequences of regulation or business trends, for example new risks for the company created by legal, technical, political and scientific developments, and the physical impacts of climate change."[29] Canada has had similar requirements since 2008. In addition, in the near future, Canada will make such disclosures obligatory rather than voluntary.[30]

7.5 Conclusion

The economics and finance of climate change show that climate change and economic and financial matters are causally linked in both directions. Chapters 6 and 7 vividly bring out this interplay. Economic and financial changes, backed by a carefully designed international legal framework, can have a significant impact on climate change mitigation and adaptation. Moreover, climate change concerns induce changes in the global economic and financial architecture.

CO2 markets have been and are being created unilaterally and regionally in both developed and developing countries. While this is a step in the right direction, and demonstrates again how unilateral action can serve as a stepping stone, CO2 markets would be more effective if they were integrated and expanded. Although a global CO2 market would be ideal, even a large CO2 market, made up of several major players, could serve as an effective catalyst for change. There are precedents at

[29] Securities and Exchange Commission Investor Advisory Committee, Possible Refinements to the Disclosure Regime, July 27, 2009 <http://www.sec.gov/spotlight/invadvcomm/iacmeeting072709-briefingpaper.pdf> (accessed January 14, 2013).

[30] Ontario Securities Commission Notice 51-717: Corporate Governance and Environmental Disclosure, December 18, 2009 <www.osc.gov.on.ca/documents/en/Securities-Category5/sn_20091218_51-717_corp-gov-enviro-disclosure.pdf> (accessed January 14, 2013).

the WTO for the negotiation of sectoral agreements, in which Members accounting for over 90 percent of trade in a particular product or service reach a trade liberalization agreement that other Members can join later. This type of negotiation model has proved effective in the liberalization of trade in financial services and information technology goods, for example. A similar approach might work well to integrate and expand CO2 markets.

The insurance industry has a valuable role to play in mitigation and adaptation and is perhaps the best example of how the private sector can play a critical role with respect to climate change. The liberalization of trade in insurance services is thus a critical component of addressing climate change. In this regard, the WTO again has a key role to play, through the liberalization of trade in financial services under the GATS. We discuss the role of GATS further in Chapter 8.

8

Climate Finance

8.1 Introduction

There are two fundamental questions regarding climate finance: where will we get the money and how should we spend it? In this chapter, we argue that the money spent on fossil fuel subsidies should be reallocated to mitigation via the development and dissemination of clean energy technologies and to adaptation via financing for subsistence farmers to acquire the inputs needed to switch to genetically modified (GM) crops that can tolerate a changing climate. Eliminating fossil fuel subsidies kills two birds with one stone, since it reduces GHG emissions and frees up funding for other mitigation projects and for adaptation projects.

A related issue is how the funding will be distributed. To stabilize GHG emissions (mitigation) would require a significant reduction of emissions both in the developed and the developing world. This requires large-scale investment in energy infrastructure and other large-scale mitigation projects. Multilateral funding channels may be the best option for such large-scale financing in developing countries. On the multilateral side, we will focus on the Clean Development Mechanism and the Green Climate Fund. We also examine the three-part assessment of climate change and the World Bank, which addresses the role of the World Bank in energy projects and reforming fossil fuel subsidies, investment in clean energy technology, and adaptation. Our analysis in this chapter will complement the World Bank assessment by analyzing the fossil fuel subsidy issue, examining the relationship between climate finance and WTO law, and looking at one adaptation topic and one mitigation topic.

Our mitigation topic is the financing of the adoption of GMO technology by subsistence farmers. In contrast to large-scale mitigation projects, financing adaptation for subsistence farmers can be local and small-scale. It also has the benefit of focusing funding on the poor, in contrast to fossil fuel subsidies, which primarily benefit the rich. Financing adaptation for subsistence farmers can be carried out at the national level, by simply redirecting fossil fuel subsidies in national budgets. Our analysis expands on the analysis needed to address this adaptation topic, which was mentioned in the World Bank assessment as a priority.

National economic policy can play a central role in climate finance, but so does global economic governance. As we saw in Chapter 1, intellectual property rights can raise the cost of GM seeds beyond the capacity of subsistence farmers to pay. Thus, in Chapter 5 we argued that developing countries, in particular, should seek to maintain policy flexibility regarding intellectual property rights for plant varieties. Another way around this problem is to provide financing for subsistence farmers to afford GM seeds and the pesticides and fertilizers that accompany their use. However, as we noted in Chapter 3, subsidies may be inconsistent with the WTO SCM Agreement or the WTO Agreement on Agriculture. Therefore, in this chapter we analyze how national financing for subsistence farmers might be structured to avoid violating these international obligations.

Financing for clean energy projects also needs to be structured to avoid violating international obligations. As we have seen in Chapter 3, Ontario's FIT program is the subject of both WTO complaints and claims filed by foreign investors under NAFTA Chapter 11. These cases highlight the importance of designing clean energy projects, whether financed locally, bilaterally, regionally, or multilaterally, with these international obligations in mind.

This chapter is organized as follows. First, we provide an overview of multilateral financing mechanisms. Second, we analyze the issue of fossil fuel subsidies. Third, we analyze how to structure financing for adaptation by subsistence farmers in accordance with the WTO Agreement on Agriculture. Fourth, we analyze financing for clean energy projects. Finally, we analyze the scope for private sector participation and the importance of international trade in services.

8.2 Multilateral Financing Mechanisms

Under the United Nations Framework Convention on Climate Change, the Green Climate Fund is the main vehicle of climate financing. In the Kyoto Protocol, the main vehicles of climate financing include the CDM and Official Development Assistance.[1] It is important to distinguish between UNFCCC parties and Kyoto Protocol parties when discussing climate financing mechanisms. While 192 parties ratified the Kyoto Protocol, the United States did not, due to the absence of binding targets for developing countries and concerns regarding its economic impact.[2] Canada withdrew from the Kyoto Protocol, effective 15 December 2012. Japan and the Russian Federation formally notified the secretariat that they do not intend to make commitments after 2012.[3] At the 2012 COP, Australia and European countries extended their Kyoto commitments. The second commit-

[1] Little Climate Finance Book <http://www.odi.org.uk/resources/download/4619.pdf> (accessed December 28, 2012).

[2] Fabio Morosini, "Trade and Climate Change: Unveiling the Principle of Common but Differentiated Responsibilities from the WTO Agreements" (2010) 42 George Washington International Law Review 713.

[3] Amendment to the Kyoto Protocol pursuant to its Article 3, paragraph 9, Report of the Ad Hoc Working Group on Further Commitments for Annex I Parties under the Kyoto Protocol, Draft decision proposed by the President, Conference of the Parties serving as the meeting of the Parties to the Kyoto Protocol, Eighth session, Doha, November 26–December 7, 2012, FCCC/KP/CMP/2012/L.9, December 8, 2012 <http://unfccc.int/resource/docs/2012/cmp8/eng/l09.pdf> (accessed December 28, 2012).

ment period runs from 1 January 2013 to 31 December 2020.[4] Thus, there is now a very limited number of Kyoto parties with emissions reduction commitments, which limits the effectiveness of Kyoto climate financing mechanisms.

As we noted in Chapter 2, Kyoto Protocol Article 3(14) requires each Annex I party to "strive to implement" its commitments in such a way as to minimize adverse social, environmental and economic impacts on developing country parties, and to consider issues such as the establishment of funding, insurance and transfer of technology. The Kyoto parties can meet their emissions targets through three market-based mechanisms: the carbon emissions trading market, joint implementation, and the clean development mechanism. The CDM allows industrialized countries to invest in projects that reduce emissions in developing countries as an alternative to domestic reductions. Such projects earn saleable certified emission reduction (CER) credits, each equivalent to one ton of CO2, which can be counted towards meeting Kyoto targets. In 2007, in Bali, the Kyoto parties launched the Adaptation Fund. It was decided that 2 percent of the CER would be earmarked for the Adaptation Fund.

At the Copenhagen Convention of the Parties (COP), the UNFCCC developed country parties promised to provide USD 30 billion in "fast start" climate finance in 2010–2012 and to mobilize USD 100 billion per year by 2020 to address the adaptation and mitigation needs of developing countries. At the Cancun COP in 2010, the UNFCCC parties established a "Green Climate Fund" to finance mitigation and adaptation to climate change in developing countries.[5] UNDP has developed an Adaptation Policy Framework.[6] The UNFCCC parties have established the Cancun Adaptation Framework, an Adaptation Committee, a work program on loss and damage and a national adaptation plan process for developing countries.[7]

The UNFCCC parties agreed that the USD 100 billion may come from public and private, bilateral and multilateral sources and that a "significant" share of new multilateral funding for adaptation should flow through the Green Climate Fund. The Green Climate Fund is to be an operating entity of the financial mechanism of the Convention under Article 11, governed by a Board of 24 members, comprising an equal number of members from developing and developed country parties. The trustee for the Green Climate Fund will manage its financial assets in accordance

[4] Amendment to the Kyoto Protocol; Roger Harrabin, "UN Climate Talks Extend Kyoto Protocol, Promise Compensation" BBC News, December 8, 2012 <http://www.bbc.co.uk/news/science-environment-20653018> (accessed December 28, 2012).

[5] The Cancun Agreements: Outcome of the work of the Ad Hoc Working Group on Long-term Cooperative Action under the Convention, 1/CP.16, Report of the Conference of the Parties on its sixteenth session, held in Cancun from 29 November to 10 December 2010, FCCC/CP/2010/7/Add.1, March 15, 2011 <http://unfccc.int/resource/docs/2010/cop16/eng/07a01.pdf#page=2> (accessed December 29, 2012); Robert N. Stavins, "Why Cancun Trumped Copenhagen: Warmer Relations on Rising Temperatures" The Christian Science Monitor, December 20, 2010 <http://theenergycollective.com/robertstavins/49196/why-cancun-trumped-copenhagen> (accessed December 2, 2012).

[6] Morosini, "Trade and Climate Change."

[7] Draft decision 1/CP.18, Agreed outcome pursuant to the Bali Action Plan (Advance version) <http://unfccc.int/resource/docs/2012/cop18/eng/08a01.pdf> (accessed April 10, 2013).

with the relevant decisions of the Green Climate Fund Board; the World Bank is to serve as the interim trustee. An independent secretariat is to support the operation of the Fund.[8] The Green Climate Fund is to balance the allocation of the resources between adaptation and mitigation activities. As of December 2012, the Green Climate Fund was not yet operational, but had received cumulative contributions of USD 4.298 million and cumulative pledges of USD 4.554 million for the administrative budget of the secretariat in South Korea.[9]

The Report of the High-level Advisory Group on Climate Change Financing recommended a carbon price of USD 20–25 per ton of CO2e as a key element of reaching the climate finance goal of USD 100 billion per year. The focus of the Advisory Group on revenue raised by developed countries for developing countries raised issues regarding whether to count some sources of financing, because they might imply financial contributions from developing countries. The Advisory Group did not agree on whether private and non-concessional financing (i.e. loans provided with a market-based interest rate) should count towards the USD 100 billion per year, but did agree that the size of such flows is likely to be greater the better the investment climate in the developing countries. They also disagreed on whether carbon offset flows should count, since these mechanisms are designed to reduce the cost of mitigation in developed countries. The Advisory Group notes that the benefit of carbon pricing is that it both raises revenue and provides incentives for mitigation, but did not decide whether this should be achieved via taxes or carbon markets. The Report estimates that allocating 10 percent of total revenues from auctions of emission allowances and domestic carbon taxes in developed countries for international climate action could mobilize around USD 30 billion annually. A further USD 10 billion annually could be raised from allocating 25–50 percent of total revenues from carbon pricing international transportation, but the Advisory Group disagreed on whether such carbon pricing measures should apply to developing countries. USD 10 billion could come from other sources, such as the redeployment of fossil fuel subsidies in developed countries or a financial transaction tax, but there is political resistance to applying the latter in developing countries. International private investment flows in carbon markets could generate USD 10 billion of net transfers. The Report notes that international private investment flows are essential for mitigation and adaptation. Net flows from multilateral development banks could be USD 11 billion. Finally, the Report notes that the domestic fiscal situation of many developed countries is now under extreme pressure, but projects that direct contributions from developed country governments will play a key role in the long term. The Report recommends

[8] Decisions adopted by the Conference of the Parties, Report of the Conference of the Parties on its sixteenth session, held in Cancun from November 29 to December 10, 2010, Addendum, Part Two: Action taken by the Conference of the Parties at its sixteenth session, FCCC/CP/2010/7/Add.1, March 15, 2011, paras. 98–112.

[9] Draft decision 1/CP.18, Agreed outcome pursuant to the Bali Action Plan para. 72; Draft decision 6/CP.18, Report of the Green Climate Fund to the Conference of the Parties and guidance to the Green Climate Fund (Advance unedited version) <http://unfccc.int/resource/docs/2012/cop18/eng/08a01.pdf> (accessed April 10, 2013).

prioritizing funding for adaptation for the most vulnerable developing countries (the least developed countries, small island developing States, and Africa) and stopping and reversing the destruction of rainforests as a cost-effective abatement solution.[10]

Seventeen years after the UNFCCC was concluded in Rio de Janeiro, a 2009 World Bank report by the Independent Evaluation Group (IEG) recommended that the World Bank more systematically address energy efficiency and climate change in its energy projects. As the report notes, it is not just that fossil fuel subsidies are expensive, inflate GHG emissions, and disproportionately benefit the middle and upper classes. Their reduction also encourages energy efficiency, increases the relative attractiveness of renewable energy, and frees up resources for poverty alleviation, including investments in clean energy for those who do not have electricity. At the same time, the social and economic benefits of providing power to the poorest greatly outweigh the social costs. The World Bank has mobilized concessional finance from the Global Environment Facility and carbon finance from the Clean Development Mechanism to promote renewable energy and other GHG-reducing activities, and has used Global Environment Facility funds to stimulate the development of non-commercial technologies. It has also assisted developing countries and economies in transition with energy pricing reforms. However, energy efficiency and climate change have not been the focus of the overwhelming majority of World Bank energy projects, which have tended to prefer investments in power generation. The 2009 report recommends that the World Bank: (1) systematically promote the removal of fossil fuel subsidies and provide technical assistance and policy advice to address related social and political economy concerns; (2) emphasize energy efficiency policies as a way to transition to market-based energy prices; (3) provide incentives to address climate change across different sectors through system-wide energy planning at the national level; (4) collaborate with the International Energy Agency on energy efficiency indicators, including information on energy prices and subsidies; and (5) do more rigorous economic and environmental assessments for energy investments that release or prevent carbon emissions. Thus, the report indicates that the UNFCCC process and World Bank energy projects have been working at cross purposes, with the UNFCCC seeking to reduce GHG emissions while the World Bank financed energy projects. World Bank management's response to the report identified over 20 percent of energy projects as supporting supply-side energy efficiency. Management also noted that the World Bank has committed to accelerate lending for new renewable energy and energy efficiency projects, and funding for low-carbon technologies when the funds become available. Its response also notes the lack of World Bank resources to maintain a comprehensive and reliable database on energy policies, prices, subsidies, and energy efficiency at the national level. The report was

[10] Report of the Secretary-General's High-Level Advisory Group on Climate Change Financing, November 5, 2010 <http://www.un.org/wcm/webdav/site/climatechange/shared/Documents/AGF_reports/AGF%20Report.pdf> (accessed January 1, 2013).

the first phase of a three-part IEG evaluation on Climate Change and the World Bank Group.[11]

The second phase of the IEG evaluation examined the World Bank Group's project-level experience in promoting technologies for renewable energy, energy efficiency, and transport. The transport element addressed issues such as financing for public transportation, such as the Metrobus system in Mexico City. From 2003 to 2008, the World Bank increased annual investments in renewable energy and energy efficiency from USD 200 million to USD 2 billion and helped to mobilize over USD 5 billion in concessional funds for GHG reduction. In 2008, it adopted the Strategic Framework on Development and Climate Change. The report argues that the World Bank can leverage its resources through advice and support for policies to remove energy subsidies and of other obstacles to renewable energy and energy efficiency. It also can support technology transfer and adaptation to local conditions by investing in pilot projects, in order to identify development solutions that can be deployed on a large scale as climate finance expands. It should refocus on high-impact sectors and instruments, particularly energy efficiency, distributing compact fluorescent light bulbs for example. Energy efficiency combines high economic returns and GHG benefits. Other projects individually have high carbon returns (forestry) or economic returns (solar home photovoltaic systems). Carbon finance needs to be redirected away from hydropower, which is more vulnerable to the impact of climate change on water flows, to applications with more leverage, such as long-duration loans for support of renewable energy. Technical assistance can help overcome barriers to capital-intensive investments in less-proven clean energy technologies. When countries pay price premiums for renewable energy, World Bank and Multilateral Investment Guarantee Agency (MIGA) guarantees against breach of contract and other political risks could catalyze investment.[12] This is a useful addition to our analysis of impact of international investment agreements on political risks in renewable energy subsidization programs, which we addressed in Chapter 4.

The third phase of the IEG evaluation examined adaptation issues. This report notes that the World Bank lacks a comprehensive, outcome-oriented results framework for guiding and tracking its adaptation efforts. The Independent Evaluation Group recommended modifying the results framework to better track resilience outcomes and to promote learning; developing operational guidelines on screening projects for climate risk; investing more in hydrometeorological systems and promoting their use, especially in Sub-Saharan Africa; and devoting more attention to learning how to promote resilient land uses. Climate adaptation comprises adaptation to current climate variability and to future climate change. The World Bank has extensive experience with efforts to address climate variability, but new efforts explicitly designated as climate adaptation are mostly framed as

[11] World Bank, *Climate Change and the World Bank Group, Phase I: An Evaluation of World Bank Win-Win Energy Policy Reforms* (World Bank, Washington 2009).

[12] World Bank, *Climate Change and the World Bank Group, Phase II: The Challenge of Low-Carbon Development* (World Bank, Washington 2010).

current climate variability. The World Bank Group is just beginning to examine when and how to invest to insure against future risks, for example with the Pilot Program on Climate Resilience to integrate resilience into development activities. This report focuses on disaster risk management, agriculture, and hydrometeorological services to deal with adverse climate, climate variability, and climate extremes. In agriculture, World Bank attention has focused on rain-fed areas, which face the most climate variability and have high poverty rates. World Bank support for watershed management and sustainable land and water management has increased agricultural yields and incomes, but there is little or no project-based evidence of direct biophysical impacts on agricultural resilience. The report considers expansion of irrigation as a potentially important avenue of adaptation, as well as financial products for risk management, where there is a major gap in international climate resilience risk management. One of the areas that the report identified as requiring anticipatory adaptation is the development of a portfolio of new crop varieties to be ready for emerging pests and climate patterns. The World Bank has supported this goal indirectly by funding the Consultative Group on International Agricultural Research. However, with respect to conservation of agrobiodiversity, which could be an important input into new crop and animal varieties, the World Bank has supported only a few projects.[13] The report does not analyze the international legal constraints on multilateral investments in new crop varieties, in particular regarding intellectual property rights and subsidies, which we analyze in Chapter 5 and in this chapter, respectively.

There are a number of other multilateral initiatives such as the Forest Carbon Partnership Facility, the Climate Investment Funds established at the World Bank, Reduced Emissions from Deforestation and Degradation (REDD) and the proposal for using the Special Drawing Rights of the International Monetary Fund (IMF) for mitigating and adapting to climate change. They all have the same problem: financing comes from the developed countries and a substantial part of the money will be spent in the developing countries.[14]

The World Bank has played a leading role in the provision of fast-start climate finance through the Climate Investment Funds, which include adaptation. However, institutional developments on adaptation finance will become more complex and more diverse, with new organizations emerging with core mandates in adaptation support (such as the Green Climate Fund). The World Bank will have to adapt to such changes.[15]

[13] World Bank, *Climate Change and the World Bank Group, Phase III: Adapting to Climate Change: Assessing World Bank Group Experience* (World Bank, Washington 2013), advance copy <http://ieg.worldbankgroup.org/content/dam/ieg/climate_change3/cc3_full_eval.pdf> (accessed January 4, 2013).

[14] Sarah A. Mason-Case and Marie-Claire Cordonier Segger, "International Law and Climate Finance", Working Paper, The Center for International Sustainable Development Law and The International Development Law Organization, 2010, <http://cdm266901.cdmhost.com/cdm/singleitem/collection/p266901coll4/id/3368/rec/4> (accessed April 10, 2013); Katherine Serra, "Designing the International Green Climate Fund: Focusing on Results" Global Economy and Development at Brookings, Policy Paper 2011-5, April 2011 <http://www.brookings.edu/~/media/Files/rc/papers/2011/0425_green_climate_fund_sierra/0425_green_climate_fund_sierra.pdf> (accessed December 8, 2012).

[15] World Bank, *Phase III.*

REDD will both reduce emissions and lead to new development finance. The World Bank's Prototype Carbon Fund is important in this context.[16] Indigenous and protected areas that permit sustainable use have reduced tropical deforestation. Forest loss generates a quarter of developing countries' emissions. The local and global values of forests often greatly exceed the gains from their destruction. The Forest Carbon Partnership Facility is a pilot program for monetizing the value of standing forests. However, the mechanisms to use the funds to conserve forests are still being planned. These findings support the feasibility of the REDD initiative.[17]

The current state of climate finance has been criticized for three main reasons: (1) for not achieving sufficient scale; (2) for the relatively low share of private-sector investment; and (3) for insufficient institutional framework. This section has provided an overview of current multilateral efforts for scaling up climate finance. In the next section, we analyze the consistency of such finance mechanisms with WTO law. We then highlight the importance of reallocating fossil fuel subsidies to free up funding for climate finance.

This section has highlighted the role that the World Bank can play in creating incentives for private-sector investment, by investing in pilot projects and lowering political risk for private investors. In Chapter 4, we have highlighted the importance of international investment agreements in this regard. Later in this chapter, we will examine the role of international trade in services private-sector investment.

This section also has provided an overview of the work that is being done to improve the institutional framework, both at the World Bank and with respect to the Green Climate Fund. In Chapters 3 and 5, we have provided an analysis of WTO issues that are relevant to climate change mitigation and adaptation. Later in this chapter we add a further analysis of WTO subsidies law in the context of climate finance, analyzing the Agreement on Agriculture and agricultural adaptation and the Agreement on Subsidies and Countervailing Measures and climate finance.

8.3 Climate Finance and WTO Subsidies Law

Are bilateral and multilateral foreign aid subsidies consistent with the SCM Agreement? Regarding bilateral foreign aid, it is common practice to make the aid conditional on the use of suppliers from the donor country. However, such practices also can occur in the context of CDM projects that include bilateral official development assistance from Kyoto Annex 1 parties.[18] A UNFCCC study

[16] World Bank, *Phase I*, xxxiv.

[17] World Bank, *Phase II*, xvi–xvii.

[18] Public funding for clean development mechanism projects from parties in Annex I is not to result in the diversion of official development assistance and is to be separate from and not counted towards the financial obligations of parties included in Annex I (17/CP.7, preamble). To this end, 3/CMP.1, Annex, Appendix B requires that any public funding for a CDM project activity be disclosed in the project design document (PDD) <http://cdmrulebook.org/758> (accessed January 3, 2013). For example, for the CDM Project for the Mexico City Metrobus, the International Bank for

of technology transfer under the CDM found that there is a strong tendency for countries to receive CERs from projects for which they are technology suppliers. In the study's sample, 91 percent of the CERs received by Denmark came from projects for which it was a technology supplier. The percentages for Spain, Germany, and Japan were 50 percent, 40 percent, and 37 percent, respectively. However, it is not known whether the CERs purchases were contractually linked with the technology supply.[19] However, there is circumstantial evidence of making climate finance aid conditional on the use of suppliers from the donor country. More convincing evidence might come to light in the event of WTO litigation over this issue. These practices may divert scarce resources away from more cost-effective suppliers. They might also be inconsistent with the SCM Agreement.

SCM Agreement Article 1.1(a)(1) defines "subsidy" as "a financial contribution by a government or any public body *within* the territory of a Member" (emphasis added). It is unclear whether the phrase "within the territory of a Member" applies to "a government," "public body," or "financial contribution." If the financial contribution must take place within the territory of a Member, "a Member" could be interpreted to mean the Member that makes the financial contribution or could mean any Member. The former interpretation might exclude foreign aid from the definition of subsidy and, thus, from the application of the SCM Agreement. Otherwise, the SCM Agreement would apply.

Let us assume that the SCM Agreement would apply. In accordance with Article 1.1(a)(1), the financial contribution may involve a direct transfer of funds, provision of goods or services other than general infrastructure, purchases of goods, payments to a funding mechanism, or any form of income or price support in the sense of Article XVI of GATT 1994. In the case of bilateral climate financing, there could be a direct transfer of funds or some other form of income or price support, depending on the terms of the aid package. In the case of multilateral climate financing, it might take the form of payments to a funding mechanism. This would require an affirmative demonstration of the link between the government and the specific conduct of the funding mechanism (see Chapter 3). Thus, this first element is likely to be met.

There must also be a benefit conferred. In *Canada—Aircraft*, the Appellate Body interpreted of the term "benefit" under Article 1.1(b) as follows: "a financial contribution will only confer a 'benefit', i.e., an advantage, if it is provided on terms that are more advantageous than those that would have been available to the

Reconstruction and Development (IBRD) acted as Trustee of the Spanish Carbon Fund (SCF). Thus, while the PDD lists Spain as a project participant, the PDD also states, "There is no Official Development Assistance in this project and the project will not receive any public funding from Parties included in Annex I." Project 4945: BRT Metrobus Insurgentes, Mexico <http://cdm.unfccc.int/Projects/DB/AENOR1309257514.77/view> (accessed January 3, 2013).

[19] UNFCCC, *The Contribution of the Clean Development Mechanism under the Kyoto Protocol to Technology Transfer* (2010) 32 <http://cdm.unfccc.int/Reference/Reports/TTreport/TTrep10.pdf> (accessed January 3, 2013).

recipient on the market."[20] One of the rationales behind foreign aid, and the CDM in particular, is that the investment would not occur in countries with limited resources in the absence of the funding. Thus, if the funding creates the necessary financing for the participation of the donor country's suppliers, thereby creating an opportunity that would not have existed otherwise, it likely would qualify as a benefit. Alternatively, if, in the absence of the aid program, there is no market that can be used as an appropriate benchmark against which to measure whether a benefit is conferred then the subsidy would not be proved.[21]

The person or entity receiving the benefit does not have to be the same as the one who received the financial contribution.[22] Thus, the recipient country could receive the financial contribution but the supplier would still qualify as the recipient of the benefit within the meaning of the SCM Agreement. The assessment of benefit must examine the terms and conditions of the challenged transaction at the time it is made and compare them to the terms and conditions that would have been offered in the market at that time.[23] Thus, depending on the facts of each case, bilateral or multilateral climate financing could confer a benefit within the meaning of the SCM Agreement.

Export subsidies are deemed to be specific and are prohibited.[24] An export subsidy is defined as contingent, in law or in fact, upon export performance.[25] The standard of contingency in fact is met when the facts demonstrate that the granting of a subsidy, without having been made legally contingent upon export performance, is in fact tied to actual or anticipated exportation.[26] In the case of climate financing, the subsidy is tied to the exportation to the recipient country. Thus, climate financing could be found to be a prohibited export subsidy under the SCM Agreement.

However, this issue has not been addressed in the WTO jurisprudence. A conflict between multilateral agreements on climate finance and the SCM Agreement might be avoided by interpreting the latter so as to avoid such a conflict, in accordance with the presumption against conflict in international law, as noted above with respect to the definition of subsidy and the issue of benefit. Alternatively, if such an interpretation is not considered possible, the conflict might be resolved in favor of multilateral agreements on climate finance by applying the

[20] Appellate Body Report, *Canada—Measures Affecting the Export of Civilian Aircraft* (*Canada—Aircraft*), WT/DS70/AB/R, adopted 20 August 1999, para. 149.

[21] Panel Reports, *Canada—Certain Measures Affecting the Renewable Energy Generation Sector (Canada—Renewable Energy)*, WT/DS412/R, and *Canada—Measures Relating to the Feed-in Tariff Program (Canada—Feed-In Tariff Program)*, WT/DS426/R, adopted May 24, 2013, paras. 7.276–7.277. See also Appellate Body Report, WT/DS/426/AB/R, paras. 5.79, 5.219.

[22] Appellate Body Report, *United States—Countervailing Measures Concerning Certain Products from the European Communities*, WT/DS212/AB/R, adopted January 8, 2003, para. 110.

[23] Appellate Body Report, *European Communities—Measures Affecting Trade in Large Civil Aircraft (EC and Certain Member States—Large Civil Aircraft)*, WT/DS316/AB/R, adopted June 1, 2011, para. 838; Appellate Body Report, *US—Large Civil Aircraft (2nd complaint)*, WT/DS353/AB/R, adopted March 23, 2012, para. 636.

[24] SCM Agreement art. 2.3, 3.1(a).

[25] SCM Agreement art. 3.1(a).

[26] SCM Agreement fn. 5.

international rules of *lex specialis* (the specific prevails over the general) or *lex posterior* (the later treaty prevails over the earlier treaty). Another possibility would be to apply the exception in GATT Article XXI(c) regarding compliance with obligations under the United Nations Charter for the maintenance of international peace and security, which serves to divide jurisdiction between the security interests addressed by the UN and the commercial interests addressed by the GATT.[27] However, while climate change is likely to qualify as a matter of international peace and security,[28] it is not clear that GATT Article XXI could be applied to justify violations of the SCM Agreement. A final alternative would be to seek a waiver of the relevant WTO obligations, in accordance with Article IX:3 of the WTO Agreement.

8.4 Fossil Fuel Subsidies

Estimates of fossil fuel production and consumption subsidies vary, due to differences in calculation methodology and availability of data.[29] As the G20 notes, "[i]nefficient fossil fuel subsidies encourage wasteful consumption, reduce our energy security, impede investment in clean energy sources and undermine efforts to deal with the threat of climate change."[30] However, different G20 countries have interpreted this language to minimize their reporting and action on fossil fuel subsidies, based on what counts as "subsidy," an "inefficient subsidy" and a subsidy that is both "inefficient" and "encourage[s] wasteful consumption."[31] While the WTO SCM Agreement requires Members to report their subsidies, incomplete reporting on subsidies makes WTO data on fossil fuel subsidies unreliable as well.[32] The WTO reporting obligation only applies to subsidies that meet the definition in Article 1.1 and are specific within the meaning of Article 2.[33] Notification of a measure does not prejudge its legal status under GATT 1994 and the SCM Agreement.[34] Nevertheless, compliance with this reporting requirement would have the effect of bringing potentially illegal subsidies to the attention of other WTO Members, which could lead to WTO complaints, countervailing duties, or both, as we discuss below.

[27] Analytical Index, Guide to GATT Law and Practice (World Trade Organization, Geneva 1995) vol. I, 600.

[28] For example, the *Draft Climate Assessment Report* recognizes that US security interests are affected by climate changes experienced in other parts of the world. National Climate Assessment Development Advisory Committee, *Draft Climate Assessment Report* (January 11, 2013) 12 <http://ncadac.globalchange.gov/> (accessed January 12, 2013).

[29] G-20, Report to Leaders on the G20 Commitment to Rationalize and Phase Out Inefficient Fossil Fuel Subsidies (26 June 2010) 1–2 <www.g20.org/documents/> (accessed January 3, 2013).

[30] G-20, Leader's Statement, The Pittsburgh Summit (September 25, 2009) para. 24 <http://www.g20.org/documents/> (accessed January 3, 2013).

[31] Doug Koplow, *Phasing Out Fossil-Fuel Subsidies in the G20: A Progress Update* (Earth Track, Inc. and Oil Change International, Cambridge, MA 2012).

[32] Koplow, *Phasing Out Fossil-Fuel Subsidies in the G20*; SCM Agreement art. 25.

[33] SCM Agreement art. 25.2.

[34] SCM Agreement art. 25.7.

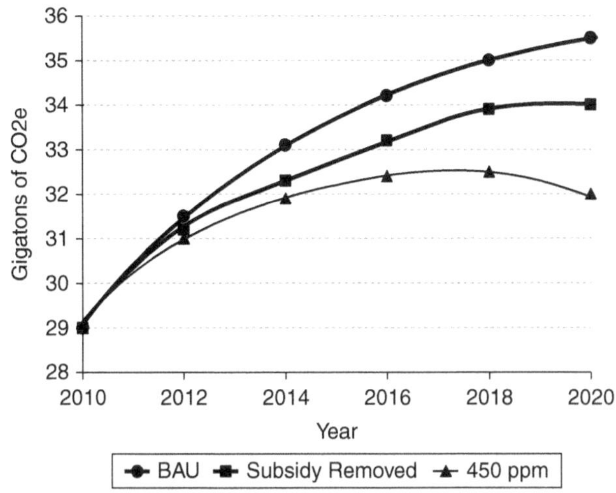

Figure 8.1 GHG emissions under different scenarios
Source: International Energy Agency, World Energy Outlook (2010) 585

Fossil fuel subsidies create incentives to use fossil fuels, thereby running counter to the objective of climate change mitigation. As Figure 8.1 shows, merely eliminating fossil fuel subsidies would achieve over forty percent of the reduction of concentrations of GHGs in the atmosphere required to achieve a concentration of 450 ppm by 2020, compared with the business-as-usual scenario.

As Figure 8.2 shows, developed countries spend over six times more on fossil-fuel subsidies than on aid pledged to help developing nations to mitigate and to adapt to climate change. At USD 57.664 billion, in 2010 these fossil fuel subsidies total more than half of the USD 100 billion of annual climate financing promised by 2020. At USD 66.79 billion, the combined total of fossil fuel subsidies and finance pledges is enough to achieve two-thirds of the 2020 climate finance goal.

As Figure 8.3 shows, a large number of developing countries also subsidize the consumption of fossil fuels. The world price of gasoline was slightly over 120 US cents in 2009. Venezuela sold it for two cents per gallon, and Iran for ten cents per gallon. In fact, all the OPEC countries sold gasoline below production cost. Since they produce oil themselves, it has been justified as a subsidy that does not "cost" anything. Of course, in terms of opportunity cost, it does *cost.* A better way of spending the money would have been to give general grants to the poor, if this was meant to be a poverty alleviation program. In the current discussion, it simply raises the question of why these resources are not dedicated instead to the acquisition or development of clean energy technologies (keeping in mind the limitations on subsidies in the SCM Agreement).

As Table 8.1 shows, many non-OECD countries spend significant amounts of money on fuel subsidies. Some large countries with over 100 million people spend the bulk of the money (such as Brazil, China, India, Indonesia, and Russia). Among them, India and China stand out for two reasons. First, both of them are large importers of oil and, second, each of them has over a billion people. They are

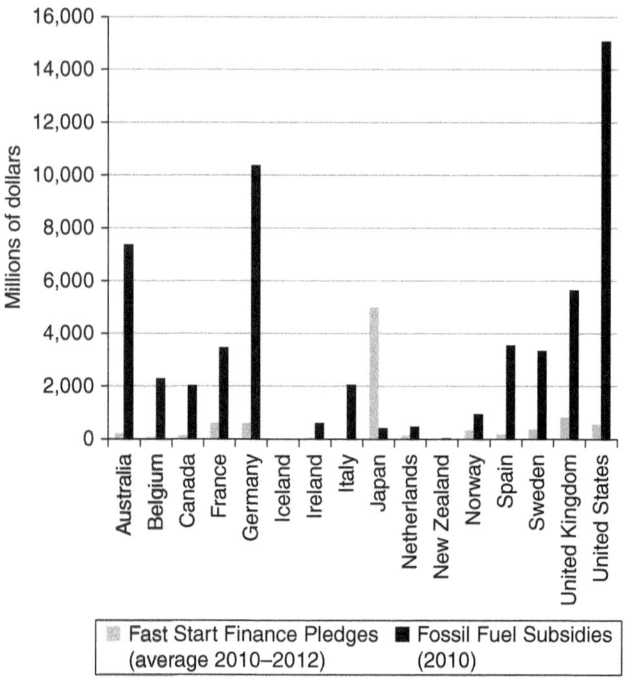

Figure. 8.2 Fossil fuel subsidies versus climate finance pledges (USD millions)

becoming significant sources of GHG emissions on a global scale. During the course of this century, the other three, Brazil, Indonesia, and Russia, net exporters of oil, will also contribute significant amounts of GHG emissions as they become developed countries. At USD 408.8 billion, the total fossil fuel subsidies in these developing countries in 2010 represent over four times the 2020 climate finance goal.

In order to discuss the fossil fuel subsidy issue at a disaggregate level we need to understand the living conditions of the poor in developing countries and their dependence on fossil fuel. At the end of 2010, there were 1.3 billion people in the world who did not have electricity at all. In addition, there were another two billion people who did not have a reliable supply of electricity. Almost all of them live in the developing countries. In addition, in these countries, there are 2.7 billion people who still rely on the traditional use of biomass for cooking.

The fundamental justification for fossil fuel subsidies (either at the level of production or at the level of consumption) in the developing world is the following: subsidies are needed to help the poor gain or maintain access to minimum living standards. Unfortunately, studies have found that fossil-fuel subsidies are regressive, that is, they benefit higher income groups who can afford consumption.[35] The

[35] Javier Arze del Granado, David Coady, and Robert Gillingham, "The Unequal Benefits of Fuel Subsidies: A Review of Evidence for Developing Countries" (2010) IMF Working Paper WP/10/202.

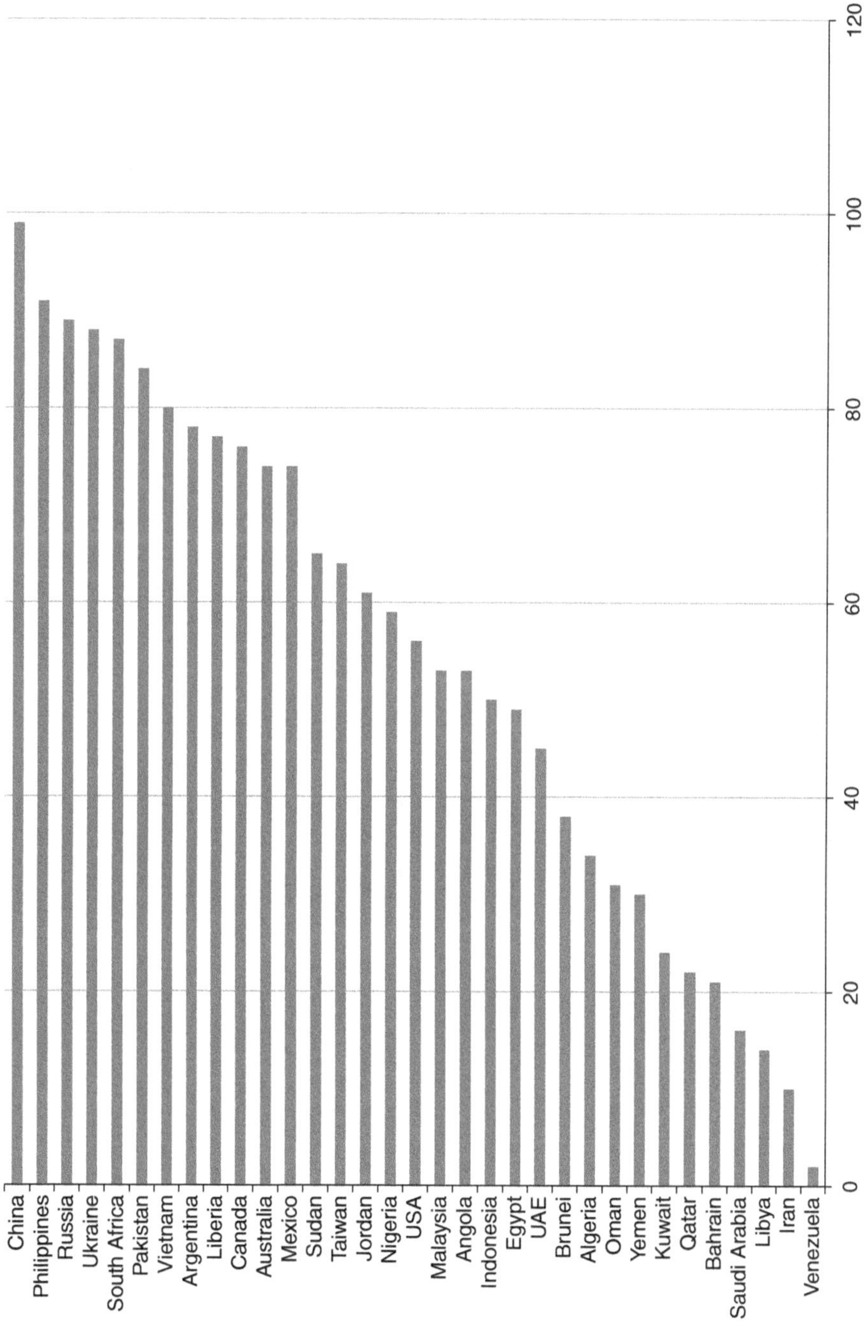

Figure 8.3 Price of a liter of gasoline in US cents in 2009

Source: GTZ Gasoline Price, 2009

Table 8.1 Developing country fossil fuel consumption subsidies (USD billions)

Country	2007	2008	2009	2010
Iran	64.56	101	64.63	80.84
Saudi Arabia	32.31	48.51	32.54	43.52
Russia	33.33	51.5	32.97	39.21
India	24.31	44.1	20.42	22.29
China	16.31	43.9	16.52	21.32
Egypt	19.39	27.93	14.86	20.28
Venezuela	18.02	24.21	14.08	19.97
UAE	8.03	15.02	11.16	18.15
Indonesia	13.17	19.02	12.56	15.94
Uzbekistan	8	15.21	11.5	11.9
Iraq	10.17	15.72	6.75	11.31
Algeria	5.6	8.83	5	10.59
Mexico	17.61	22.51	3.43	9.5
Thailand	2.82	7.38	3.8	8.47
Ukraine	6.09	9.78	6.35	7.67
Kuwait	6.44	9.52	6.19	7.62
Pakistan	7.62	12.74	5.38	7.3
Argentina	12.02	18.1	5.87	6.5
Malaysia	4.6	9.78	3.85	5.67
Bangladesh	2.08	4.4	3.6	5.03
Turkmenistan	3.55	4.82	3.11	5.01
Kazakhstan	1.75	2.71	1.27	4.32
Libya	3.35	4.32	2.21	4.21
Qatar	2.47	4.16	2.83	4.15
Ecuador	3.18	4.58	1.62	3.74
Vietnam	2.1	3.56	1.2	2.93
Nigeria	2.37	3.51	0.54	2.91
South Africa	5.16	5.74	2.96	2.12
El Salvador	0	0	0	1.19
Angola	0.64	1.18	0.27	1.12
Philippines	0.16	0.12	0.03	1.1
Azerbaijan	1.34	2.44	0.79	0.83
Taiwan	2.11	5.03	1.14	0.58
Sri Lanka	0.44	0.9	0.05	0.51
Colombia	0.68	1.01	0.25	0.48
Brunei	0.21	0.37	0.23	0.34
Korea	0	0.21	0.18	0.18
Peru	0.16	0.62	0	0
Total	342.15	554.44	300.14	408.8

Source: International Energy Agency, 2012

reason is simple. Poor households in developing countries not have access to subsidized energy directly. They do not have electricity, cooking facilities with natural gas, nor do they have vehicles that use gasoline or diesel. Moreover, low income households spend less (in absolute terms) on energy than higher-income households. Subsidizing fossil fuel consumption is an extremely inefficient way to alleviate poverty.

The World Energy Outlook of 2011 estimated "that out of the USD 409 billion spent on fossil-fuel consumption subsidies in 2010, only USD 35 billion, or 8 percent of the total, reached the poorest income group (the bottom 20 percent). This finding is based on a survey of 11 of the 37 economies identified as having fossil-fuel consumption subsidies and does not take into account subsidies specifically provided to extend access to basic energy services."[36]

The countries in Figure 8.4 have a population of over 3 billion. Suppose the fossil fuel consumption subsidies are evenly spread across the population. The bottom 20 percent of the population would have received 20 percent of the total subsidy. However, that is not the case. The best case scenario is for Pakistan—the share of the total fossil fuel consumption subsidy that goes to the bottom 20 percent of the population reaches 10 percent. Among the countries surveyed, only 2 percent of the subsidy reaches to the bottom 20 percent in South Africa—the worst case of the distribution of the subsidy (see also Table 8.2).

Subsidies to kerosene seem to be best targeted on the poor—despite being sold in the black market not just in the targeted countries but also in the neighboring countries.[37] Subsidies for electricity and natural gas were in the middle of the range,

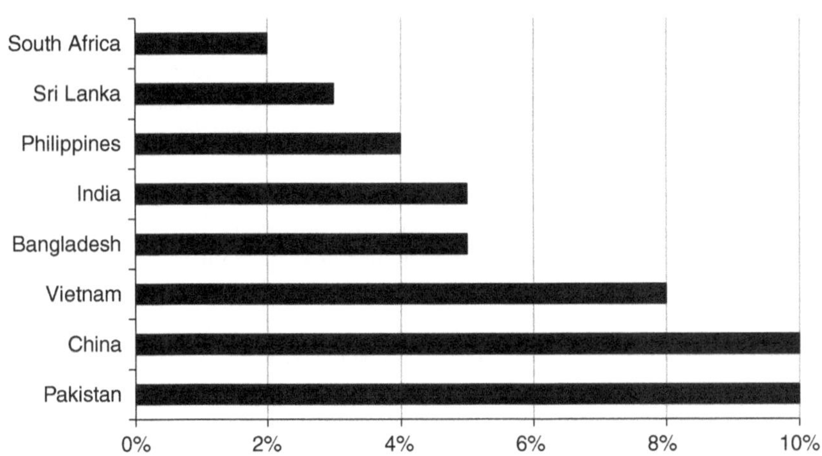

Figure 8.4 Percentage of subsidy going to the bottom 20 percent of the population

[36] International Energy Agency, *World Energy Outlook* (International Energy Agency, Paris 2011) ch. 13.

[37] International Energy Agency, *World Energy Outlook* ch. 13.

Table 8.2 Share of fossil fuel subsidies received by the lowest 20 percent income group by fuel in surveyed economies, 2010

Diesel	6%
Kerosene	15%
Gasoline	6%
Natural Gas	10%
LPG	5%
Electricity	9%

Note: Countries surveyed were Angola, Bangladesh, China, India, Indonesia, Pakistan, Philippines, South Africa, Sri Lanka, Thailand and Vietnam

Source: International Energy Agency, Recent Developments in Energy Subsidies (International Energy Agency, Paris 2012), page 103.

with shares of 9 percent and 10 percent to the bottom 20 percent of the income group. Subsidies to Liquified Petroleum Gas, gasoline, and diesel perform the worst—only 5 to 6 percent reach the bottom 20 percent of the population. A much more efficient way of implementing social welfare programs for the poorest households is conditional cash transfer programs.[38]

How much investment do we need to achieve GHG concentrations of 450 ppm? Table 8.3 uses 2010–2030 figures estimated by McKinsey. We have added up those numbers. The bottom line is that, even if the euros are converted into USD, the financial needs of the developing countries are still smaller than their own energy subsidy expenditures.

Financial difficulties in developed countries, such as the United States and the European Union, reduce the likelihood that they will be willing to finance the acquisition of technology in developing countries to the degree that is needed to achieve sufficient climate change mitigation and adaptation. However, economic growth in major developing countries will increase the financial resources available to acquire technologies over time. Reallocating resources from fuel subsidies to the acquisition and development of clean energy technologies and adaptation technologies would increase the availability of financing for these ends.

At the WTO, generally available fuel subsidies are unlikely to be subject to multilateral action (dispute settlement to address the adverse effects of subsidies on the interests of other Members) or unilateral action (countervailing duties) under Parts III and V of the SCM Agreement, respectively, since they are not specific to one enterprise or industry. They may be prohibited under Part II if they are contingent upon export performance or upon the use of domestic over imported goods. However, subsidies that are targeted specifically to develop clean energy industries are more vulnerable to multilateral and unilateral action, as evidenced by recent WTO complaints. The United States has complained to the WTO about

[38] Conditional Cash Transfer programs provide cash payments to poor households that meet certain "family behavioral requirements." Typically, these requirements are tied to health care and education of the children in the family.

Table 8.3 Investment needs per year for the world and the developing countries (in billions of euros of 2010) for 450 ppm

	Global Investment needs	Global Investment needs	Developing Countries' needs	Developing Countries' needs
	2010–2020	2020–2030	2010–2020	2020–2030
Buildings (mainly energy efficiency)	€125	€155	€25	€45
Transportation (mainly energy efficiency)	€70	€215	€25	€100
Industry (mainly energy efficiency)	€75	€80	€40	€50
Power	€65	€125	€30	€70
Waste	€10	€10	€5	€5
Forestry and agriculture (terrestrial carbon)	€5	€5	€5	€5
Total	€350	€590	€130	€275

Source: Marcel Brinkman, "Incentivizing Private Investment in Climate Change Mitigation," Chapter 14, Table 1, Richard B. Stewart, Benedict Kingsbury, and Bryce Rudyk (eds.), *Climate Finance: Regulatory and Funding Strategies for Climate Change and Global Development* (New York University Press, New York 2009)

Chinese and Indian subsidies, including those for clean energy technologies.[39] The United States also initiated a WTO dispute against Chinese wind power products and applied countervailing duties on Chinese solar panel imports. China initiated a WTO dispute against the United States regarding the countervailing duties on Chinese solar panel imports.[40]

As we noted in Chapter 3, while the SCM Agreement initially contained an exception for environmental subsidies, the exception expired. Thus, the current subsidies rules create perverse incentives for WTO Members to grant generally available fossil fuel consumption subsidies, but not subsidies to develop and disseminate clean energy technology. Countervailing duties increase the cost of clean energy technology, when the objective should be to lower the cost. Countervailing duties

[39] Keith Bradsher, "200 Chinese Subsidies Violate Rules, U.S. Says" New York Times, October 6, 2011 <http://www.nytimes.com/2011/10/07/business/us-says-some-chinese-subsidies-violate-trade-rules.html?ref=business&pagewanted=all> (accessed December 15, 2012); Office of the United States Trade Representative, "United States Details China and India Subsidy Programs in Submission to WTO" October 2011 <www.ustr.gov/about-us/press-office/press-releases/2011/october/united-states-details-china-and-india-subsidy-prog> (accessed December 15, 2012).

[40] *China—Measures Concerning Wind Power Equipment*, DS419, <http://www.wto.org/english/tratop_e/dispu_e/cases_e/ds419_e.htm> (accessed December 15, 2012); *United States—Countervailing Duty Measures on Certain Products from China*, DS437, <http://www.wto.org/english/tratop_e/dispu_e/cases_e/ds437_e.htm> (accessed December 15, 2012); United States—Countervailing and Anti-dumping Measures on Certain Products from China, WT/DS449, <http://www.wto.org/english/tratop_e/dispu_e/cases_e/ds449_e.htm>, in which China requested consultations regarding US antidumping and countervailing duties on September 17, 2012.

also create a barrier to international technology transfer. WTO complaints and countervailing duty investigations also can hamper international trade in clean energy technology regardless of the outcome, by increasing commercial uncertainty.

While the current state of national and international policy on energy subsidies is deplorable, there is good news. First, the foregoing figures demonstrate that there is ample energy financing available, both in developed and in developing countries. It simply is misallocated. Second, while multilateral negotiation paralysis at the WTO makes reforms to the SCM Agreement unlikely at the moment, it may be possible to structure clean energy subsidies in a way that minimizes the risk of violating the SCM Agreement, depending on how the issue of "benefit" is resolved.[41] It is also possible to structure adaptation financing for subsistence farmers in a manner that is consistent with the Agreement on Agriculture. We analyze these issues further in the following two sections.

8.5 Financing Adaptation by Subsistence Farmers

As we noted in Chapter 1, climate change will have a greater impact on the viability of traditional plant varieties in developing countries in the tropics than in developed countries in temperate zones. This means that the need for GMO seeds will be greater in developing countries. However, a larger percentage of the population depends on agriculture in developing countries (50 percent in India, for example) and the poorest in developing countries depend on subsistence agriculture. Their poverty means that they rely on collecting seeds from traditional plant varieties to sow future crops. However, climate change will make these varieties increasingly untenable. As production in these varieties decline, developing countries will require greater access to GM varieties that can raise yields and adapt to climate change.

In theory, farmers in developing countries could adapt to a changing climate by purchasing GMO seeds. In many developing countries, especially in Latin America and the Middle East, the majority of the population has moved to urban areas. Nevertheless, in the rest of the developing countries, principally in Asia, the majority of the population is rural and poor. The world's poor will live in rural areas for many decades to come.[42] There are 1.4 billion people living in extreme poverty (defined as those living on less than USD 1.25 per day), of which 70 percent live in rural areas. Half of the developing-country rural population farms 3 hectares or less of crop land.[43]

[41] Appellate Body Report, *Canada—Certain Measures Affecting the Renewable Energy Generation Sector (Canada—Renewable Energy)*, WT/DS412/R, and *Canada—Measures Relating to the Feed-in Tariff Program (Canada—Feed-In Tariff Program)*, WT/DS/426/AB/R, adopted May 24, 2013.

[42] Martin Ravallion, Shaohua Chen, and Prem Sangraula, "New Evidence on the Urbanization of Global Poverty" (2007) Working Paper 4199, World Bank, Washington, DC <http://siteresources.worldbank.org/INTWDR2008/Resources/2795087-1191427986785/RavallionMEtAl_UrbanizationOfGlobalPoverty.pdf> (accessed March 15, 2013).

[43] International Fund for Agricultural Development, "New Realities, New Challenges: New Opportunities for Tomorrow's Generation" Rural Poverty Report (2011) 47 <http://www.ifad.org/rpr2011/report/e/rpr2011.pdf> (accessed March 15, 2013).

Would farmers with 3 hectares with less than USD 1.25 per day income be able to afford to cultivate a crop like genetically modified corn? Three hectares is equivalent to 8 acres. One acre needs 20 lbs of seeds. Therefore, 8 acres will need 160 lbs of seeds. We know that 40 lbs of seeds cost USD 125.[44] Therefore, 8 acres will need USD 500 worth of seeds. However, genetically modified seeds need a fertilizer and herbicides. The cost of such inputs would be at least USD 1,400. Therefore, the total cost of inputs would be at least USD 2,000 per year for such farmers. Per capita income of such farmers is USD 450 a year. Thus, even with a family of four, an income of USD 1,800 would still be below what these farmers will be able to afford even if they plough their entire income into farming. What this calculation demonstrates is that it is impossible for such farmers to be availing themselves of genetically modified corn. The story is similar for other major crops like rice or wheat.

Suppose that some agency is willing to loan the farmer all of these input costs. Would it then be possible for the same farmer to generate profits? The answer is yes. The output of an acre of genetically modified corn is 100 bushels per acre.[45] In 8 acres of land, a farmer would be able to produce 800 bushels. The global price of a bushel of corn averaged at least USD 7 per bushel during 2006–2010. Hence the total value of the output of the farmer with 8 acres of land would be USD 5,600. Even with a total input cost of USD 2,500 (seeds plus fertilizer and herbicides costs USD 2,000) the farmer would be able to generate a profit of USD 2,900 per year.

This calculation does not take into account the risk faced by the farmer. For example, there is risk of crop failure (due to floods, severe drought, or other natural calamities). In addition, there is a risk of low market prices as the price of agricultural products fluctuates. Nevertheless, this calculation shows that with proper available credit, it is possible for a poor farmer to mitigate the risk of severe weather that genetically modified crops can withstand. However, it requires the availability of credit.

The WTO Agreement on Agriculture prohibits subsidies to domestic producers in excess of each Member's commitment levels.[46] However, agricultural input subsidies generally available to low-income or resource-poor producers in developing country Members are exempt from domestic support reduction commitments.[47] Least-developed country Members are not required to undertake any reduction commitments.[48] These exemptions should suffice to permit financing to subsistence farmers to acquire GM seeds, fertilizer, and pesticides. However, if

[44] Sam McNeill, "Control Seed Costs to Manage Profits" <http://www.bae.uky.edu/ext/grain_storage/calculators/Control_Seed_Costs_to_Manage_Profits.pdf> (accessed December 15, 2012).

[45] Commodity Costs and Returns, United States Department of Agriculture, document available at <http://www.ers.usda.gov/data-products/commodity-costs-and-returns.aspx> (accessed April 6, 2013).

[46] Agreement on Agriculture art. 3.2.

[47] Agreement on Agriculture art. 6.2.

[48] Agreement on Agriculture art. 15.2.

these subsidies are made contingent upon the use of domestic over imported goods, they could violate the SCM Agreement.[49]

In addition to the foregoing exemptions, WTO Members can claim exemptions from the reduction commitments for domestic support measures under Annex 2.[50] Annex 2 measures must have no, or at most minimal, trade-distorting effects or effects on production. Accordingly, these measures must conform to three basic criteria: (1) the support must be provided through a publicly-funded government program not involving transfers from consumers; (2) the support must not have the effect of providing price support to producers; and (3) the support must meet the policy-specific criteria and conditions set out in Annex 2 according to the category of support.[51] There are two categories that are relevant to climate-change related domestic agricultural support payments: natural disasters and environmental programs.

Governments can make payments for relief from natural disasters that have occurred or are occurring. The term "natural disaster" is not defined, but includes disease outbreaks, pest infestations, nuclear accidents, and war on the territory of the Member concerned.[52] Climate change and its effects are likely to qualify as natural disasters that have occurred or are occurring. Eligibility for payments is subject to two requirements: (1) prior formal recognition by government authorities that a natural or like disaster has occurred or is occurring; and (2) a production loss which exceeds 30 percent of the average of production in the preceding three-year period or a three-year average based on the preceding five-year period, excluding the highest and the lowest entry.[53] In addition, the payments must be applied, following a disaster, only in respect of losses of income, livestock, land, or other production factors due to the natural disaster in question and, during a disaster, to prevent further losses of this nature.[54] Payments must not compensate for more than the total cost of replacing such losses and must not require or specify the type or quantity of future production.[55] Where a producer receives in the same year payments under this category and under the category of income insurance and income safety-net programs, the total of such payments must be less than 100 percent of the producer's total loss.[56]

Payments under environmental programs are subject to fewer requirements. Eligibility for payments must be determined as part of a clearly-defined government environmental or conservation program and be dependent on the fulfillment of specific conditions under the government program, including conditions related to production methods or inputs.[57] In addition, the amount of payment must be

[49] SCM Agreement art. 3.1(b); Appellate Body Report, *United States—Subsidies on Upland Cotton* (*US—Upland Cotton*), WT/DS176/AB/R, adopted March 21, 2005, para. 550.

[50] Agreement on Agriculture art. 6.1, Annex 2 para. 1.

[51] Agreement on Agriculture Annex 2 para. 1.

[52] Agreement on Agriculture Annex 2 para. 8.

[53] Agreement on Agriculture Annex 2 para. 8(a).

[54] Agreement on Agriculture Annex 2 para. 8(b) and (d).

[55] Agreement on Agriculture Annex 2 para. 8(c).

[56] Agreement on Agriculture Annex 2 para. 8(e).

[57] Agreement on Agriculture Annex 2 para. 12(a).

limited to the extra costs or loss of income involved in complying with the government program.[58] In contrast to the category of natural disasters, which would allow support in response to the effects of climate change, the focus in the category of environmental programs is on support to comply with environmental requirements. Thus, the first category is relevant for adaptation measures, while the second is relevant for mitigation measures. Once again, if these subsidies are made contingent upon the use of domestic over imported goods, they could violate the SCM Agreement.[59]

8.6 Financing Clean Energy Projects

As we argued in Chapter 1, with respect to clean energy technologies, while IPRs may have a negative impact on innovation, competition, and affordable access, we argue that the focus of the debate regarding access to clean energy technologies should not be on IPRs. First, achieving reforms to the international intellectual property regime is likely to prove difficult. Second, IPRs do not represent the main obstacle to innovation, competition, and affordable access for clean energy technologies. For example, the fact that the United States has applied countervailing duties on imports of solar panels from China indicates that IPRs are not a sufficient barrier to competition in this sector. Otherwise, countervailing duties would not be necessary to protect the US solar panel industry from Chinese competition. With respect to clean energy technologies, the debate on IPRs is misplaced and distracts from the real issues: creating incentives for and removing obstacles to clean energy development and dissemination.

Fossil fuel subsidies need to be reallocated to the development and the dissemination of clean energy technologies. This is a policy that many developing countries can adopt without the need for multilateral agreements or foreign aid, by reallocating their budgets for fossil fuel subsidies. This may require reforms to WTO subsidies law, depending on the direction that WTO jurisprudence takes. Countries also need to remove barriers to trade in clean energy technologies, rather than erect such barriers. This, too, may require reforms to WTO subsidies law, although reforms to national countervailing duty law could be undertaken without reforming WTO law. WTO law does not require countries to apply countervailing duties; it merely permits this practice as long as it is done in accordance with the requirements of the SCM Agreement.

Clean energy projects make economic sense for the rural population who do not have electricity. In 2010, 1.4 billion people lived without electricity with 85 percent living in rural areas that are far from electricity grids. Without any additional help, the number of people without electricity will fall to 1.2 billion

[58] Agreement on Agriculture Annex 2 para. 12(b).

[59] SCM Agreement art. 3.1(b); Appellate Body Report, *United States—Subsidies on Upland Cotton* (*US—Upland Cotton*), WT/DS176/AB/R, adopted March 21, 2005, para. 550.

by 2030. The number of people relying on traditional biomass as fuel was 2.7 billion in 2010 and projected to rise to 2.8 billion in 2030.[60]

As a result, it is more practical to consider decentralized, renewable energy technologies for rural areas that are far from the electricity grid. In these circumstances, it is feasible to leapfrog fossil-fuel generated electricity and go directly to clean energy sources. We illustrate this principle with an example from India with three options: (1) traditional coal fired power grid, (2) biomass fuel, and (3) hybrid solar/wind powered electricity.

Solar and wind powered village-scale mini-grids can provide electricity for lighting and other small electric appliances such as radio, television, and cellular phones. In addition, solar cookers can provide clean cooking. Solar crop dryers and solar water heaters can be used for crops, other agricultural processing and for hot water needs. Mechanical wind powered pumps or solar photovoltaic pumps can replace diesel generators for pumping water. Biogas from household scale digester and small-scale biomass gasifiers can provide for village-level lighting and other small-scale electric needs. Biomass gasifiers can be fed into small electricity grid systems to power small industry. They can also pump water for drinking and agriculture. Biogas digesters can also help with heating, cooking, heating water, and commercial stoves and agricultural processing.

Figure 8.5 shows us an experiment conducted in India to demonstrate the following. Coal-fired power plants are cost-effective when the power source is less

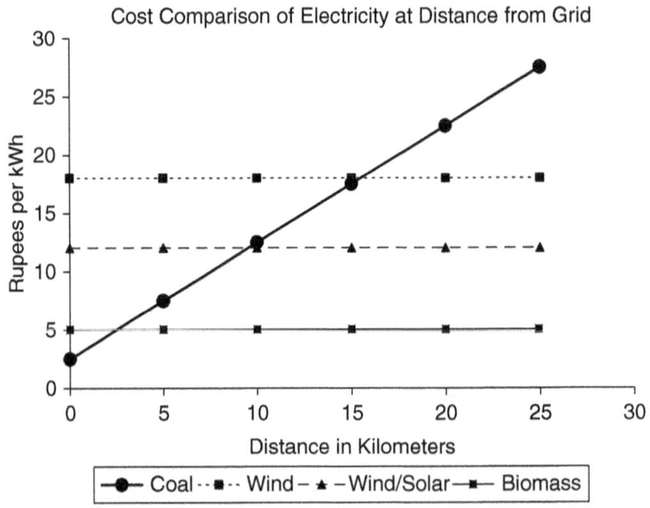

Figure 8.5 Cost comparison of coal powered electric grid and biomass and solar/wind hybrid generators in India

Source: Access to Energy for the Poor: The Clean Energy Option. Oil Change International, June 2011, Figure 2.

[60] International Energy Agency, *Energy Poverty: How to Make Modern Energy Access Universal* (IEA/OECD, Paris 2010) 7.

than 5 kilometers from the place of consumption. The cost of consumption keeps rising with the distance because of loss of power over longer distance. The advantage of wind, solar, and biomass fuels comes from their local nature. The cost does not depend on distance because they are always locally produced. For more than a billion people, coal-fired power plants are not really an option. Figure 8.5 clearly shows that it is inefficient to try to connect these people to the power grid. It is better to assist them financially to get them to clean energy sources like biomass or wind or solar power.

We should also note that the cost calculation for coal powered electricity here did not take into account the externality that coal powered plants produced (discussed in detail in Chapter 6). Neither does it take into account the subsidy that the government in India provides for all the power plants (fixed cost). It simply takes into account the marginal cost of producing a kWh of electricity in 2010.

Since 1992, the World Bank has contributed USD 790 million to solar home system components in 34 countries, mostly with Global Environmental Facility subsidies and relying on microfinance for consumers. These projects can have economic rates of return of 30 to 90 percent, but have little impact on GHG reductions because off-grid households use so little energy. Nevertheless, climate change is a serious threat to development, particularly for the poor.[61] Promoting the electrification of poor, rural communities with clean energy technology is a cost-effective means of economic development that avoids increasing GHG emissions and can benefit a large number of poor.

Dissemination of clean energy technologies can also be facilitated by removing barriers to foreign investment and international trade in services, a topic that we discuss in the following section.

8.7 Private Sector Finance and Technology Transfer

In Chapter 4 we have addressed the role of international investment agreements in creating incentives for foreign direct investment to serve as a vehicle for financing investment in climate change mitigation and adaptation. In this regard, we also saw that subsidies for these purposes need to be designed with international investment agreements in mind, to minimize the risk of litigation with and compensation to foreign investors for breaches of the relevant obligations. Here, we discuss the importance of international trade in services to facilitate the participation of the private sector in climate finance and technology transfer.

One entry mode for international trade in services is via a commercial presence, which implies foreign direct investment. Cross-border trade, consumption abroad and the presence of natural persons are other modes of delivery that are relevant to both climate finance and technology transfer. In this regard, the market access commitments of WTO Members under the General Agreement on Trade in

[61] World Bank, *Phase II*, xi, xxx.

Services, or the lack thereof, may imply obstacles to trade in services related to climate finance and technology transfer. The effort by some WTO Members to negotiate an International Services Agreement makes this one area where progress may be possible, on a plurilateral basis. Trade in environmental services will have an important role to play in disseminating the know-how and technology to address climate change, as will trade in financial services with respect to climate finance.

According to IPCC (2007), GHG mitigation activities relate to enhancing energy efficiency, increasing reliance on renewable energy, deploying carbon-dioxide capture and storage (CCS), taking forest-related actions, and undertaking specific solutions needed to mitigate emissions of non-CO2 GHGs. The deployment of GHG-mitigation technologies is dependent on the availability of a wide range of services, including those that are imported, particularly business services, telecommunications services, and construction and related engineering services. In certain countries and regions, financial services also play an important role, particularly for the implementation of climate-policy instruments, such as emissions trading schemes, and training is another component of the services that are sometimes provided as part of GHG-mitigation projects.[62] Initially, a project to mitigate GHG emissions involves consultancy services, R&D services, engineering, architectural, or design services. Other services become involved if a facility is built, particularly construction.[63]

Kim identifies services that are directly linked to climate-friendly goods, analyzes specific commitments of WTO Members that trade the most in these services, and examines offers made in the Doha Round. The paper covers mitigation sectors identified by the IPCC: energy supply, transport, buildings, industry, agriculture, forestry, and waste.[64]

Technology transfer requires trade in services, because expertise and capacity in GHG-mitigation technologies vary across nations. Several countries have little or no domestic capacity to design projects based on low-carbon power sources, such as wind turbines, geothermal energy, or concentrated solar power. Firms with the necessary expertise might not exist, and energy engineers with relevant expertise might not be sufficiently available in the host country. Products and technologies related to climate change are connected with the provision of services. In most developing countries, a great deal of technologically sophisticated equipment must be imported, but many construction materials can be procured locally.[65]

Trade liberalization could make an important contribution to the mitigation of GHG emissions, particularly by facilitating the adoption of energy-efficient and

[62] Ronald Steenblik and Massimo Geloso Grosso, "Trade in Services Related to Climate Change: An Exploratory Analysis" OECD Trade and Environment Working Papers, 2011/03 (OECD Publishing 2011) 7 <http://dx.doi.org/10.1787/5kgc5wtd9rzw-en> (accessed October 10, 2012).

[63] Steenblik and Geloso Grosso, "Trade in Services Related to Climate Change."

[64] Joy A. Kim, "Facilitating Trade in Services Complementary to Climate-friendly Technologies" Environmental Goods and Services Series, Issue Paper 16, International Centre for Trade and Sustainable Development, Geneva (2011) <www.ictsd.org> (accessed October 10, 2012).

[65] Steenblik and Geloso Grosso, "Trade in Services Related to Climate Change" 4.

low-carbon technologies.[66] However, barriers to trade in services often hamper trade in related technologies.[67] A document submitted to the WTO in December 2007 and cosponsored by Australia, Canada, the European Communities, Iceland, Japan, Korea, Norway, Kingdom of Saudi Arabia, Singapore, The Separate Customs Territory of Taiwan, Penghu, Kinmen and Matsu, and the United States (JOB(07)/208), proposes a guide to scheduling commitments in energy services, drawing on services classified under ten separate groups. This approach could be used to identify services sub-sectors to liberalize in order to improve quality and competitive pricing of services needed for GHG mitigation.[68]

Cross-border trade taking place over the Internet is becoming more commonplace, often complementing movement of personnel. Consumption abroad typically involves training of a client's personnel. Commercial presence is critical for the provision of services related to climate change. Many of these services entail construction and operation of production facilities, which makes establishing a commercial presence necessary for trade to take place. The temporary movement of natural persons is also common, particularly when expert judgment or supervision is required for a short period of time.[69] In addition, contractors for projects related to climate change depend on imports of technologically sophisticated equipment to the project site from other countries, which highlights the importance of standards affecting the mobility of goods and technologies, not just people.[70]

Carbon-market services also have emerged to help implement Kyoto finance mechanisms and emissions trading schemes.[71] Expertise and capacity in GHG-mitigation technologies varies across nations particularly in the case of low-carbon power sources based on renewable energy technologies. Even in developed countries with private sector expertise and capacity in the services needed for mitigation-technology projects, project sponsors often choose to import more cost-effective services from an international supplier.[72] Trade in climate-related services also can facilitate capacity-building to support technology transfer.[73] Finally, carbon trading and trade in CERs under the CDM may constitute financial services that would be

[66] Steenblik and Geloso Grosso, "Trade in Services Related to Climate Change" 5; WTO and UNEP, *Trade and Climate Change* (WTO, Geneva 2009).

[67] Steenblik and Geloso Grosso, "Trade in Services Related to Climate Change" 5; Ronald Steenblik and Joy Kim, "Facilitating Trade in Selected Climate Change Mitigation Technologies in the Energy Supply, Buildings, and Industry Sectors" OECD Trade and Environment Working Papers, No. 2009-02, OECD, Paris (2009) <http://www.oecd.org/tad/42738759.pdf> (accessed March 15, 2013); Ronald Steenblik et al., "Facilitating Trade in Selected Climate Change Mitigation Technologies in the Electricity Generation and Heavy-Industry Sectors" OECD Trade and Environment Working Papers, No 2010-02, OECD, Paris (2010).

[68] Steenblik and Geloso Grosso, "Trade in Services Related to Climate Change" 6.

[69] Steenblik and Geloso Grosso, "Trade in Services Related to Climate Change" 38.

[70] Steenblik and Geloso Grosso, "Trade in Services Related to Climate Change" 39.

[71] Steenblik and Geloso Grosso, "Trade in Services Related to Climate Change" 33.

[72] Steenblik and Geloso Grosso, "Trade in Services Related to Climate Change" 37.

[73] Steenblik and Geloso Grosso, "Trade in Services Related to Climate Change" 37–8.

regulated by the GATS and the Annex on Financial Services, which allows prudential measures.[74]

8.8 Conclusion

The money spent on fossil fuel subsidies should be reallocated to climate change mitigation and adaptation. Eliminating fossil fuel subsidies would reduce GHG emissions and increase climate change funding. The difficulty of achieving adequate multilateral climate financing appears likely to continue for several reasons. The lack of agreement evident in the Report of the High-level Advisory Group on Climate Change Financing, a group charged with advising on rather than deciding on these issues, is symptomatic of the larger problem of making developed countries responsible for financing climate change mitigation and adaptation in developing countries. Moreover, while fossil fuel subsidy reallocation seems a logical place to find financing, the G20 avoidance of compliance with their joint commitment on this issue is an indication of how difficult this task may prove to be. The slow response of the World Bank to climate change is another indication of how difficult change can be. At some point, developing countries will have to assume greater responsibility for financing their national climate change strategies, but should draw upon the growing expertise of institutions such as the World Bank. Financial difficulties in developed countries, combined with growing wealth in developing countries and the increasing risk of climate change causing damage, in developing countries in particular, means that the current approach will have to change. Developing countries could start by reallocating their fossil fuel subsidies to finance their mitigation and adaptation efforts, with a particular focus on helping the poorest. In this chapter, we have provided two examples: financing adaptation for subsistence farmers and financing access to clean energy in poor rural communities.

Multilateral cooperation and financing will continue to be an essential element in climate finance. The WTO will have to adapt subsidies rules, via judicial interpretations on a case-by-case basis or via a negotiated response to agreements reached in other multilateral organizations. The increasing number of actors in climate finance activities will require more coordination and adaptation of roles among institutions such as the World Bank and the Green Climate Fund. The activities of these institutions will have to be coordinated with those of the private sector participants as well.

The United States and China, in particular, need to take on the leadership roles that are necessary to promote multilateral change and to use unilateral measures judiciously to create incentives for multilateral progress. The US and China have a large share of world CO2 emissions. China's growing economic and technological clout makes it well placed to take on a leadership role, particularly with respect to

[74] Robert Howse and Antonia Eliason, "Carbon Trading and the CDM in WTO Law" in Stewart et al. (eds.), *Climate Finance: Regulatory and Funding Strategies for Climate Change and Global Development* (New York University Press, New York 2009) 254, 256–7.

the approach of developing countries to climate finance and technology transfer. The United States plays a key role in various spheres of international affairs, where its leadership is essential to move forward.

The United States is the largest donor country (even though per capita it is near the bottom end of the OECD countries). It plays a large role in CO2 emissions. The current US contribution to global emissions is about 20 percent.[75] It has the biggest share of voting rights at the World Bank, at 16 percent of the total votes. At the IMF, the United States has 17 percent of the votes—roughly the same as all of Asia combined (whose population is more than 12 times that of the United States).

However, the United States is a country where less than half the population believes that climate change is caused by human activities, in contrast to neighboring Canada, where 61 percent believe that climate change is caused by human activities. As a result, a large number of US politicians deny a human role in climate change. This, in turn, has an impact on (1) the foreign aid directed to climate change, (2) activities to reduce emissions, and (3) funding activities of large international agencies such as the World Bank and the IMF. Nevertheless, in the aftermath of Hurricane Sandy, and with the release of the 2013 Draft Climate Assessment Report, there are signs that the United States is preparing to take on a leadership role with respect to climate change. US leadership on climate change could serve as an important catalyst for not just climate finance, but wider actions on climate change as well.

[75] *Draft Climate Assessment Report* 6.

9

Creating Incentives and Removing Obstacles

This book has analyzed the role of climate change in global economic governance against the backdrop of two central themes: technology diffusion and unilateral responses to multilateral negotiation failure. It has not been possible to reach multilateral agreements with respect to climate change finance, intellectual property rights for plant varieties, a multilateral investment agreement, or international trade in environmental goods and services. Multilateral progress in all of these areas would facilitate technology diffusion and diminish the need for unilateral action.

In the case of the WTO, deadlock at the multilateral level has led to unilateral, bilateral, and regional policy responses. The same may happen with respect to climate change negotiations. Unilateral measures may be taken to address local or global concerns and may be used to create incentives for multilateral action. They may be consistent with international obligations or not, depending on the circumstances of each case. Unilateral measures can serve as catalysts for multilateral action on climate change, by prompting the affected economic actors to pressure their governments to seek a solution, through litigation or negotiation. Climate change agreements should either comply with or prompt modifications to international economic law and global models of economic governance.

Existing international economic law places limitations on the right of national and sub-national governments to regulate to address climate change. Given the current difficulty in reaching multilateral agreements, we assume for the most part that countries will have to develop climate change policy and law within the constraints of the existing legal, economic, and financial framework. The shifting fortunes of developed and emerging economies have altered the dynamics of global governance. The ensuing multilateral negotiation paralysis means that unilateral action will be necessary to create incentives to address climate change. However, the risk of regulatory capture needs to be addressed to ensure that these unilateral measures are consistent with international law and economically viable.

The failure to advance in the UNFCCC negotiations prevents advances at the WTO with respect to the compatibility of multilateral and unilateral responses to climate change. Moreover, the climate of uncertainty leads to risk aversion, making it less likely that financing of green technologies will come from the private sector. However, public financing of green technologies will be uneven across countries and could lead to serious market distortions that will create further political and economic instability. As a result, it is important to identify policy issues and options and ways to overcome negotiation obstacles.

One proposal, with respect to WTO negotiations, is to make negotiations less ambitious, by abandoning the rule that "nothing is agreed until everything is agreed," and to abandon decision-making by consensus. There are precedents for this approach at the WTO, in which a limited number of Members agreed to liberalize specific sectors once enough Members were on board to cover ninety percent of trade in the sector. The most-favored-nation rule extends concessions to all WTO Members and the resulting agreement is left open for other Members to join. The same approach to environmental goods and services would reduce barriers to technology diffusion for climate change. A similar approach could be taken with respect to GHG emissions, by seeking agreement among the countries that account for the overwhelming majority of emissions, and by leaving it open for other countries to join.

Countries may choose to apply taxes that discriminate between different products based on differences in national climate change policies or differences in the carbon footprints of products. Concerns regarding the impact of carbon taxes on competitiveness can lead to the implementation of accompanying border tax adjustments or export subsidies to compensate for a lack of carbon taxes or equivalent measures in trade partners. While some have argued in favor of imposing countervailing duties against products from countries that do not require emissions reductions, the definition of subsidy would likely preclude such actions unless, for example, a country applied a general carbon tax but then subsidized a specific industry by not collecting the tax. Similarly, antidumping duties cannot be used to counteract so-called "environmental dumping," since differences in domestic environmental regulation cannot be used as a basis for applying antidumping duties.

However, the Antidumping Agreement and the SCM Agreement's rules on countervailing duties might serve as a model for regulating this type of unilateral climate change measure. Antidumping duties and countervailing duties are unilateral trade actions. WTO law regulates their use, by imposing substantive and procedural rules on the national investigating authorities that investigate dumping and subsidies allegations and calculate the antidumping and countervailing duties. A limited number of like-minded countries could develop a similar approach to regulate the use anti-carbon duties between them. While such investigations create commercial uncertainty that can disrupt trade, if countries begin to take such measures it would be better to regulate their use. Measurement of carbon footprints may be difficult, but investigating authorities are already set up to measure complex economic and legal issues.

Mitigation and adaptation deal with different aspects of the risks imposed by climate change, but they are interrelated. Since mitigation would prevent the worst-case scenario from occurring, it increases the chances that the remaining climate risks can be successfully managed through adaptation. The benefits of mitigation are global, but the cost of reductions is primarily local. In contrast, adaptation costs and benefits are both primarily local. Mitigation benefits are long-term, whereas adaptation benefits are short and medium-term. Because of these differences, national and international climate change response efforts have followed separate mitigation and adaptation paths, with a major focus on mitigation. However, some climate change impacts are unavoidable now, so adaptation efforts have begun. Nevertheless, the technological and financial capacity of countries to adapt to climate change differs significantly, as does their vulnerability to the effects of climate change. These differences between countries are part of the reason for incorporating the principle of common but differentiated responsibilities into the UNFCCC.

Policy responses that depend on the maturity of financial markets or the robustness of national regulatory institutions may not work in all countries. Differences in levels of economic development influence the design of global solutions both at the negotiation stage and at the implementation stage. However, simply dividing countries into developed and developing is an outdated approach that is not working.

Incentives to take mitigation and adaptation measures are asymmetrical, since the costs and benefits of these measures are not the same for all countries. Discussing the costs and benefits of mitigation and adaptation in terms of developed and developing countries ignores the global nature of the problem, the potential seriousness of the consequences, and the great variation among countries in vulnerability to and capacity to adapt to the effects of climate change.

Dividing the world into developed and developing countries, and using that categorization as a basis for addressing climate change, whether through adaptation or mitigation, is too simplistic. The financial and technological endowments of countries are not frozen in time. It makes more sense to assign responsibility on a scale, based on objective criteria that are in accordance with the governing principles of international environmental law and determine responsibility for the cost of mitigation and adaptation, respectively. These criteria could be used to create an index of countries that evolves as conditions change, what we call the "Climate Sensitivity Index."

With respect to clean energy technologies, the debate over intellectual property rights, and technology transfer from developed to developing countries is similarly misplaced. This type of North–South debate distracts from the real issues: creating incentives for and removing obstacles to clean energy development and dissemination. For example, fossil fuel subsidies need to be reallocated to clean energy technologies. Developing countries' fossil fuel subsidies are four times greater than the financing they seek for climate change mitigation and adaptation. The World Bank is prepared to provide technical assistance to developing countries to change these policies. Moreover, developing countries are becoming an important source of clean energy technology, particularly China. Dissemination of clean energy technologies can be facilitated by removing barriers to foreign investment and international trade in clean energy technology and related services. For these reasons, the analysis of clean energy technology transfer should focus on these issues, rather than on an outdated North–South debate over intellectual property rights.

While countries have been debating intellectual property rights and technology transfer in the context of the UNFCCC negotiations, some developing countries have been signing free trade agreements in which they commit to stronger intellectual property rights for plant varieties.[1] This is one area where intellectual property rights are likely to create obstacles to climate change adaptation and where a North–South debate over intellectual property rights would be appropriate. While the World Bank has discussed the importance of new plant varieties for climate change adaptation in developing countries, this discussion has taken place without any mention of the implications of intellectual property rights for the dissemination of these biotechnologies. This is an example of where existing WTO law strikes the right balance, by maintaining policy flexibility. But this flexibility is being eroded in bilateral and regional negotiations.

Subsidies law is an area where WTO law does not strike the right balance, by facilitating fossil fuel subsidies, discouraging clean energy subsidies and facilitating the application of countervailing duties against imports of clean energy technologies.

International jurisprudence has become an increasingly important source of international law and plays an important role in the interpretation of treaties. However, there are limits to what WTO jurisprudence can do to correct deficiencies in the SCM Agreement. Moreover, international climate change regulation involves many different fields of international law, and WTO tribunals are limited to WTO law. In addition, the multiple sources of international climate change regulation make the process of interpreting and applying international climate change law particularly complex.

General international law can sometimes be used to fill in the gaps in treaties. However, general principles cannot displace clear treaty language. Moreover, while some principles have been recognized as forming part of customary international law, others have not. A treaty interpreter must take care not to exaggerate the importance of non-binding principles, which can guide actions without obligating governments, and to carefully consider the role that they might play in the interpretative process. At the same time, a treaty interpreter must avoid conflicting interpretations of obligations in different fields of international law, and general principles may prove useful in this regard. In general, the current status of the principles of international environmental law that are relevant to climate change are compatible with WTO law: sustainable development, the precautionary principle, common but differentiated responsibility, environmental impact assessment, responsibility for transboundary environmental damage, and the duty to cooperate to address transboundary and global environmental problems.

The polluter-pays principle has not been applied effectively in either climate change law or WTO law, which also avoids incompatibility between these two areas of law, though not in a positive way. The polluter-pays principle, which forms part of international environmental law, has encountered political resistance in international climate change law. In particular, the wide use of fossil fuel subsidies and the disincentives to subsidize clean energy technology reveal a need to change the

[1] Members of the International Union for the Protection of New Varieties of Plants, Status on December 5, 2012 <http://www.upov.int/export/sites/upov/members/en/pdf/pub423.pdf> (accessed January 9, 2013).

multilateral regime, particularly the WTO SCM Agreement. This represents a weak point, where multilateral consensus is lacking and unilateral measures might begin to fill the gap.

International environmental law and WTO law both permit unilateral trade measures to conserve transboundary or global resources. They should be preceded by good faith efforts to reach a negotiated solution, be applied flexibly to take into account different conditions among countries (but not a simplistic division between developed and developing countries), and comply with transparency and procedural fairness. However, the degree of urgency may be a factor that requires much greater or lesser effort to reach a negotiated solution, particularly with respect to the time frame involved. The need for climate change mitigation and adaptation is now urgent, which buttresses arguments in favor of unilateral action.

While the precautionary principle might be used to persuade countries that they should reach international agreements to address effectively the risks from climate change, it is far less important than the actual terms of those agreements. Nevertheless, the precautionary principle may serve as an additional argument to justify unilateral and multilateral measures taken to mitigate or adapt to climate change measures based on limited or insufficient scientific proof, both in WTO law and in international investment law. In both areas of international economic law, it would be relevant to arguments regarding the nature of the measure. In the case of international investment law, it could support an argument that the measure relates to environmental protection, not investment, and is not subject to the provisions of the international investment agreement. In the case of WTO law, it could be used to support the same argument, that the measure is an environmental measure, and that any discrimination is neither arbitrary nor unjustifiable.

The EU Aviation Directive demonstrates how a unilateral climate change measure can create incentives to make progress on climate change multilaterally. The same is true with respect to WTO law. WTO Members are free to eliminate trade barriers unilaterally, as long as the MFN rule is observed, and bilaterally or plurilaterally, as long as they comply with the exceptions for regional trade agreements in GATT Article XXIV and GATS Article V. In the case of climate change, however, the regime is less developed and there is greater uncertainty regarding the compatibility of unilateral and regional approaches to mitigation and adaptation. It might be useful to add explicit provisions to the UNFCCC, to reduce the uncertainty regarding the consistency of unilateral, bilateral, and plurilateral approaches to climate change regulation. As with the WTO, unilateral, bilateral, and plurilateral approaches can complement multilateral approaches, by pushing countries to follow suit multilaterally. The UNFCCC already incorporates the language of GATT Article XX regarding arbitrary or unjustifiable discrimination between countries, but this is insufficient to address this issue.

Regulatory capture creates risks that unilateral measures will serve as disguised restrictions on international trade rather than legitimate efforts to combat climate change. For this reason, unilateral measures should be designed and applied in accordance with GATT Article XX, to minimize the risk of unilateral measures that constitute arbitrary or unjustifiable discrimination or disguised restrictions on international trade. In this regard, it is helpful that this same language has been incorporated into international environmental law and the UNFCCC. The political and economic context that has led to multilateral negotiation paralysis means that unilateralism may represent the future of climate change regulation, at least in the short to medium term. However, this does not mean that we cannot use the multilateral consensus that has been achieved so far to regulate the use of unilateral measures.

The ongoing implementation of climate change policies could raise several unresolved issues in WTO law. GATT Article XX will play an important part in determining the WTO consistency of climate change measures. The scope of paragraphs (b) and (g) in GATT Article XX still needs to be defined in many aspects, as does the relationship between these two paragraphs. Multilateral environmental agreements on climate change will probably be relevant to determining the consistency of climate change measures with GATT Article XX and the provisions of the TBT Agreement. However, it is unlikely that GATT Article XX will be applied to the SCM Agreement, the Agreement on Agriculture, or the TBT Agreement. Its application to provisions in other agreements in Annex 1A will have to be analyzed on a case-by-case basis.

If processing and production methods are relevant to determining the issue of "like products" in GATT Articles I and III, the SCM Agreement, the Antidumping Agreement, and the TBT Agreement, then this may provide an alternative analytical approach to determine the WTO consistency of climate change measures. Again, this will have to be analyzed on a case-by-case basis in light of specific climate change measures. However, if environmental subsidies are designed so that they are not specific to certain enterprises, they will be not be subject to multilateral action under Part III or unilateral action under Part V. If the subsidies apply to agricultural products, they will have to comply with the commitments of Members under the Agreement on Agriculture. In the case of export subsidies, compliance with the Agreement on Agriculture may shield subsidies on agricultural products from action under SCM Agreement Article 3.1(a). However, opinion differs on this issue. In the case of subsidies contingent on the use of domestic products, it will be necessary to comply with both the SCM Agreement and the Agreement on Agriculture.

With respect to WTO non-discrimination obligations, "less favorable treatment" requires a determination of whether the contested measure modifies the conditions of competition to the detriment of imported products. However, the existence of such a detrimental effect is not sufficient to demonstrate less favorable treatment if the detrimental impact on imports stems exclusively from a legitimate regulatory distinction, provided that it is evenhanded. Thus, the "legitimate regulatory distinction" test serves as a defense to allegations of WTO-inconsistent discrimination, where risks are addressed in an evenhanded way, for example where distinctions in treatment are based on evidence that the risks are different in different situations and therefore the different situations need to be addressed in different ways to achieve the ultimate policy goal. With respect to climate change, emissions from different fuels could be subject to different taxes where the different emissions pose different risks, for example due to the nature and quantity of GHG emissions for each fuel or the GHG emissions from their productions processes. Different treatment of products, based on their processing and production methods, also might not constitute less favorable treatment, for example due to differences in their carbon footprint. The difficulty is that carbon footprints may be difficult to measure and the design of carbon labeling programs runs the risk of being distorted to benefit domestic industry lobbies. *De facto* discrimination, which creates incentives for private actors to choose domestic inputs over imported ones, could be incorporated into some element of the design of a regime of carbon taxes and border tax adjustments, for example where the taxes themselves do not discriminate but the reporting or filing requirements are more burdensome for the imported products.

At the end of the day, multilateral climate change regulation will likely prove insufficient to tackle climate change effectively. This gives WTO Members an argument to adopt unilateral technical regulations, since the international standards, if any exist, will be ineffective, in light of the growing scientific evidence of the urgency of addressing climate change. Multilateral negotiation paralysis, and the dramatic changes in the economic growth, technological capacity and GHG emissions of developing countries since 1992, has made the UNFCCC approach outdated and ineffective to address climate change adaptation and mitigation. Moreover, new evidence indicates that the climate is changing faster than expected.

International investment agreements could prove problematic, not because of the substance but due to the lack of predictability in the outcomes of arbitrations. The language of international investment agreements is sufficiently flexible to accommodate climate change regulation. While international investment tribunals do not create precedent that is binding upon other tribunals, this jurisprudence does influence other tribunals. However, the approach of different international investment arbitrators to similar issues can vary considerably, which creates a degree of uncertainty regarding the outcome of international investment litigation.

Governments must strike a balance between regulation that discourages foreign investment and foreign investment protection that discourages regulation. If countries implement international agreements or domestic climate change policies in a way that violates the rights of foreign investors, they may have to pay compensation to the foreign investors. This risk can create disincentives to regulation, particularly in countries where the responsible government officials are unsure of the scope of their obligations to foreign investors. At the same time, foreign investment is an important source of knowledge and technology diffusion, together with trade in goods and services. Thus, it is important to create adequate incentives for foreign investors to transfer best practices and technologies that can address climate change adaptation and mitigation. This means that governments must provide adequate protection to foreign investors. International investment agreements can lower regulatory and political risks for foreign investors, and thus lower the cost of and create incentives for foreign investment in clean energy or other carbon mitigation technologies.

Traditionally, debates regarding climate change and financial and technology flows have been framed as North–South. However, this has changed, with the emergence of new low-carbon technology companies in developing countries, such as China and India, which could diffuse clean technology nationally and internationally. In addition, the financial crisis has limited the financial capacity of major developed countries, notably the United States, European Union, and Japan. Moreover, investment banks (including the World Bank) can help to mobilize capital for low-carbon technology, making their own investments and structuring investments for classes of investors with different risk-reward profiles and return expectations. This paradigm shift in financial and technology flows should inform debates regarding both.

In relation to environmental regulation, there are some important differences between international trade agreements and international investment agreements. In particular, only governments have access to the dispute settlement system of the WTO (and comparable dispute settlement systems in free trade agreements). This filters out challenges to some measures. In contrast, private investors can file claims directly against host governments under international investment agreements. Private investors may be less likely to question the wisdom of challenging governments' right to regulate than governments themselves. Moreover, the majority of international investment agreements do not contain references to environmental concerns. This highlights the importance of determining the extent to which environmental regulation is subject to the disciplines contained in these agreements.

International investment agreements do not eliminate a State's right to regulate in the public interest. Legitimate environmental measures should not be subject to international investment agreements, since they would not qualify as measures "relating to" investments. A key issue is the legitimacy of the disputed environmental measure. One way to define legitimacy is by asking whether the measure serves the public interest or a private interest. Of course, a measure can simultaneously serve both public and private interests. The real question here is whether the evidence demonstrates bad faith, protectionist intent, or intent to harm foreign investors on the part of the legislator or the judiciary. The majority of State practice is consistent with the view that international investment agreements do not negate the right to regulate climate change. Moreover, customary international law does not require a State to maintain a stable legal and business environment for investments. The customary international law standard does not prevent a public authority from changing the regulatory environment to take account of new policies and needs, even if some of those changes may have far-reaching consequences and effects, and even if they impose significant additional burdens on an investor. It does not provide a guarantee against regulatory change or entitle an investor to expect no material changes to the regulatory framework within which an investment is made. Governments can change, and policies and rules can change. The rules of customary international law only protect against egregious behavior and do not require a legal and business environment to be set in stone. The minimum standard of treatment of foreign investors under customary international law has to be interpreted in accordance with evolving customary international environmental law. The obligation to avoid activities causing significant damage to the environment of another State is likely to encompass regulations to address climate change. Thus, legitimate climate change regulation would not be inconsistent with the minimum standard of treatment. To conclude otherwise would create a conflict between customary international investment law and customary international environmental law.

The general body of precedent usually does not treat regulatory action as amounting to expropriation, because expropriations tend to involve the deprivation of ownership rights and regulations a lesser interference. Under customary international law, where economic injury results from *bona fide* regulation within the police powers of a State, compensation is not required. Thus, as a general matter, States are not liable to compensate foreign investors for economic loss incurred as a result of a nondiscriminatory action to protect the public interest. However, once an expropriation has taken place, compensation is due even if it is for an environmental purpose. In the context of NAFTA Article 1110, if there is a finding of expropriation, compensation is required, even if the taking is for a public purpose, non-discriminatory and in accordance with due process of law and Article 1105(1). However, not all government regulatory activity that makes it difficult or impossible for an investor to carry out a particular business is an expropriation. Given the economic and environmental consequences of climate change, it seems that *bona fide* climate change regulation should take precedence over investors' rights, though the correct balance likely will have to be decided on a case-by-case basis. It is also important to protect foreign investors from unfair or arbitrary treatment by governments who are motivated by short-term political interests rather than long-term environmental risks. For this reason, we emphasize that we are referring to *bona fide* climate change regulation.

Striking the right balance between the regulatory risks that investors face and the litigation risk that governments face is not the same for all markets. Larger markets can have a greater degree of regulatory risk and still attract foreign investors. In contrast, smaller, less economically attractive markets may need to strike a balance that is more in favor of investors' rights and reduces regulatory risk to a greater degree, in order to attract foreign investment. Larger markets are also a greater source of GHG emissions, so the balance should favor climate change regulations over compensation to foreign investors, in order to limit the risk of regulatory chill and to enhance the right to regulate. Science-based regulatory decisions should withstand scrutiny, even if the science was preliminary, when there is sufficient scientific evidence of the potentially serious environmental effects to support the regulation. There is sufficient scientific evidence of the potentially serious effects of climate change to justify climate change regulation, even if it also has the effect of diminishing the value of some foreign investments.

International investment law has the potential to have a chilling effect on climate change regulation, by raising issues regarding the risk that climate change regulation will expose host states to claims from foreign investors. Our analysis shows that legitimate climate change regulation should not trigger liability to compensate foreign investors. However, this may not eliminate the chilling effect, since it is costly for States to defend against such claims even if they do not succeed. Awards of costs against investors who file such claims may discourage such claims, but this may not be sufficient to overcome the chilling effect in the short term. Moreover, the lack of a system of precedents for tribunals permits tribunals to reach different conclusions on similar issues, which increases the uncertainty regarding the outcome of litigation. Nevertheless, there is room in international investment law for striking an appropriate balance between the right to regulate climate change and rights of foreign investors to seek compensation for arbitrary and discriminatory governmental actions. Striking the right balance will facilitate the mobilization of foreign investment as a source of climate finance and technology dissemination.

However, to stabilize GHG emissions (mitigation) would require a significant reduction of emissions both in the developed and the developing world. This calls for large-scale investment in energy infrastructure and other large-scale mitigation projects. Multilateral funding channels may be the best option for such large-scale financing in developing countries. However, private, bilateral, and multilateral funding sources are not mutually exclusive. For example, the World Bank can play a role in creating incentives for private-sector investment, by investing in pilot projects and lowering political risk for private investors. However, bilateral and multilateral climate financing could be found to be a prohibited export subsidy under the SCM Agreement.

Multilateral cooperation and financing will continue to be an essential element in climate finance. The WTO will have to adapt subsidies rules, via judicial interpretations on a case-by-case basis or via a negotiated response to agreements reached in other multilateral organizations. The increasing number of actors in climate finance activities will require more coordination and adaptation of roles among institutions such as the World Bank and the Green Climate Fund. The activities of these institutions will have to be coordinated with those of the private sector participants as well.

In the current political and economic context, multilateralism is not working in the UNFCCC and WTO systems. This requires unilateralism, but in a well-considered approach that creates economic incentives to engage the private sector and to push governments into effective multilateral agreements. For example, funding for adaptation and technology transfer should be made conditional on recipients implementing PPP-based mitigation measures, in order to channel funding and technology to combat climate change. This can be accompanied by unilateral trade measures on goods and services to create political will in developing countries; the private industry that opposes climate change action based on competitive concerns might change their stance if unilateral trade measures begin to affect market access and competitiveness. However, unilateral measures to combat climate change need to be taken in a manner that is consistent with existing obligations and principles of international environmental and economic law, as far as possible, in order to more effectively address this urgent global issue.

A single country or a single region cannot reduce the greenhouse gas emissions of other regions or countries. It may be possible to create incentives for other countries, for example with unilateral trade restrictions or foreign aid that is conditional upon emissions reductions. However, trade restrictions do not just impose costs on the exporting country. They also impose costs on the importing country, where importers are affected by the increased cost of inputs. In addition, few domestic markets are large enough for trade barriers to have an economic impact that would be sufficient to create an adequate incentive to reduce emissions. Foreign aid that is conditioned upon the use of inputs from the donor country also may violate WTO law. Moreover, foreign aid costs money for donor countries, too. Even if we resolve the problems of cost and WTO consistency of trade barriers and foreign aid, they remain partial solutions only because they will not achieve the desired level of emissions reductions.

We can estimate the future concentrations of greenhouse gases and we can estimate the probability of a range of temperature increases, but we cannot say precisely what the temperature increase will be, how that temperature increase will vary from one part of the planet to another, and what the ecological and economic effects will be. What we do know is that the risks are grave. Debating whether we have underestimated or overestimated the proximity and severity of those risks misses the point. We need to address those risks through mitigation and adaptation. To do so effectively will require creating incentives for multilateral action and removing obstacles to financing the dissemination of the technologies needed for mitigation and adaptation.

Selected Bibliography

BOOKS AND MONOGRAPHS

Abdalati, Waleed et al. (Panel on Advancing the Science of Climate Change), *Advancing the Science of Climate Change* (National Academies Press, Washington, DC 2010).

Aldy, Joseph E. et al., *Beyond Kyoto: Advancing the International Effort against Climate Change* (Pew Center on Global Climate Change, Arlington, VA 2003).

Aldy, Joseph E. and Robert N. Stavins, *Architectures for Agreement: Addressing Global Climate Change in the Post-Kyoto World* (Cambridge University Press, Cambridge 2007).

Amiraslany, Afshin, *The Impact of Climate Change on Canadian Agriculture: A Ricardian Approach*, PhD thesis, University of Saskatchewan, 2010.

Andersen, Stephen O., K. Madhava Sarma, and Kristen N. Taddonio, *Technology Transfer for the Ozone Layer: Lessons for Climate Change* (Global Environment Facility and Earthscan Publications, Oxford 2007).

Andersen, Stephen O., K. Madhava Sarma, and Kristen N. Taddonio, *Technology Transfer For the Ozone Layer, Lessons For Climate Change* (Earthscan Publications, Oxford 2007).

Anderson, K. and R. Blackhurst (eds.), *The Greening of World Trade Issues* (University of Michigan Press, Ann Arbor, MI 1992).

Arnoldus, Michiel and Roger Bymolt, *Demystifying Carbon Markets: A Guide to Developing Carbon Credit Projects* (KIT, Amsterdam 2011).

Arrow, K.J, W.R. Cline, K.-G. Maler, M. Munasinghe, R. Squitieri, J. E. Stiglitz, "Intemporal Equity, Discounting, and Economic Efficiency" in James P. Bruce et al. (eds.), *Climate Change 1995—Economic and Social Dimensions of Climate Change* (Cambridge University Press, Cambridge 1996).

Aust, Anthony, *Modern Treaty Law and Practice* (Cambridge University Press, Cambridge 2000).

Barkley, P. and D. Seckler, *Economic Growth and Environmental Decay* (Harcourt Brace Jovanovich, New York 1972).

Birnie, Patricia W. and Alan E. Boyle, *International Law and the Environment* (Oxford University Press, Oxford 1992).

Blakeney, Michael, *Intellectual Property Rights and Food Security* (CABI, Wallingford, MA 2009).

Bowman, Michael and Alan Boyle (eds.), *Environmental Damage in International and Comparative Law* (Oxford University Press, Oxford 2002).

Brenner, Carliene, *Integrating Biotechnology in Agriculture: Incentives, Constraints, and Country experiences* (OECD, Paris 1996).

Buckley, Ross P. (ed.), *The WTO and the Doha Round: The Changing Face of World Trade* (Kluwer Law International, The Hague 2003).

Burroughs, William J., *Climate Change in Prehistory. The End of the Reign of Chaos* (Cambridge University Press, Cambridge, MA 2005).

California Council on Science and Technology, *California's Energy Future: The View to 2050 Summary Report* (CCST, May 2011).

Calvello, Angelo A., *Environmental Alpha* (John Wiley & Sons, Hoboken, NJ 2009).

Chandrappa, Ramesha, Sushil Gupta, and Umesh Chandra Kulshrestha, *Coping With Climate Change: Principles and Asian Context* (Springer, Berlin 2011).

Charnovitz, Steve, Gary Clyde Hufbauer, and Jisun Kim, *Global Warming and the World Trading System* (Peterson Institute for International Economics, Washington, DC 2009).

Condon, Bradly J., *El Derecho de la Organización Mundial de Comercio: Tratados, Jurisprudencia y Práctica* (Cameron May, London 2007).

Condon, Bradly J., *Environmental Sovereignty and the WTO: Trade Sanctions and International Law* (Transnational Publishers, Ardsley 2006).

Condon, Bradly J., *NAFTA, WTO and Global Business Strategy: How AIDS, Trade and Terrorism Affect Our Economic Future* (Quorum Books, Westport, CT 2002).

Condon, Bradly J., Joyce Sadka, and Tapen Sinha, *Insurance Regulation in North America: Integrating American, Canadian and Mexican Markets* (Kluwer Law International, the Hague 2003).

Condon, Bradly J. and Tapen Sinha, *Global Lessons from the AIDS Pandemic. Economic, Financial, Legal and Political Implications* (Springer Verlag, Berlin 2008).

Cordonier Segger, Marie-Claire, Markus W. Gehring, and Andrew Newcombe (eds.), *Sustainable Development in World Investment Law* (Kluwer Law International, The Hague 2010).

Daly, H. and J. Cobb, *For the Common Good* (Beacon Press, Boston, MA 1989).

De La Torre, Augusto, Pablo Fajnzylber, and John Nash, *Low Carbon, High Growth: Latin American Responses to Climate Change* (World Bank, Washington, DC 2009).

Esty, Daniel C., *Greening the GATT: Trade Environment, and the Future* (Institute for International Economics, Washington, DC 1994).

Faure, Michael and Marjan Peeters (eds.), *Climate Change and European Emissions Trading: Lessons for Theory and Practice* (Edward Elgar, Cheltenham 2008).

Food and Agriculture Organization, *"Climate-Smart" Agriculture Policies, Practices and Financing for Food Security, Adaptation and Mitigation* (FAO, Rome 2010).

Ford, James D. and Lea Berrang-Ford, Lea (eds.), *Climate Change Adaptation in Developed Nations, Advances in Global Change Research 42* (Springer, Berlin 2011).

Freestone, David and Charlotte Streck (eds.), *Legal Aspects of Carbon Trading: Kyoto, Copenhagen, and Beyond* (Oxford University Press, Oxford 2009).

GATT Secretariat, *Trade and the Environment* (GATT, Geneva 1992).

Gore, Al, *Earth in the Balance: Ecology and the Human Spirit* (Houghton Mifflin, Boston, MA 1992).

Gurjar, Bhola R., et al., *Air Pollution Health and Environmental Impacts* (CRC Press, Boca Raton, FL 2010).

Hamilton, Clive, *Requiem For a Species. Why We Resist the Truth about Climate Change* (Allen & Unwin, Sydney 2010).

Hoekman, Bernard and Beata Smarzynska Javorcik (eds.), *Global Integration and Technology Transfer* (Palgrave Macmillan and the World Bank, Washington, DC 2006).

Intergovernmental Panel on Climate Change, *Managing the Risks of Extreme Events and Disasters to Advance Climate Change Adaptation, Special Report of the Intergovernmental Panel on Climate Change* (Cambridge University Press, Cambridge 2012).

International Energy Agency, *Energy Poverty: How to Make Modern Energy Access Universal* (IEA/OECD, Paris 2010).

International Energy Agency, *Recent Developments in Energy Subsidies* (International Energy Agency, Paris 2012).

International Energy Agency, *World Energy Outlook* (International Energy Agency, Paris 2011).

International Fund for Agricultural Development, *Rural Poverty Report: New Realities, New Challenges: New Opportunities for Tomorrow's Generation* (2011).

Karl, Thomas R., Jerry M. Melillo, and Thomas C. Peterson (eds.), *Global Climate Change Impacts in the United States* (Cambridge University Press, Cambridge 2009).

Karmali, Abyd, "Observations from the Carbon Emission Markets: Implications for Carbon Finance" in Cary Krosinsky and Nick Robins (eds.), *Sustainable Investing: the Art of Long-Term Performance* (Earthscan, London 2008) 59.

Koplow, Doug, *Phasing Out Fossil-Fuel Subsidies in the G20: A Progress Update* (Earth Track, Inc. and Oil Change International, Cambridge, MA 2012).

Lang, Winfried (ed.), *Sustainable Development and International Law* (Springer, Berlin 1995).

Low, Patrick (ed.), *International Trade and the Environment*, World Bank Discussion Papers No. 159 (World Bank, Washington, DC 1992).

Makarczyk, Jerzy (ed.), *Theory of International Law at the Threshold of the 21st Century* (Springer, Berlin 1996).

Markell David L. and John H. Knox (eds.), *Greening NAFTA: the North American Commission for Environmental Cooperation* (Stanford University Press, Stanford, CA 2003).

Maskus, Keith E. and Jerome H. Reichman (eds.), *International Public Goods and Transfer of Technology under a Globalized Intellectual Property Regime* (Cambridge University Press, Cambridge 2005).

National Climate Assessment Development Advisory Committee, *Draft Climate Assessment Report* (January 11, 2013) <http://ncadac.globalchange.gov/> (accessed January 12, 2013).

Metz, Bert, Ogunlade R. Davidson, Jan-Willem Martens, Sascha N. M. van Rooijen, and Laura van Wie McGrory (eds.), *Methodological and Technological Issues in Technology Transfer: A Special Report of the Intergovernmental Panel on Climate Change* (Cambridge University Press, Cambridge 2000).

Newell, Peter and Paterson, Matthew, *Climate Capitalism Global Warming and the Transformation of the Global Economy*, (Cambridge University Press, Cambridge 2010).

Nordhaus, William, *A Question of Balance: Weighing the Options on Global Warming Policies* (Yale University Press, New Haven, CT 2008).

Nordhaus, William, *The Efficient Use of Energy Resources* (Cowles Foundation for Research in Economics, Yale University Press, New Haven, CT 1979).

Nordstrom, Hakan and Scott Vaughan, *Trade and Environment*, WTO Special Studies, (WTO, Geneva 1999).

OECD, *Biotechnologies in Agricultures and Related Natural Resources to 2015* (OECD, Paris 2009).

OECD, *The Bioeconomy to 2030: Designing a Policy Agenda* (OECD, Paris 2009).

OECD, *Environmental Labelling in OECD Countries* (OECD, Paris 1991).

OECD, *OECD and the Environment* (OECD, Paris 1976).

OECD, *Processes and Production Methods (PPMs): Conceptual Framework and Considerations on Use of PPM-based Trade Measures* (OECD, Paris 1997).

OECD-FAO, *Agricultural Outlook 2008–2017* (OECD-FAO, Paris 2008).

Owen, Anthony D. and Nick Hanley (eds.), *Economics of Climate Change* (Routledge, London 2004).

Pauwelyn, Joost, *Conflict of Norms in Public International Law: How WTO Law Relates to Other Rules of International Law* (Cambridge University Press, Cambridge 2003).

Petersmann, Ernst-Ulrich, *International and European Trade and Environmental Law after the Uruguay Round* (Martinus Nijhoff, Dordrecht 1995).

Pinkse, Jonatan and Ans Kolk, *International Business and Global Climate Change* (Routledge, London 2009).

Potsdam Institute for Climate Impact Research and Climate Analytics, *Turn Down the Heat: Why a 4° Warmer World Must be Avoided* (World Bank, Washington, DC 2012).

Richardson, Katherine, Will Steffen, and Diana Liverman, *Climate Change: Global Risks, Challenges and Decisions* (Cambridge University Press, Cambridge 2011).

Roberts, J. Timmons and Bradley C. Parks, *A Climate of Injustice: Global Inequality, North-South Politics, and Climate Policy* (MIT Press, Cambridge, MA 2006).

Sagoff, Mark, *The Economy of the Earth: Philosophy, Law, and the Environment* (2nd edn., Cambridge University Press, Cambridge 2008).

Sandor, Robert, *Good Derivatives* (John Wiley & Sons, Hoboken, NJ 2012).

Saunders, J.O. (ed.), *The Legal Challenge of Sustainable Development* (Canadian Institute of Resources Law, Calgary 1990).

Schwarze, R., John O. Niles, and Eric Levy Law, *Economics in International Climate Change Policy* (Springer, Berlin 2001).

Solow, Robert, *Growth Theory: An Exposition* (Oxford University Press, Oxford 1988).

Steinberg, Richard H. (ed.), *The Greening of Trade Law: International Trade Organizations and Environmental Issues* (Rowman & Littlefield, Boulder, CO 2002).

Stern, Nicholas, *Stern Review on the Economics of Climate Change* (H.M. Treasury, London 2006).

Stern, Nicholas, *The Economics of Climate Change* (Cambridge University Press, Cambridge 2007).

Stewart, Richard B. et al. (eds.), *Climate Finance: Regulatory and Funding Strategies for Climate Change and Global Development* (New York University Press, New York 2009).

UNCTAD, *World Investment Report 2010: Investing in a Low-Carbon Economy* (United Nations, Geneva 2010) <http://unctad.org/en/Docs/wir2010_en.pdf> (accessed November 2, 2012).

UNFCCC, *The Contribution of the Clean Development Mechanism under the Kyoto Protocol to Technology Transfer* (2010) <http://cdm.unfccc.int/Reference/Reports/TTreport/TTrep10.pdf> (accessed January 3, 2013).

Vanderheiden, Steve, *Atmospheric Justice. A Political Theory of Climate Change* (Oxford University Press, Oxford 2008).

Victor, David G., *Global Warming Gridlock* (Cambridge University Press, Cambridge 2011).

Vranes, Erich, *Trade and the Environment: Fundamental Issues in International and WTO Law* (Oxford University Press, Oxford 2009).

Weiss, Edith Brown and John H. Jackson (eds.), *Reconciling Trade and Environment* (Transnational Publishers, Ardsley 2001).

Werksman, Jacob, Kevin A. Baumert, and Navroz K. Dubash, *Will International Investment Rules Obstruct Climate Protection Policies?* (World Resources Institute, Washington, DC 2001).

Wilkins, Gill, *Technology Transfer for Renewable Energy* (The Royal Institute of International Affairs and Earthscan Publications, Oxford 2002).

World Bank, *Climate Change and the World Bank Group, Phase I: An Evaluation of World Bank Win-Win Energy Policy Reforms* (World Bank, Washington, DC 2009).

World Bank, *Climate Change and the World Bank Group, Phase II: The Challenge of Low-Carbon Development* (World Bank, Washington, DC 2010).

World Bank, *Climate Change and the World Bank Group, Phase III: Adapting to Climate Change: Assessing World Bank Group Experience* (World Bank, Washington, DC 2013), advance copy <http://ieg.worldbankgroup.org/content/dam/ieg/climate_change3/cc3_full_eval.pdf> (Accessed January 4, 2013).

World Bank, *International Trade and Climate Change: Economic, Legal, and Institutional Perspectives* (World Bank, Washington, DC 2007).

World Bank, *World Development Report 2010, Development and Climate Change* (The International Bank for Reconstruction and Development/the World Bank, Washington, DC 2010).

World Commission on Environment and Development, *Our Common Future* (Oxford University Press, Oxford 1987).

World Trade Organization, Committee on Trade and Environment Special Session, Note by the Secretariat, *Multilateral Environmental Agreements (MEAs) and WTO Rules: Proposals Made in the Committee on Trade and Environment (CTE) from 1995–2002*, WTO Doc. TN/TE/S/1 (WTO, Geneva 2002).

World Trade Organization, *GATT/WTO Dispute Settlement Practice relating to GATT Article XX, Paragraphs (b), (d) and (g)*, WT/CTE/W/203 (WTO, Geneva 2002).

World Trade Organization, *Guide to GATT Law and Practice* (6th edn., WTO, Geneva 1995).

World Trade Organization, *Matrix on Trade Measures Pursuant to Selected MEAs*, WTO Doc. WT/CTE/W/160 (WTO, Geneva 2000).

World Trade Organization, *Trade and Environment in the GATT/WTO (Background Note by the Secretariat)* (WTO, Geneva 1999).

World Trade Organization, *Trade and Environment Report* (WTO, Geneva 1999).

World Trade Organization and UNEP, *Trade and Climate Change* (WTO, Geneva 2009).

Yamin, Farhana, *Climate Change and Carbon Markets: A Handbook of Emissions Reduction Mechanisms* (Earthscan, Oxford 2005).

Yandle, Bruce, *The Political Limits of Environmental Regulation: Tracking the Unicorn* (Praeger, New York 1989).

JOURNAL ARTICLES

Adcock, J. and J. Kildow, "Environment and the Trading System" (1992) 16 Fletcher Forum of World Affairs 55.

Alvarez, Jose E., "Critical Theory and the North American Free Trade Agreement's Chapter Eleven" (1997) 28 University of Miami Inter-American Law Review 303.

Arrow, Kenneth J., "Global Climate Change: A Challenge to Policy" (2007) 4(3) Economist's Voice <http://dx.doi.org/10.2202/1553-3832.1270> (accessed March 15, 2013).

Atik, Jeffery, "Repenser NAFTA Chapter 11: A Catalogue of Legitimacy Critiques" (2004) 3 Asper Review of International Business and Trade Law 215.

Barker, Terry, "The Economics of Avoiding Dangerous Climate Change. An Editorial Essay on the Stern Review" (2008) 89 Climatic Change 173.

Barpujari, Indrani, "Facilitating Access or Monopoly: Patent Pools at the Interface of Patent and Competition Regimes" (2010) 15 Journal of Intellectual Property Rights 345.

Barron, Jason A., "Genetic Use Restriction Technologies: Do the Potential Environmental Harms Outweigh the Economic Benefits?" (2008) 20 Georgetown International Environental Law Review 271.

Bergeijk, P., "International Trade and the Environmental Challenge" (1991) 25 Journal of World Trade 105.

Berger, Joseph R., "Unilateral Trade Measures to Conserve the World's Living Resources: An Environmental Breakthrough for the GATT in the WTO Sea Turtle Case" (1999) 24 Columbia Journal of Environmental Law 355.

Berman, Paul Schiff, "The Globalization of Jurisdiction" (2002) 151 University of Pennsylvania Law Review 311.

Bernazani, Jennifer A., "The Eagle, the Turtle, the Shrimp and the WTO: Implications for the Future of Environmental Trade Measures" (2000) 15 Connecticut Journal of International Law 207.

Bonds, Matthew H., Andrew P. Dobson, and Donald C. Keenan, "Disease Ecology, Biodiversity, and the Latitudinal Gradient in Income" (2012) 10 PLOS Biology e1001456.

Bordwin, Harold J., "The Legal and Political Implications of the International Undertaking on Plant Genetic Resources" (1985) 12 Ecology Law Quarterly 1053.

Boute, Anatole, "Combating Climate Change through Investment Arbitration" (2012) 35 Fordham International Law Journal 613.

Bree, Axel, "Article XX GATT—Quo Vadis? The Environmental Exception After the Shrimp/Turtle Appellate Body Report" (1998) 17 Dick. Journal of International Law 99.

Brower, Charles H. II, "Structure, Legitimacy, and NAFTA's Investment Chapter" (2003) 36 Vanderbilt Journal of Transnational Law 37.

Carrico, Amanda R. et al., "Energy and Climate Change: Key Lessons for Implementing the Behavioral Wedge" (2011) 2 Journal of Energy and Environmental Law 61.

Chan, Gabriel, Robert Stavins, Robert Stowe, and Richard Sweeney, "The SO2 Allowance-Trading System and the Clean Air Act Amendments" (2012) 65 National Tax Journal 419.

Chang, Howard, "Toward a Greener GATT: Environmental Trade Measures and the Shrimp-Turtle Case" (2000) 74 Southern California Law Review 31.

Charnovitz, Steve, "Exploring the Environmental Exceptions in GATT Article 20" (1991) 25 Journal of World Trade 37.

Charnovitz, Steve, "The Law of Environmental 'PPMs' in the WTO: Debunking the Myth of Illegality" (2002) 27 Yale Journal of International Law 59.

Charnovitz, Steve, "Transnational and Supranational Democracy: The Emergence of Democratic Participation in Global Governance" (2003) 10 Indiana Journal of Global Legal Studies 45.

Christensen, L. and W. Green, "Economies of Scale in U.S. Electric Power Generation" (1976) 84 Journal of Political Economy 655.

Condon, Bradly J., "Climate Change and Unresolved Issues in WTO Law" (2009) 12 Journal of International Economic Law 895.

Condon, Bradly J., "GATT Article XX and Proximity of Interest: Determining the Subject Matter of Paragraphs b and g" (2004) 9 UCLA Journal of International Law and Foreign Affairs 137.

Condon, Bradly J., "Lost in Translation: Plurilingual Interpretation of WTO Law" (2010) 1 Journal of International Dispute Settlement 191.

Condon, Bradly J., "Multilateral Environmental Agreements and the WTO: Is the Sky Really Falling?" (2002) 9 Tulsa Journal of International and Comparative Law 533.

Condon, Bradly J., "NAFTA and the Environment: A Trade-Friendly Approach" (1994) 14 Northwestern Journal of International Law and Business 528.

Condon, Bradly J., "Reconciling Trade and Environment: A Legal Analysis of European and North American Approaches" (2000) 8 Cardozo Journal of International and Comparative Law 1.

Condon, Bradly J. and Tapen Sinha, "Chronicle of a Pandemic Foretold: Lessons from the 2009 Influenza Epidemic" (2010) 22 Florida Journal of International Law 1.

Condon, Bradly J. and Tapen Sinha, "Global Diseases, Global Patents and Differential Treatment in WTO Law: Criteria for Suspending Patent Obligations in Developing Countries" (2005) 26 Northwestern Journal of International Law and Business 1.

Cottier, Thomas and Mario Panizzon, "Legal Perspectives on Traditional Knowledge: The Case for Intellectual Property Protection" (2004) 7 Journal of International Economic Law 371.

Cottier, Thomas, "The Protection of Genetic Resources and Traditional Knowledge: Towards More Specific Rights and Obligations in World Trade Law (1998) 1 Journal of International Economic Law 555.

Dailey, Virginia, "Sustainable Development: Reevaluating the Trade vs. Turtles Conflict at the WTO" (2000) 9 Transnational Law and Policy 331.

de La Fayette, Louise, "Case Report: United States—Import Prohibition of Certain Shrimp and Shrimp Products" (2002) 96 American Journal of International Law 685.

de Lima-Campos, Aluisio, "Políticas de Comercio y Medio Ambiente: En Busca de un Alineamiento" (2012) 2 Revista de Derecho Económico Internacional 35.

Deere-Birkbeck, Carolyn, "Global governance in the context of climate change: the challenges of increasingly complex risk parameters" (2009) 85 International Affairs 1173.

Desmedt, A., "Proportionality in WTO Law" (2001) 4 Journal of International Economic Law 441.

DiMascio, Nicholas and Joost Pauwelyn, "Nondiscrimination in Trade and Investment Treaties: Worlds Apart or Two Sides of the Same Coin?" (2008) 102 American Journal of International Law 48.

Dutton, A. and K. Lambeck, "Ice Volume and Sea Level During the Last Interglacial" (2012) 337 Science 216.

Fletcher, Charles R., "Greening World Trade: Reconciling GATT and Multilateral Environmental Agreements within the Existing World Trade Regime" (1996) 5 Journal of Transnational Law and Policy 341.

Fox, Sean, "Responding to Climate Change: The Case for Unilateral Trade Measures to Protect the Global Atmosphere" (1996) 84 Georgetown Law Journal 2499.

Gaines, Sanford E., "Processes and Production Methods: How to Produce Sound Policy for Environmental PPM-Based Trade Measures?" (2002) 27 Columbia Journal of Environmental Law 383.

Green, Andrew James, "Climate Change, Regulatory Policy and the WTO: How Constraining are Trade Rules?" (2005) 8 JIEL 143.

Gronewold, Nathanial, "Chicago Climate Exchange Closes Nation's First Cap-and-Trade System But Keeps Eye to the Future" New York Times, January 3, 2011 <www.nytimes.com/cwire/2011/01/03/03climatewire-chicago-climate-exchange-closes-but-keeps-ey-78598.html> (accessed March 15, 2013).

Hansen, J., M. Sato, and R. Ruedy, *Climate Variability and Climate Change: The New Climate Dice*, November 2011 <http://www.columbia.edu/~jeh1/mailings/2011/20111110_NewClimateDice.pdf> (accessed April 5, 2013).

Hardin, Garrett, "The Tragedy of the Commons" (1968) 162 Science 1243.

Hart, Michael and Bill Dymond, "Special and Differential Treatment and the Doha Development Round" (2003) 37 Journal of World Trade 395.

Hodges, Brian Trevor, "Where the Grass is Always Greener: Foreign Investor Actions Against Environmental Regulations Under NAFTA's Chapter 11, *S.D. Myers, Inc. v. Canada*" (2001) 14 Georgetown International Environmental Law Review 367.

Holland, K., "Exploitation on Porpoise: The Use of Purse Seine Nets by Commercial Tuna Fishermen in the Eastern Tropical Pacific Ocean" (1991) 17 Syracuse Journal of International Law and Commerce 267.

Howse, Robert and Donald Regan, "The Product/Process Distinction—An Illusory Basis for Disciplining 'Unilateralism' in Trade Policy" (2000) 11 European Journal of International Law 249.

Howse, Robert, "The Appellate Body Rulings in the Shrimp/Turtle Case: A New Legal Baseline for the Trade and Environment Debate" (2002) 27 Columbia Journal of Environmental Law 491.

Hudnall, Shannon, "Towards a Greener International Trade System: Multilateral Environmental Agreements and the World Trade Organization" (1996) 29 Columbia Journal of Law and Social Problems 175.

Jackson, John H., "Dolphins and Hormones: GATT and the Legal Environment for International Trade after the Uruguay Round" (1992) 14 University of Arkansas at Little Rock Law Journal 429.

Jackson, John H., "World Trade Rules and Environmental Policies: Congruence or Conflict?" (1992) 49 Washington and Lee Law Review 1227.

Janis, Mark D. and Stephen Smith, "The Protection of Rights in Plant Varieties: Technological Change and the Design of Plant Variety Protection Regimes" (2007) 82 Chicago-Kent Law Review 1557.

Johnson, Lise, "International Investment Agreements and Climate Change: The Potential for Investor-State Conflicts and Possible Strategies for Minimizing It" (2009) 39 Environmental Law Reporter 11147.

Joskow, Paul L. and Richard Schmalensee, "The Political Economy of Market-Based Environmental Policy: The U.S. Acid Rain Program" (1998) 41 Journal of Law and Economics 37.

Kriebaum, Ursula, "Partial Expropriation" (2007) 8 Journal of World Investment and Trade 69.

Kuei-Jung Ni, "Contemporary Prospects for the Application of Principle 12 of the Rio Declaration" (2001) 14 Georgetown International Environmental Law Review 1.

Lamont, Owen A. and Richard H. Thaler, "Law of one Price in the Financial Markets" (2003) 17 Journal of Economic Perspectives 191.

Lang, W., "Environmental Protection: The Challenge for International Law" (1986) 20 Journal of World Trade Law 489.

Lomas, O., "Environmental Protection, Economic Conflict and the European Community" (1988) 33 McGill Law Journal 506.

Lones, L., "The Marine Mammal Protection Act and International Protection of Cetaceans: A Unilateral Attempt to Effectuate Transnational Conservation" (1989) 22 Vanderbilt Journal of Transnational Law 997.

M'Gonigle, Michael, "Between Globalism and Territoriality: The Emergence of an International Constitution and the Challenge of Ecological Legitimacy" (2002) 15 Canadian Journal of Law and Jurisprudence 159.

Marceau, Gabrielle and Joel Trachtman, "The Technical Barriers to Trade Agreement, the Sanitary and Phytosanitary Measures Agreement, and the General Agreement on Tariffs and Trade" (2002) 36 Journal of World Trade 811.

Masarek, Adam, "Treetop View of the Cathedral: Plant Variety Protection in South and Southeast Asian Least-Developed Countries" (2010) 24 Emory International Law Review 433.

Mavroidis, Petros C., "Trade and Environment after the Shrimp-Turtles Litigation" (2000) 34 Journal of World Trade 73.

McDorman, Ted, "The 1991 U.S.–Mexico GATT Panel Report on Tuna and Dolphin: Implications for Trade and Environment Conflicts" (1992) 17 North Carolina Journal of International Law and Commercial Regulation 461.

McDorman, Ted, "The GATT Consistency of U.S. Fish Import Embargoes to Stop Driftnet Fishing and Save Whales, Dolphins and Turtles" (1991) 24 George Washington Journal of International Law and Economics 477.

Mills, Eric, "Weighing the Risks of Climate Change Mitigation Strategies" (2012) 68 Bulletin of the Atomic Scientists 67.

Mills, Evan, "A Global Review of Insurance Industry Responses to Climate Change" (2009) 34 The Geneva Papers 323.

Morosini, Fabio, "Trade and Climate Change: Unveiling the Principle of Common but Differentiated Responsibilities from the WTO Agreements" (2010) 42 George Washington International Law Review 713.

Nordhaus, William, "An Optimal Transition Path for Controlling Greenhouse Gases" (1992) 258 Science 1315.

Ortino, Federico, "Legal Reasoning of International Investment Tribunals: A Typology of Egregious Failures" (2012) 3 Journal of International Dispute Settlement 31.

Parker, Richard W. "The Use and Abuse of Trade Leverage to Protect the Global Commons: What We Can Learn from the Tuna-Dolphin Conflict" (1999) 12 Georgetown International Law Review 1.

Pauwelyn, Joost, "Bridging Fragmentation and Unity: International Law as a Universe of Inter-Connected Islands" 25 Michigan Journal of International Law 903 (2004).

Pauwelyn, Joost, "The Role of Public International Law in the WTO: How Far Can We Go?" (2001) 95 American Journal of International Law 535.

Perez, Oren, "Reconstructing Science: the Hormone Conflict between the EU and the United States" (1998) 3 European Foreign Affairs Review 563.

Petersmann, E.-U., "Trade Policy, Environmental Policy and GATT: Why Trade Rules and Environmental Rules Should be Mutually Consistent" (1991) 46 Aussenwirtschaft 197.

Quiggin, John, "Equity between Overlapping Generations" (2012) 14 Journal of Public Economic Theory 273.

Reuveny, Rafael, "Climate change-induced migration and violent conflict" (2007) 26 Political Geography 656.

Richards, Eric L. and Martin A. McCrory, "The Sea Turtle Dispute: Implications for Sovereignty, the Environment, and International Trade Law" (2000) 71 University of Colorado Law Review 295.

Rose, Gregory, "International Law of Sustainable Agriculture in the 21st Century: The International Treaty on Plant Genetic Resources for Food and Agriculture" (2003) 15 Georgetown International Environmental Law Review 583.

Rosendahl, R., "The Development of Mexican Fisheries and Its Effect on United States-Mexican relations" (1984) 3 UCLA Pacific Basin Law Journal 1.

Ruiz Euler and Nadja Dorothea, "El Trato Nacional y la Nación Más Favorecida en el Acuerdo General sobre el Comercio de Servicios de la OMC" (2012) 2 Revista de Derecho Económico Internacional 5.

Runge, C.F., "Trade Protectionism and Environmental Regulations: The New Nontariff Barriers" (1990) 11 Northwestern Journal of International Law and Business 47.

Safrin, Sabrina, "Hyperownership in a Time of Biotechnological Promise: The International Conflict to Control the Building Blocks of Life" (2004) 98 American Journal of International Law 641.

Samson, Jason, D. Berteaux, B.J. McGill, and M.M. Humphries, "Geographic Disparities and Moral Hazards in the Predicted Impacts of Climate Change on Human Populations" (2011) 20 Global Ecology and Biogeography 532.

Sands, Philippe, "The 'Greening' of International Law: Emerging Principles and Rules" (1994) 1 Indiana Journal of Global Legal Studies 293.

Schill, Stephan W., "Do Investment Treaties Chill Unilateral State Regulation to Mitigate Climate Change?" (2007) 24 Journal of International Arbitration 469.

Schoenbaum, Thomas J., "International Trade and Protection of the Environment: The Continuing Search for Reconciliation" (1997) 91 American Journal of International Law 268.

Schoenbaum, Thomas J., "Trade and Environment: Free International Trade and Protection of the Environment: Irreconcilable Conflict?" (1992) 86 American Journal of International Law 700.

Simmons, Benjamin, "In Search of Balance: An Analysis of the WTO Shrimp/Turtle Appellate Body Report" (1999) 24 Columbia Journal of Environmental Law 413.

Skilton, Thomas E., "GATT and the Environment in Conflict: The Tuna-Dolphin Dispute and the Quest for an International Conservation Strategy" (1993) 26 Cornell International Law Journal 455.

Somvanchi, Vishal S., "Patenting Drought Tolerance in Organisms" (2009) 3 Recent Patents on DNA & Gene Sequences 16.

Spiegel Feld, Danielle and Stephanie Switzer, "Whither Article XX? Regulatory Autonomy Under Non-GATT Agreements After China—Raw Materials" (2012) 38 Yale Journal of International Law 16.

Stewart, Terence P. and Mara M. Burr, "Trade and Domestic Protection of Endangered Species: Peaceful Co-existence or Continued Conflict? The Shrimp-Turtle Dispute and the World Trade Organization" (1998) 23 William and Mary Environmental Law and Policy Review 109.

Stone, Christopher D., "Common but Differentiated Responsibilities in International Law" (2004) 98 American Journal of International Law 276.

Strauss, Debra M., "The Application of TRIPS to GMOs: International Intellectual Property Rights and Biotechnology" (2009) 45 Stanford Journal of International Law 287.

Tarasofsky, Richard G., "Heating Up International Trade Law: Challenges and Opportunities Posed by Efforts to Combat Climate Change" (2008) 2 Carbon & Climate Law Review 7.

Titus, James G., "Rising Sea Levels: The Impact They Pose" (1986) 12 EPA Journal 17.

Tobey, J., "The Effects of Domestic Environmental Policies on the Patterns of World Trade: An Empirical Test" (1990) 43 Kyklos 191.

Tol, Richard S.J., "The Economic Effects of Climate Change" (2009) 23 Journal of Economic Perspectives 29.

Urgese, Joseph J., "Dolphin Protection and the Mammal Protection Act have Met their Match: The General Agreement on Tariffs and Trade" (1998) 31 Akron Law Review 457.

Verdirame, Guglielmo, "The Definition of Developing Countries under GATT and Other International Law" (1996) 39 German Yearbook of International Law 164.

Voon, Tania and Andrew Mitchell, "Time to Quit? Assessing International Investment Claims against Plain Tobacco Packaging in Australia" (2011) 14 Journal of International Economic Law 515.

Weitzman, Martin, "A Review of *The Stern Review on the Economics of Climate Change*" (2007) 45 Journal of Economic Literature 703.

Weitzman, Martin, "On Modeling and Interpreting the Economics of Catastrophic Climate Change" (2009) 91 Review of Economics and Statistics 1.

Winter, Ryan L., "Reconciling the GATT and WTO with Multilateral Environmental Agreements: Can We Have Our Cake and Eat It Too?" (2000) 11 Colorado Journal of International Environmental Law 223.

Wirth, D., "A Matchmaker's Challenge: Marrying International Law and American Environmental Law" (1992) 32 Virginia Journal of International Law 377.

Wisthoff-Ito, Rita M, "The United States and Shrimp Import Prohibitions: Refusing to Surrender the American Goliath Role in Conservation" (1999) 23 Maryland Journal of International Law and Trade 247.

Wold, Chris, "Multilateral Environmental Agreements and the GATT: Conflict and Resolution?" (1996) 26 Environmental Law 841.

Index